X

Benton MacKaye

Creating the
North American
Landscape

GREGORY CONNIFF
EDWARD K. MULLER
DAVID SCHUYLER
Consulting Editors

GEORGE F. THOMPSON
Series Founder and Director

Published in cooperation with
the Center for American Places,
Santa Fe, New Mexico, and
Harrisonburg, Virginia

.

Benton MacKaye

Conservationist, Planner, and Creator
of the Appalachian Trail

Larry Anderson

The
Johns Hopkins
University Press
Baltimore & London

© 2002 The Johns Hopkins University Press
All rights reserved. Published 2002
Printed in the United States of America
on acid-free paper

2 4 6 8 9 7 5 3 1

The Johns Hopkins University Press
2715 North Charles Street
Baltimore, Maryland 21218-4363
www.press.jhu.edu

Library of Congress Cataloging-in-Publication Data
Anderson, Larry, 1950–
Benton MacKaye, creator of the Appalachian Trail / Larry Anderson.
p. cm.— (Creating the North American landscape)
Includes bibliographical references (p.).
ISBN 0-8018-6902-1 (hardcover : alk. paper)
1. MacKaye, Benton, 1879–1975. 2. Conservationists—United States—
Biography. 3. Regional planners—United States—Biography.
4. Appalachian Trail—History. I. Title. II. Series.
QH31.M137 A83 2002
333.7'2'092—dc21
2001007182

A catalog record for this book is available
from the British Library.

The article beginning on p. 371 and, unless otherwise specified,
all photographs and maps are reproduced by permission
of the Dartmouth College Library.

TO
Frances H. Anderson
and the memory of
Albert E. Anderson

CONTENTS

Photographs follow page 210.

ACKNOWLEDGMENTS

Work on this book extended over many years, during which time I accumulated a large debt of gratitude to the many individuals who provided assistance and encouragement. They all helped to improve the final product; I alone bear responsibility for any of its shortcomings.

My wife, Nan Haffenreffer, has provided support, patience, and love beyond measure. Our son, Sam, has for most of his life shared his father with the slowly emerging biographical character of Benton MacKaye. I am blessed to belong to such a family.

I am grateful to staff members at all the research institutions named in the Note on Sources, as well as at the Arnold Arboretum, the Harvard Forest, the Henry S. Graves Memorial Library at the Yale University School of Forestry and Environmental Studies, the Brown University Library, and the Brownell Library in Little Compton, Rhode Island. But I am especially beholden to the staff of the Rauner Special Collections Library at Dartmouth College, who were unfailingly courteous, friendly, and helpful. Philip N. Cronenwett, Special Collections Librarian, provided every possible measure of support. With apologies to those whose names I may have omitted, I offer special thanks to some of his current and former colleagues: Stanley W. Brown, John Schwoerke, Suzy Schwoerke, Barbara Krieger, Bonnie Wallin, Joyce Pike, and Patti Houghton. I am grateful to the Dartmouth College Library for permission to quote manuscript materials and reproduce images from the MacKaye Family Papers.

Other individuals who were particularly helpful in providing assistance on behalf of their organizations include Meredith Marcinkewicz of the Shirley Historical Society, Bennett Beach and the late Tom Watkins of the Wilderness Society, and Brian King, Robert Rubin, and Jean Cashin of the Appalachian Trail Conference.

I owe thanks to Harley P. Holden, Benton MacKaye's literary executor and Harvard University Archivist, both for his permission to use certain materials included in this book and for his encouragement. I am also grateful to Robert

Wojtowicz of Old Dominion University, Lewis Mumford's literary executor, for permission to quote from Mumford's correspondence and manuscripts; and to Wellington Huffaker of the Aldo Leopold Foundation, Inc., for permission to quote from Leopold's correspondence.

David M. Sherman, of the U.S. Forest Service, unstintingly shared materials from his own remarkable archive documenting the history of the Appalachian Trail and related conservation matters. He also suggested numerous avenues for research, introduced me to many individuals familiar with MacKaye and the Appalachian Trail, and offered constant encouragement. Laura Waterman and her late husband Guy shared with me some of their own extensive research into the history of hiking in the mountains of the Northeast. The example of their work and lives has been truly moving and inspiring.

Other individuals who generously shared correspondence, manuscripts, photographs, publications, recollections, and other materials included Robert Adam, Christy MacKaye Barnes, David Bates, Marion Bridgman Billings, Harriet Bridgman Blackburn, Howard Bridgman, Keller Easterling, Hermann Field, Frederick Gutheim, Cynthia Holmes, Robert M. Howes, Mable Abercrombie Mansfield, Joan Pifer Michaels, James E. Moorhead, Margaret E. Murie, Michael Nadel, Robert O'Brien, Paul Oehser, Betsy Pifer Rush, Jamie Sayen, Daniel Schaffer, Marion Stoddart, Carroll A. Towne, Donald Theoe, and Ed Zahniser.

John L. Thomas of Brown University offered encouragement and helpful commentary during the early stages of my work. Paul S. Sutter of the University of Georgia provided a particularly useful critical reading of a later draft of my manuscript. I am also grateful to Mark W. T. Harvey of North Dakota State University for sharing thoughts and ideas about the wilderness movement in general and Howard Zahniser in particular. Terry Tempest Williams provided timely encouragement at the Bread Loaf Writers' Conference.

Sheila and Bill Mackintosh, Kathleen Cushman, David and Serena Mercer, and Ransom and Mary Lee Griffin have been the most faithful of friends. My brother, Robert Charles Anderson, has provided a model of dedicated and exemplary scholarship.

The late Kermit C. Parsons of Cornell University and his former student Robert McCullough, now of the University of Vermont, were instrumental in bringing my work to the attention of the Center for American Places and the Johns Hopkins University Press. I owe Bob McCullough special thanks for numerous acts of generosity, including sharing some of his own research materials and ideas concerning MacKaye and Clarence Stein.

I am grateful to George F. Thompson, Randall Jones, and Christine Hoepfner of the Center for American Places, who patiently shepherded the book toward completion. I also owe thanks to Anne M. Whitmore of the Johns Hopkins University Press for applying her editing skills to the improvement of my work.

This book is dedicated to my parents, without whose support my work on the project would not have been possible. I regret that my father did not live to see its publication.

❖

Several periodicals published my articles about Benton MacKaye and Steele MacKaye. These included *Appalachia, Appalachian Trailways News, Chicago History, Harvard Magazine,* the *Harvard (Mass.) Post,* the *Providence Journal-Bulletin,* and *Wild Earth.* I also presented early versions of some sections of this book at the 1993 conference of the American Society for Environmental History and a 1996 conference, "Benton MacKaye and the Appalachian Trail: A Seventy-Fifth Anniversary of Vision, Planning, and Grassroots Mobilization," at the University at Albany, State University of New York.

Introduction

"Expedition 9"

"Hunting Hill is situated at the southeast point of a plateau about three miles square on the north side of Mulpus Brook," wrote fourteen-year-old Benton MacKaye in his self-designed "Geographical Hand Book" on June 12, 1893. "The hill itself is a drumling." The earnest teenager grappled with spelling, but the geological term *drumlin* accurately described the area's swarm of gentle hills, which had been rounded off by the retreating glacier many millennia earlier. Already, Benton MacKaye knew where he stood on Earth and in the broad sweep of time. "As I sit looking off this drumling only 542 feet high, taking in the beauty of the scenery," he continued, "I have the country spread out like a map before me."[1]

MacKaye (the name rhymes with "high") was embarked on the ninth of his "expeditions" to explore the country within walking distance of his home in Shirley Center, Massachusetts. Benton's tramps across his hometown landscape had been inspired in part by such celebrated explorers as Alexander von Humboldt, John Wesley Powell, and Robert Peary, the latter two of whom he had heard lecture in Washington, D.C., several years earlier. Always restless in New York City, where the family often spent its winters, he returned to the small north-central Massachusetts town in the spring of 1893 determined to pursue some adventures of his own. "Why shouldn't I, my own self, be an explorer?" he wondered, "up there in my own homeland, containing the 'canyons' of the Squannacook?" Thus began the "first stunt" of his future careers in forestry, regional planning, and a field he called "geotechnics," "the applied science of making the earth more habitable."[2]

Vivid, stimulating maps; colorful, high-minded prose; horizon-sweeping viewpoints—such were the methods the adolescent Benton already employed as he explored his neighborhood's gentle hills and rivers. They were the same techniques he would apply throughout a long and productive career that spanned three-quarters of a century and significantly influenced the evolving American conservation and environmental movements.

He had pasted into his notebook sections of topographical maps depicting the region surrounding Shirley. But he also drafted his own colorful pen-and-ink maps of the local terrain, documenting his daily expeditions. "By 'No. 9' I had reached the top of the divide (between Mulpus and Squannacook)—Hunting Hill," he later recalled.[3] Surveying the exhilarating panorama from Hunting Hill that day in 1893, he could see "the Mulpus Valley breaking through the hills" to the southeast, widening as it neared the Nashua River. Across Mulpus Brook, he wrote, "stands Shirley Ridge 422 feet high on which the little town of Shirley Center is situated hemed [sic] in by the woods so that high as it is very little can be seen of the outside world. A pretty place but over-run with Gossip."[4]

Benton's tongue-wagging neighbors in Shirley Center (a village in the northern part of the town of Shirley) were not the only objects of the young philosopher's lofty, censorious concerns:

> Why is it that the beauty of nature must be spoiled so by Man? Man, though the highest of beings, is, in one sence [sic], the lowest, never contented until he has spoiled all the beauty of Nature in his power by cutting down vegetation, killing animals, and even cutting down hills when he has the power to do so. . . .
>
> How much more beautiful the surrounding country would be, in every way, if men were not such fools.
>
> Yes, a very true saying is "What fools these Mortals be."[5]

Turning his gaze toward 2,000-foot Mount Wachusett in the southwest, then to the even humbler Mount Watatic in the northwest, he pondered the relationship between human endeavor and the natural landscape. "As I sit here, taking in the glory of Nature, the wind blowing through the trees, the cows bellowing now and then, I wish only that man was as peaceful as Nature," he opined. "When I hear the train whistle blow, I think that doesn't sound so very bad if only man who built it did not spoil Nature and corrupt himself, by building his railroad and other infernal signs of civilazition [sic]."[6]

When James MacKaye, one of Benton's older brothers, came across the record of this Shirley excursion, he teasingly described the young pathfinder's habits as "expedition nining."[7] Benton used the phrase proudly thereafter as a metaphor for his new explorations, both physical and philosophical. Mountain summits, even on so modest a scale as Hunting Hill, again and again provided the vantage points from which he visualized humankind's proper place in relation to the natural environment. No matter on which prominence he stood, MacKaye always envisioned hopefully "a land in which to live."[8]

❖

I was brought up within view of Hunting Hill's summit. For most of my early years, in fact, I lived only a few miles from MacKaye's Shirley Center home. Though I never met him during the quarter-century that our lives overlapped in time and space, I had become dimly aware of MacKaye's identity, perhaps when a local newspaper described the latest award or recognition he had received—for conceiving the Appalachian Trail, helping create the Wilderness Society, or promoting other environmental and community causes.

MacKaye was an obsessive recordkeeper. For most of his adult life he kept a terse pocket diary. From this record I would learn that on at least one day, in 1957, he and I had been in the same place at the same time. We both attended a parade celebrating the 225th anniversary of the founding of Townsend, the town where I had recently lived. As a boy racing across Townsend's classic New England town common that festive day, perhaps I had bumped into a thin, dignified man well into his seventies, the brim of his battered felt hat tilted to block the breeze as he lit his pipe.[9]

The Algonquian names for the local natural features—the Squannacook, Nashua, and Nissitissit rivers, the mountains called Monadnock, Wachusett, and Watatic—were as much a part of my own native consciousness as they had been of MacKaye's. My first and abiding awareness of the contours of the world had been formed in the same New England watershed that Benton MacKaye called home: that of the Nashua River, a tributary of the grander Merrimack. Out of curiosity, fate, and circumstance, I gradually came to learn more about MacKaye's life and career beyond our common locale.

Identification with one's subject is perhaps the biographer's greatest potential pitfall, but as I proceeded with a biography of Benton MacKaye, I hoped that my own life experience might provide a useful viewpoint from which to recount the story of this remarkable American personality. On the same campus where MacKaye spent his undergraduate and graduate years, I was a college student during the tumultuous years of the late 1960s and early 1970s. At this highwater mark of the "ecology" movement and other countercultural enthusiasms, I perused the *Whole Earth Catalog* as closely as the college course catalogue. My own callow quest for meaning and "relevance" occasionally struck pay dirt. I took an eye-opening course on the man-made American environment from the legendary landscape historian J. B. Jackson, who revealed new ways of observing the commonplace. Somewhere along the way, I began reading my way through Lewis Mumford's books. Mumford's prose, passion, and range of interests, though not entirely fashionable at the time, somehow struck a chord in me. Along with many others of my generation, during the era's backpacking boom, I began to hike New England's north country, sometimes toting Aldo Leopold's *A Sand County Almanac* or Henry David Thoreau's *Walden.* My first job after college was in an ancient Vermont sawmill, working beside a tough seventy-five-year-old woodsman who felled trees on his own 4,000 surrounding acres. Later, as a small-town journalist, then as a freelance writer and local environmental activist, my own eclectic interests and skills seemed, fortuitously, to provide some suitable tools for pursuing the story of MacKaye's life, a story I followed not as an academic specialist but as a generalist. My goal has been to provide an account of MacKaye's life that will appeal not only to scholars in such fields as environmental history, American intellectual and cultural studies, architecture, and planning but also to readers interested in the Appalachian Trail, hiking, wilderness preservation, and community activism.

MacKaye is best known for his conception of the Appalachian Trail (AT), the 2,160-mile mountain footpath that stretches between Maine and Georgia. "Perhaps it is unrivaled by any other single feat in the development of American outdoor recreation," he wrote in 1972, just over a half-century after he publicly proposed the idea.[10] If a ninety-three-year-old might be excused for boasting, the claim could still be defended a generation later. Following the natural course of the Appalachian ridgeline, the trail traverses and links fourteen states, more than two hundred counties and municipalities, and some seventy-five public-land areas, including eight national forests and six units of the national park system. It is managed by the National Park Service, in cooperation with the U.S. Forest Service, nearly one hundred state agencies, and the thirty-one trail-maintaining clubs coordinated by the Appalachian Trail Conference. By various estimates, three to four million hikers set foot on the trail every year. In recent years, some two thousand backpackers have started out each spring, most of them headed from south to north, moving with the season, to travel the full length of the trail in one continuous hike. As the chill days of autumn take hold at the summit of Maine's Katahdin, months later, no more than three hundred to four hundred of those trampers are likely to have endured the rigors of the AT to complete the trek as "thru-hikers."[11]

But statistics only hint at the trail's unique significance as an American cultural institution and social invention, described by MacKaye's friend Lewis Mumford in 1927 as "one of the fine imaginative works of our generation."[12] In that most essential and democratic of American publications, the annual *Rand McNally Road Atlas,* the only route depicted in the eastern United States for people traveling by foot is the thin broken line representing the Appalachian National Scenic Trail. And, just as MacKaye hoped in 1921 when he envisioned the project as "a retreat from profit," the AT remains a physical and symbolic wilderness bulwark against the inroads of modern industrialism, urbanism, and commercialism. "It's a folk product pure and simple," MacKaye continued to insist after the trail was completed.[13] There are no mandatory admission fees or tollbooths along the Appalachian Trail's route. A hiker can still walk the trail from Maine to Georgia free of charge, joining a community of hikers linked across generations and across the mountain landscape.

But the Appalachian Trail has come to represent not just another popular recreational destination for adventure and retreat. For some, the trail has also become a spiritual domain, the setting for a rite of American cultural passage and pilgrimage. Indeed, AT chronicles have become a popular American literary genre. These trail narratives range from the laconic and reflective, such as *Walking with Spring*, Earl V. Shaffer's account of the first "thru-hike" in 1948, to the humorous and irreverent, like Bill Bryson's *A Walk in the Woods*, which became a bestseller a half-century after Shaffer's legendary trek (even as the seventy-eight-year-old Shaffer was completing his third thru-hike).[14]

The successful example of the Appalachian Trail has also inspired the creation of many other trails and recreational corridors throughout the country. In the southern Appalachians, trail activists in the 1980s began creating a 250-mile Benton MacKaye Trail, originating at Springer Mountain and circling west to rejoin the AT in Great Smoky Mountains National Park. A decade later, northern enthusiasts were promoting an International Appalachian Trail, which would continue along the Appalachians from Katahdin, across Maine, New Brunswick, and Quebec, to terminate at Mount Jacques Cartier on the Gaspé Peninsula. In the same decades, a nationwide "greenways" movement was taking root. During an era when other environmental initiatives often met resistance, grassroots-inspired trails and greenways won support and funding from federal, state, and local governments. The President's Commission on Americans Outdoors, for instance, in its 1987 report *Americans Outdoors*, envisioned "a continuous network of recreation corridors which could lead across the country . . . reaching out from communities all across America to link cities, towns, farms, ranches, parks, refuges, deserts, alpine areas, wetlands, and forests into a vast and varied network of open spaces."[15] It was an idea Benton MacKaye had proposed seventy years earlier.

❖

The idea for the Appalachian Trail, and its continuing influence on the creation of other public trails, greenways, and protected lands, would constitute a substantial achievement for any lifetime. But MacKaye also played a significant role in many other noteworthy American social and intellectual causes and organizations—whether as a young forester with the fledgling U.S. Forest Service; a proponent within the Labor Department of government-sponsored new communities on federal lands; a self-created regional planner conceiving innovative approaches to land use, community building, and transportation with the Regional Planning Association of America in the 1920s; an advocate of comprehensive river-basin planning at the Tennessee Valley Authority in the 1930s; an early promoter of wilderness preservation, as a founder and leader of the Wilderness Society; or an elder statesman and prophet who inspired a new generation of environmental activists in the post–World War II era.

MacKaye's life and career sliced through many important events, movements, and organizations, but his own substantial and unique legacy has not always been well known or clearly understood. In fact, the example of MacKaye's life confounds many of the conventional criteria of American success. During a long career in government, he never attained high office. His literary and popular reputation during his own lifetime never matched that of such illustrious friends and acquaintances as Lewis Mumford, Aldo Leopold, Bob Marshall, Sigurd Olson, and Walter Lippmann. In a manner that eluded some of his better-known peers, though, MacKaye tapped the interests, desires, and energies of ordinary citizens to inspire the creation of enduring social and environmental institutions. And in contrast to the legacy of a near-contemporary, the notorious Robert Moses—the "power broker" who conceived and created the literally monumental bridges, parkways, housing developments, and parks that dramatically reshaped the physical landscape of the state and city of New York—the success of MacKaye's ideas can best be measured by the modesty of their visible impact on the physical environment. Moses represented one centralized, autocratic extreme of American land-use planning. MacKaye's long career embodied a hopeful planning alternative: he strove to reconcile somehow the grassroot endeavors of citizens, the integrity of small communities and natural landscapes, and the technical resources and legal authority of a benign federal government. As it happened, his seemingly utopian ambitions achieved some genuine success, the lessons of which are still worth heeding.

David Brower, serving as executive director of the Sierra Club, came to know MacKaye during the tumultuous conservation battles of the 1950s and 1960s. Years later, describing the distinctive nature of his friend's influence, he cited what he called "Benton MacKaye's Theory of How to Build Big by Start-ing Small."[16] MacKaye himself, in a 1922 article about the prospects for the nascent Appalachian Trail, explained his subtle technique of reform in con-crete terms comprehensible to any hiker or trailworker. "In almost every local-ity along the Appalachian ranges a greater or less amount of trail-making is going on anyhow from year to year. Various local projects are being organized, and in one way or other financed, by the local outing groups. The bright idea, then, is to combine these local projects—to do one big job instead of forty small ones."[17] The Appalachian Trail and the National Wilderness Preserva-tion System (by 2002 comprising almost 106 million acres in some 644 units of federal land) were established during MacKaye's own lifetime, partly through his efforts and influence. Such achievements demonstrate the effec-tiveness of a patient, determined, step-by-step approach to accomplishing the "one big job."

Snapshots of MacKaye's writings and activities at particular junctures of his lengthy career might place him comfortably in standard scholarly pigeonholes labeled conservationist, preservationist, environmentalist, forester, regional-ist, planner, or ecologist. Partly by virtue of his longevity, but primarily be-cause he was motivated more by intuition and personal experience than by a systematic philosophy or discipline, such stereotypes do not prove very useful in MacKaye's case. He was a complicated figure, both personally and intellec-tually. "The highly polished persona he offered his following of trailsmen and wilderness-seekers, the grizzled, craggy New Englander, cast in the mould of Thoreau, is something less than the whole man—and may, indeed, not be the essential one," observed planner and writer Frederick Gutheim shortly after the death of his often circumspect friend.[18] A self-styled "amphibian as be-tween urban and rural life," MacKaye sometimes lived near the brink of pov-erty, punctuating his forestry and planning assignments with long stretches of solitary reflection and writing in Shirley.[19] His colorful, even eccentric per-sonality attracted a legion of friends and followers. But "although he had a genius for friendship," observed one longtime acquaintance, "he did not care to be too friendly."[20]

MacKaye's habit of compartmentalizing, even concealing, aspects of his personal life has hindered the emergence of a complete account of his life, ca-reer, and legacy. Nonetheless, a number of recent authors have recognized the comprehensive, multifaceted, and even somewhat ethereal nature of Mac-Kaye's wide-ranging view. "As MacKaye saw it, you changed people's mental maps by first altering their physical maps," writes journalist Tony Hiss, whose 1990 book, *The Experience of Place,* reported the fervent interest in MacKaye's pioneering ideas by a new generation of planners, preservationists, and activists.[21] New England forest historian Robert McCullough has described how MacKaye's ideas "were creatively stitched together into a quilt of humanistic-environmental planning: community, regional, continental, and, ultimately global."[22] Architect Keller Easterling has offered a provocative commentary on the "mammoth" and "gigantic" scale of MacKaye's self-appointed vocation as designer of a "global infrastructure."[23]

Others have focused on the social dimensions of MacKaye's work. Regionalist scholar Robert L. Dorman observes that MacKaye's ideas represented a "broadened vision" which "had striven to bind up the unraveling of conservationism and preservationism, two doctrines that needed each other for political and ethical completeness."[24] Paul S. Sutter, in an important and discriminating study of the development of the American wilderness movement in the twentieth century, portrays MacKaye as "one of the nation's most important and least understood environmental thinkers," notable for his "socially informed brand of wilderness advocacy."[25] Robert Gottlieb, another historian searching for the roots of social activism in the contemporary environmental movement, discerns in MacKaye's career a pioneering attempt "to define a new progressive politics consisting of rational decision making, social justice, and resource management of both city and countryside."[26] And Paul T. Bryant, whose scholarship provides the foundation for any study of MacKaye's life and work, likewise wrote that his subject's "interest primarily was in humanity. If nature had no relationship to mankind, he would have had little interest in nature."[27]

I have tried to do justice to the important intellectual and historical themes evoked by the events of MacKaye's life, in the context of his own times; but this biography does not offer the in-depth critical analysis and interpretation other scholars have brought to bear on particular aspects of MacKaye's career. Rather, my overarching theme is the example of how MacKaye lived his life. Composing a life story as a moral and inspirational tale may be a somewhat old-fashioned rationale for a biography, but in MacKaye's case I think it applies. He was a genuine idealist and visionary, who lived and worked against the American grain. His unconventional and original environmental ideas were firmly grounded in the realities of the natural landscape, yet his was also and always an inclusive social vision. And despite experiencing his share of personal tragedy and professional frustration, he remained an irrepressible optimist.

MacKaye's ideas still resonate along the unorthodox American intellectual continuum that historian John Thomas has described as an "adversary tradition."[28] The life story of Benton MacKaye, a truly distinctive American intellectual and activist, suggests what can be accomplished by proceeding, as Lewis Mumford once described his friend's purposeful approach, "dead in the face of the prevailing power system, which puts profits and prestige and machine-made productivity above the needs of life."[29]

"Speak softly and carry a big map," MacKaye urged a meeting of the Appalachian Trail Conference in 1930.[30] His exhortation to the hiking community neatly captured the expansive perspective and the singular method that characterized his own lifelong quest. The adventure was already well under way when, as a precocious young explorer ranging across the evocative Massachusetts landscape on his solitary "expedition 9," he reached the top of Hunting Hill to survey "the country spread out like a map" before his expectant, farseeing eyes.

❖ 1 ❖

The MacKaye Inheritance

1879–1896

W hen Benton MacKaye was born, in Stamford, Connecticut, on March 6, 1879, his family was engulfed in the worst yet of its recurrent financial crises. Just four days after Benton's birth, his father's aptly titled play *Through the Dark* opened at the Fifth Avenue Theatre in New York. "Our precious new-born boy has come in the midst of troubles," Steele MacKaye wrote to his wife, Mary, in Stamford after the initial performances, "but *every storm spends itself at last!*"[1] The production closed within three weeks. Determined to control every aspect of his productions, Steele leased another modest New York theater, setting theatrical circles abuzz with his proposal to share profits with playwrights and actors, but its opening was threatened by aggressive creditors. Reluctant to turn once again to his wealthy father for financial support, Steele this time was rescued by his wife. Early one morning, while visiting her husband in New York, Mary stole out from Steele's rooms at 23 Union Square to plead with an "obdurate" friend for the necessary funds. Using her "last ounce of resourcefulness," Mary was successful. MacKaye's "little theatre" opened with another play, *Aftermath*, a revival of his own *Won at Last.*[2] It was characteristic of the couple's frenetic, melodramatic life that months passed before they named their infant son, who was briefly dubbed "little Mr. Nemo." As it happened, the boy's name, like his father's, was adapted from that of Steele's mother, Emily Benton Steele: he was called Emile Benton MacKaye.[3]

Mary MacKaye's mettle in the face of calamity counterbalanced her husband's high-minded artistic ambitions, financial prodigality, and irrepressible faith in his own talents. Born in Newburyport, Massachusetts, in 1845, she was the daughter of the Reverend Nicholas Medbery, a Baptist minister, and the equally devout Rebecca Belknap Stetson, who had served as principal of the Charlestown (Mass.) Female Seminary and written several biographies of missionaries. Like her mother, Mary never confined her interests solely to domestic duties. She followed her husband's artistic and business affairs as closely as his sometimes secretive nature allowed. She coaxed and cajoled her children in their careers. But Mary pursued her own literary ambitions as well, adapting Jane Austen's *Pride and Prejudice* for the stage and sometimes lecturing about the theater.[4]

Mary's marriage to the dashing young "Jim McKaye," as he was known in 1865, was auspicious and romantic enough—even slightly scandalous. She had won the young man's affections away from his first wife and her own close friend, Jennie Spring. Jennie was the daughter of the founder of the progressive school in Eagleswood, New Jersey, where Steele was teaching art. In the tumult of those early years of the Civil War, Steele within a ten-day period in June of 1862 enlisted in the Seventh Regiment of New York and married Jennie, "an impulsive, war-time match."[5] They had a son, Arthur, the next year. Steele saw no action during his brief military career, and after several bouts with malaria he retreated to Eagleswood to recover. It was here that he met Mary, who was visiting her friend Jennie, in 1863. Two years later, Mary (known familiarly as "Molly") and Steele were married. They soon embarked for Paris, where the young aesthete hoped to make a name for himself as an art dealer and artist.

Eventually Mary bore six children—Harold (called Hal), William (Will), James (Jamie or Jack), Percy (Poog), Benton, and Hazel. They returned to America in the early 1870s, and Mary raised the growing family as her brilliant but mercurial husband pursued his endeavors as actor, playwright, producer, inventor, teacher, and promoter. Mary MacKaye remained ever hopeful that one of Steele's ventures would maintain the standard of living she had assumed would always be theirs when she married the son of the wealthy and distinguished Colonel James Morrison McKaye.

James Morrison Kay, as Benton's grandfather had been named, was born in 1805 in Argyle, New York, a small community of Scottish settlers near Glen Falls, in the Hudson River Valley uplands. The twelve-year-old James left home unannounced after his father's death and tramped across New York to the "half-Indian village" of Buffalo, which was then coming to life as work on the Erie Canal commenced. The enterprising youth soon made himself useful as an errand boy for the burgeoning town's most prominent citizens and as a clerk in the land office.[6] He enrolled in the military academy at Norwich, Vermont, with which he remained affiliated until 1829 as an instructor and as principal. At Norwich, he married his first wife, Elizabeth Partridge, and made the first of several emendations of the family name, adding *Mc* to his surname to become James Morrison McKay. (In the 1850s, he added *e* as a final letter. His son Steele, in thrall to the clan's Scottish heritage, eventually dubbed his family "MacKaye" and dropped the first name he shared with his father.)[7]

Returning to Buffalo, McKay studied law under an ambitious local attorney, Millard Fillmore, joined the New York bar, and became a law partner of the future president. As head of a local volunteer regiment, he assumed the title of Colonel. In his commercial ventures, McKaye prospered as an organizer of the American Express Company, secretary of Wells Fargo and Company, and president of the American Telegraph Company, predecessor of Western Union.[8] In the early 1850s, after the death of his second wife, Steele's mother, he moved his family east to Brooklyn Heights, New York; summers they spent in Newport, Rhode Island. Colonel McKaye's new acquaintances included the likes of writers Henry Wadsworth Longfellow, Julia Ward Howe, Horace Greeley, and Ralph Waldo Emerson; artist William Morris Hunt; reformer William Lloyd Garrison; and self-styled theologian and philosopher Henry James, Sr.[9]

An abolitionist and a leader of New York's Underground Railroad, Colonel McKaye was appointed by Secretary of War Frank Stanton in 1863 to the American Freedman's Inquiry Commission, along with Robert Dale Owen and Samuel Gridley Howe. He traveled through the lower Mississippi River valley, behind the Union Army's lines, on a regional survey of "the great changes which, at the present moment, slave society is everywhere undergoing." In his 1864 report, *The Mastership and Its Fruits,* he called for a constitutional amendment to secure the equality, civil rights, and enfranchisement of blacks.[10] After the Civil War, Colonel McKaye withdrew from business and political affairs. He spent most of the last two decades of his life in Europe, writing his memoirs—and worrying about his son Steele's erratic career.

Born in Buffalo in 1842, before the family's move to Brooklyn Heights, the young Steele was described by philosopher William James, his boyhood friend from Newport summers, as "effervescing with inco-ordinated romantic ideas of every description."[11] Colonel McKaye dispatched the sixteen-year-old Steele to the École des Beaux-Arts in Paris to study art, but the young man's interest soon turned to the theater. While living in Europe after his marriage to Mary, the talented young man became a devoted disciple of François Delsarte, a French exponent of a highly stylized and physically expressive acting technique. Upon his return to America, MacKaye lectured on the Delsarte method. For a time, he was one of the best-known, colorful, and controversial figures in American theater. He produced, wrote, and acted in several New York plays, which were greeted with mixed critical and popular response. After one more European sojourn, he returned to the United States in 1874 to pursue his American theatrical career in earnest.[12]

Steele MacKaye was involved in the design, construction, and management of several theaters in New York City during the 1870s and 1880s. But virtually every episode of his career as a theatrical impresario left him embroiled in controversy with partners and financiers. Periods of frenzy were punctuated by periods of mental and physical collapse. From the time of his youth, Steele experienced bouts of "brain fever," "nervous exhaustion," fainting spells, and other vague but distressing symptoms of the "neurasthenic" illnesses that so often afflicted the lives of the era's leisure class.[13] Members of the next generation of MacKayes, including Benton, at times suffered some of the same nervous maladies, brought on perhaps by the tension between their high expectations and the often less exalted realities of their daily lives.

At the time of Benton's birth, in 1879, the outward indications of the family's status appeared auspicious enough. The Stamford home where Benton was born and where the family had lived since 1875 was a very comfortable one, equipped with a bowling lane, workshops for the boys, a stable of horses, a yard full of exotic poultry, and paraphernalia for all manner of sports. Steele employed a motley staff, described in his son Percy's biography of his father as "itinerant tramps and dependants whom he enthusiastically appointed to the posts of coachman, gardener, nightwatchman, etc.," but who were most diligent at helping themselves to the family's belongings, including the contents of their master's wine cellar.[14]

In late 1879, strapped for funds, the family began a restless series of moves. The MacKayes took a comfortable brownstone at 107 West 44th Street in New York. The family's prospects fleetingly brightened, as Steele's romantic 1880 melodrama *Hazel Kirke* drew large audiences. But MacKaye did not benefit from the play's financial success. In his perpetual quest for funds, he had previously signed away to promoters the rights to his theatrical works.[15]

As the family's fortunes declined, Mary from season to season moved her family to farms and houses in Brattleboro, Vermont; Norton, Massachusetts; Mount Vernon, New York; and Ridgefield, Connecticut. In 1883, Colonel MacKaye stopped financing his son's ventures. By the fall of 1885, Steele had moved his family back to New York City, to 172 Lexington Avenue. And for a time, matters again seemed more hopeful. One of Steele's successes, and a showcase for his unique talents, was his collaboration with "Buffalo Bill" Cody during 1886 and 1887. In New York's Madison Square Garden, MacKaye restaged Cody's already popular Wild West Show as a "Drama of Civilization," employing a novel array of spectacular and realistic special effects to depict the settlement of the American West. Benton and the other MacKaye children had the best seats in the house during the production and met the colorful Cody.[16]

The year 1887 was for Steele MacKaye "a last serene isle of refuge in his battling and tempestuous life-passage," according to Percy MacKaye.[17] *Paul Kauvar,* his spectacle of the French Revolution, opened in New York for a successful three-month run. But the production's closing was followed by a sad and sudden event that quickly altered the fates and fortunes of all the MacKayes. On April 8, 1888, Colonel James McKaye died in Paris at the age of eighty-three. The Colonel bequeathed much of his estate to Steele's half-brother William, "doubtless with the expectation that, having no children, he would pass it on to Steele MacKaye and his children," as Percy wishfully observed. But only a few days after the death of the Colonel, William suffered a fatal stroke. His will provided that all he owned should pass to his wife, Maggie—who unexpectedly died a few weeks later. Thus, by Percy's embittered account, Colonel McKaye's considerable fortune passed to Maggie's relatives, bypassing Steele and his family altogether.[18]

It was providential, then, that one of Steele's sons had secured a modest family estate for the MacKayes. Not yet twenty years old, Will, the second oldest of the children, had begun a career as a professional actor in a touring company. When the family visited the small village of Shirley Center, Massachusetts for the first time in the summer of 1887, Will heard his "Aunt Sadie" speak longingly of a small but attractive cottage, known locally as the Aunt Betsy Kelsey place, which stood on the west side of Parker Road a few hundred yards north of Shirley Center's common. Sarah Stetson Pevear—"Aunt Sadie" —was Mary MacKaye's cousin and "adopted sister." She had been part of the migratory MacKaye clan for years, "the rock of stability in a household of exuberant artists and waxing children," Percy wrote, "a mothering foster-grandmother, adored by us all." During the winter, Will and his mother negotiated to buy the house for Sadie for $500. Will made a $100 down payment from his acting earnings, agreeing to pay the balance with interest a year later.[19]

The MacKayes had come to Shirley Center to visit the summer retreat of Sadie's brother, businessman Henry Pevear. Five years earlier, during a horse-and-buggy tour of central Massachusetts with his wife, Pevear had happened upon the attractive village. Smitten by the tranquil rural surroundings, he purchased the gracious Whitney house on the town common.[20] The Shirley Center of the 1880s was a classic New England setting—and it remains so still. Thirty-five miles northwest of Boston, it is the original village in a town inhabited in those last years of the nineteenth century by just over a thousand citizens. The Center was the focal point for five roads radiating out to the other

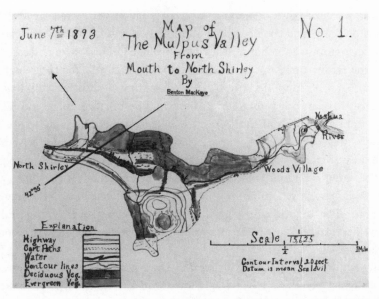

June 7th 1893

MAP of
The Mulpus Valley
From
Mouth to North Shirley
By
Benton MacKaye

No. 1.

Nashua River

North Shirley

Woods Village

42°35′

Explanation

Highway
Cart Paths
Water
Contour lines
Deciduous Veg.
Evergreen Veg.

Scale, 1/75,625

½   1 Mile

Contour Interval 20 feet
Datum is mean Sea level

One of a series of maps fourteen-year-old Benton drew to document his eleven walking "expeditions" in the countryside surrounding Shirley Center during May and June of 1893. (Courtesy Shirley Historical Society, Shirley, Massachusetts.)

parts of town. The meeting house and common were established on a height of land that fell off to Mulpus Brook and the Squannacook River in the north, Walker Brook in the east, Long Swamp in the west, and Catacoonamug Brook to the south. The Nashua River, a tributary of the Merrimack, formed the easterly border of the town. By the 1800s, the southern part of town, along the Catacoonamug, had become more populous, as various entrepreneurs dammed the brook to power their mills. And the construction of the Fitchburg Railroad through the town in 1844—the same line that ran along the southern shore of Walden Pond in Concord—had established Shirley Village, near the mills two miles south of the Center, as the town's primary outlet to the world beyond. As a result, Shirley Center was left as a sanctuary, a veritable time capsule of eighteenth-century New England life and values. Indeed, for Benton MacKaye the almost organic "starfish symmetry" of Shirley Center's tight-knit physical arrangement—town hall, church, schoolhouse, general store, homes —came to represent an idealized model of community life.[21]

In July of 1887, when eight-year-old Benton paid his first visit to Shirley Center with his sister, Hazel, and Aunt Sadie, they arrived by train at Shirley Village. They drove by wagon up the "store hill" to the Center, passing by the Longley Homestead on the west side of the road—a place that would become virtually a second boyhood home. A year later, having weathered the March 12 "Blizzard of '88" and the news of the Colonel's death, the MacKayes gave up their Lexington Avenue house. They returned to Shirley to take possession of what the family came to call "The Cottage." Henceforth, for the remaining years of Benton's long life, the Shirley Cottage that his brother Will had impetuously agreed to buy for Aunt Sadie would be his true home and refuge. "Not till I reached rurality did I start to really live," Benton wrote in his "Sky Parlor" study in the Cottage almost seventy-four years after his first night there, July 28, 1888.[22] He instantly fell in love with his new surroundings. "I like the country very much better then [sic] the city," he reported to an aunt during his first winter in Shirley.[23] In the fall, his older brothers Percy and James enrolled in a private school in nearby Groton, while Benton and his sister Hazel attended the brick one-room schoolhouse only a few steps north of the Cottage.

The family's "last and only Christmas reunion" in Shirley in 1888 proved to be an idyllic, yet ultimately a tragic, interlude for the family. The children skated on Hazen's meadow. They listened as Will read from *The Ancient Mariner, The Wandering Jew,* and other books he had illustrated with his own drawings. Steele discussed with his sons "that favourite theme of our family circle—'the universe,' its how and why and whither."[24] But within a month of returning to his New York boarding house, Will died from a respiratory illness. His surviving siblings cherished his memory in virtually hagiographic terms. And for almost a century they would hold fast to Will's legacy, the beloved Shirley Cottage.

The gently rolling hills and the bucolic valleys of the Nashua River watershed had long been the scene of a curious abundance of utopian and spiritual experiments. In the 1880s, when the MacKayes arrived in Shirley, a dwindling community of Shakers still lived in the southernmost part of town. Across the Nashua River to the east, in the neighboring town of Harvard, stood another Shaker village. And not far away from that community, on a west-facing hillside overlooking Shirley and Mount Wachusett, was the site of Fruitlands. There, Bronson Alcott, the most transcendental of the Transcendentalists, had in 1843 led his family (including eleven-year-old Louisa May) and a small band of disciples to a run-down farm to establish a short-lived "New Eden" based on vegetarian and cooperative principles.

Benton, who would go on to design a few utopias of his own, romanticized Shirley, as did many other members of his family. "More, I believe, than any other lost hamlet in New England," Percy wrote, "it was (and is) a living survival of old English serenities—a 'day-dreaming' landscape of elegy and dim folk remembrance."[25] In the late 1880s and early 1890s, the environs of Shirley Center—the Cottage, the nearby common, the streams, pastures, forests, and hills—were the focal point of Benton's personal world, the source of his education and inspiration. "The Shirley of Benton's youth was populated with the usual assortment of sages, patriarchs, eccentrics and rogues," observed his Shirley friend Harley Holden, the descendant of a town founder. In the summer months, another element was added to the community. Like the Pevears, many affluent urban families in those last years of the nineteenth century found rural retreats in pleasant Massachusetts hill towns like Shirley. The railroad afforded ready access from the city, and the decline of New England agriculture created a stock of old farmhouses and village homes that could be purchased for a modest sum. The summer residents "were generally welcomed and they added to the quality of life in the community. Sometimes, however, the natives felt that they were regarded with a missionary or explorer attitude, like Rousseau's 'noble savage,' rustics to be observed and studied like Margaret Mead's Samoans," continued Holden. "With the coming of the summer people, such aspects of culture and of gracious living as music, art and theater were encouraged. The MacKayes wrote plays and produced them in the town hall using local actors. . . . Artists, musicians, actors, dramatists—many associated with the MacKayes—found their way to Shirley, if only briefly."[26]

The MacKayes, by education and upbringing, identified most closely with these summer folk. But by economic necessity, the core of the MacKaye family —Benton, his sister Hazel, Aunt Sadie, and usually Mary MacKaye—for a time called Shirley Center their year-round home. Benton relished the rural life. With his schoolmates, he prowled the local countryside, swam in Mulpus Brook, helped out on neighborhood farms, took lessons at the town hall from dancemaster Simeon Green, and visited the local blacksmith shop. He became so familiar a figure in the little community and so identified with the place that his brother Percy titled him "The Honorable Benton MacKaye, Mayor of Shirley Center."[27] Such freedom, especially in the absence of his father, provoked in Benton a measure of adolescent willfulness. "Emil will not study more than 15 min. and says he will not go to school," his mother complained in a letter to Jamie. "When he calls me a liar it doesn't hurt me but if I whipped him, it would. He needs a master."[28] In the voluminous correspondence exchanged almost daily among family members, the mother and children alike felt free to comment on one another's foibles.

❖

By the middle of 1890, Steele's fortunes and prospects in the theater had reached their lowest ebb. His most recent play, aptly titled *Money Mad,* closed after a two-month run. The peripatetic and financially hard-pressed MacKaye family now gravitated to the point of the compass where prospects were brightest. In the winter months of 1890 and 1891, that place was Washington, D.C. Harold and James, respectively twenty-four and eighteen, both held modest government positions, Harold in the Patent Office, James with the Census Bureau. Their modest but dependable incomes then provided the foundation of the MacKayes' livelihood. In late December, the younger children—Percy, Benton, and Hazel—had accompanied their mother and Aunt Sadie by train from Shirley to Washington. Mary, Sadie, and the two youngest children took lodging at a rooming house on 13th Street, near Logan Circle, while Percy moved in with Harold and James a few blocks away. Steele was away on fruitless quest for fortune. He had recently been involved in a South Dakota real estate promotion, the development of a North Carolina gold mine, and the composition of a novel, excerpts of which he read to the family during a brief Washington visit. ("It is surely *very* daring," Percy noted in his diary.)[29] Like many of his grandiose ideas, though, these too would come to naught. However, for all its troubles, the family would never again share so intimate and happy a time as those months in Washington. And for young Benton, that winter and spring in the nation's capital revealed prospects for new explorations of America's fast-changing physical, cultural, and intellectual landscape.

December 19, 1890, the day after Benton arrived in Washington, he visited the Smithsonian Institution and the Agricultural Museum with Aunt Sadie and his brothers James and Percy. That night he went to hear the imposing Gardiner Hubbard, president of the National Geographic Society, lecture about South America. In the following weeks, Benton made the rounds of Washington's public attractions, fashioning an education that far surpassed in variety, novelty, and inspiration anything he might have gained had he been enrolled in school. "School is a wicked word," MacKaye once wrote. "It might be defined as a place that boys like to run away from. (That would be my own boyhood definition.)"[30]

The Washington in which Benton roamed (and where he would later live for many of his adult years) was a city of modest proportions, still edged by woodlands and farms, where a boy who already knew and loathed the streets of Manhattan could feel at ease. "The present Mall was partly wooded, and 'thugs' were supposed to inhabit it at night," he recalled many years later. "There were fleets of 'safety' bicycles, and girls beginning to ride them. Public transport, of course, was by horse car, though 7th Street sported cable cars. Rich people rode in carriages, with high-hatted coachmen and footmen pulled smartly by 'spanking' docked-tailed horses."[31] At the time, one of the city's several railroad terminals stood directly on the Mall, so a youngster with a passionate interest in trains could conveniently observe the busy proceedings on his way to the nearby museums.[32]

Almost daily during that winter Benton visited the Smithsonian Institution and the United States National Museum adjoining it. The museums, for those Washington months, became his classroom and fantasyland. At the Smithsonian, he spent his mornings studying the collections of mounted birds and other wildlife. During the first days of the new year, Benton sketched the deer, horses, and bison grazing outside the museums, part of the small menagerie then maintained on the grounds in those years just before the establishment of the National Zoological Park. Indeed, one striking exhibit at the Smithsonian that year was a twelve-foot-long, glass-covered topographic model of "The Proposed Rock Creek Park." The dynamic map depicting the Washington park and the site of the new national zoo was one MacKaye never forgot. "There was the whole big idea. Folks gazed upon it and said 'amen.' Automatically it came to pass."[33]

Encouraged by his brothers, Benton began to pursue his studies at the Smithsonian in a more purposeful fashion. With his usual spelling lapses, he recorded this activity in his diary. "In the afternoon I went down to the Smithsonian and National Museum to see weather it would be aloud to use my camp stool, and weather it would be in peoples way but I got, very luckly, the satisfactory answer of yes I could use it."[34] Every day he set up his stool in the museum galleries, took out his sketchpad, and began to draw and take notes on the specimens of birds and other wildlife. The intense youngster soon became so familiar a figure in the museum's exhibition halls that a few staff members befriended him. They welcomed Benton into their laboratories, libraries, and other nooks off limits to the general public, such as the towers of the Smithsonian's main building, familiarly called the Castle. One day he was allowed to handle ornithological specimens. Another staff member referred him to a book that described the birds taxonomically, "and so I am going to take note of the birds as they are in the list."[35] Benton worked his way through various orders of birds, his notations becoming more detailed and precise. Upon his completion of the "Tub-Nosed Swimmers," he was led by his friend in the galleries to meet the museum's distinguished curator of ornithology, Robert Ridgway, who showed him a collection of birds recently arrived from Africa and introduced him to the young ornithologist who had collected the specimens.[36] On another occasion, he was introduced to C. Hart Merriam, who would head the U.S. Biological Survey.

Soon Benton began to frequent the "Crab Division," where he helped assistant curator James Benedict print labels and sort specimens in the department's extensive collection of mollusks. In Benedict's laboratory he made up some cards, "on which was printed, in quite small type, 'Collection of E. B. MacKaye,' for I mean to collect something this summer in the way of Birds, perhaps what Mammals I can get, Shells etc."[37]

By early April, the family was making plans to return to Massachusetts, where James had secured a position as secretary to the eminent Harvard geologist and geographer Nathaniel Southgate Shaler. Benton, granted a Sunday pass to the Smithsonian from his friend Ridgway, vigorously continued his work at the museums, memorizing the taxonomy and "careteristicks" of the ornithological collections. And he led his mother and sister on a tour of the Smithsonian, proudly introducing them to his mentor Benedict.[38]

On April 11, Benton had joined an audience of 750 other eager and curious listeners in the lecture hall of the National Museum. The ornate building was the setting for the first public lecture sponsored by the three-year-old National Geographic Society. And the speaker that day was the already legendary Civil War hero, explorer, and scientist, Major John Wesley Powell. The "bearded smiling one-fisted rockwhacker" cut a striking figure, especially in the eyes of an imaginative boy dreaming of future expeditions of his own. Powell had lost his right arm leading troops at the battle of Shiloh, a circumstance which only magnified the drama of the well-known adventure he was describing to his enthusiastic audience—his 1869 exploration of the Colorado River and the Grand Canyon.

Powell, director of the U.S. Geological Survey and the Bureau of American Ethnology, was one of the leading figures of the American scientific establishment. A man of extraordinary experience and eclectic learning, he attempted the daunting task of depicting in mere words the magnificent spectacle of the Grand Canyon. The explorer's original ideas about the dynamic geology of the American West had sprung from the unprecedented experience of riding the raging and uncharted Colorado River through the canyon; he had sensed firsthand the water's power to slice through the millions of years represented in the chasm's brilliant geological stratigraphy. Powell's epochal journey down the Colorado had been more than just another exploration into a previously unknown recess of the North American continent; it had also been a metaphorical journey into the depths of evolutionary time.

The voyage, as MacKaye vividly recalled more than three-quarters of a century after hearing Powell speak, "appeared to consist of dashes conducted at critical sections. Each savored of experiment." Benton cheered the explorer's heart-stopping account, as "one escape followed another." For Benton MacKaye, the heroic Powell, a "man of action and of thought," personified an approach to learning that was not confined to the laboratory and the library.[39] Instead, he studied natural phenomena out-of-doors and underfoot. Powell was also a reformer—just as Benton would become; he visualized the possibilities for a genuinely democratic American civilization adapted to the unique conditions of the nation's varied and still little-known terrain.

Near the end of his Washington stay, Benton was in the audience of a thousand in the National Geographic Society's Pennsylvania Avenue lecture hall to hear the young adventurer named Robert E. Peary describe his forthcoming Arctic expedition. At the end of the talk, Benton joined in the audience's shouts of "bon voyage" for the polar explorer.[40] If the scientific and historical import of such stirring proceedings may sometimes have gone over the boy's head, the excitement and romance they imparted did not. Nor did the daily routine of Benton's education end with his visits to the Smithsonian and Washington's lecture halls. His own family maintained a domestic environment that was equally high-minded and purposeful. Percy, destined to follow in his father's footsteps as a playwright, dramatized the family members in a series of skits which he called *Half Hour Happenings in the Teeles Family*.[41] Occasionally, Benton accompanied his brothers to the theater, or family friends visited for a game of whist or to invent the not entirely frivolous game, "how to get wealth." On other evenings, as they gathered in the gaslit 13th Street rooms, members of the family took turns reading aloud. Benton also listened as family members read from the works of Charles Dickens, John Burroughs, John Fiske, Charles Darwin, Herbert Spencer, and Thomas Huxley. Indeed, the MacKaye household was a veritable seminar on natural history and evolutionary philosophy, as propounded by the era's most popular writers in those fields.[42]

The MacKaye children, following the example of their father and encouraged by their mother, all went on to pursue ambitious literary and artistic ventures. Two decades after the MacKayes' winter sojourn in Washington, at the beginning of his distinguished journalistic career, Walter Lippmann described the essence of the family's endeavors. "The best argument I ever saw for an aristocracy of birth is a family which preaches democracy. . . . Theirs is a conspiracy to further the happiness of nations," wrote Lippmann, who came to know the MacKayes during his own undergraduate years at Harvard. "That is the MacKaye inheritance—plays, novels, acting, scientific research, 'fun, fishing, and philosophy' for all the world—a creative strain, strikingly like the criminal strain. It is in their blood."[43] Benton MacKaye, drawing upon this creative inheritance, would spend a long lifetime inventing his own professions and envisioning innovative social and environmental institutions.

❖

When Benton returned to Shirley in the late spring of 1891, he and his neighborhood friends Ned Stone and Warren Brown drafted the constitution for the Rambling Boys' Club. "The first object of the R.B.C.," they wrote, "is to give the members an education of the lay of the land in which they live, also of other lands, taking in the Geography, Geology, Zoology and Botany of them. Not only to know the Science of it but also the History and Progress of the different places."[44] The "first object" of the Rambling Boys' Club—"an education of the lay of the land"—was an uncannily precise description of Benton Mac-Kaye's lifelong vocation.

Another element of Benton's Shirley education, more important, at least in his own estimation, than the red brick schoolhouse on Parker Road, was Melvin Longley's barn. The Longleys had lived on the same substantial farm just south of the common since the eighteenth century. When Benton began frequenting the place at every spare moment, its owner welcomed the boy's company and enthusiastic assistance. For a youngster whose own, unpredictable father was usually absent, the good-natured masculine presence of Melvin Whittemore Longley filled a special need. One winter day, as Benton approached Longley's barn, he heard the farmer shouting. He "ran inside and found him alone, holding forth at the hayloft. He stopped short: 'I'm practicing a speech,' he said and went right on, roaring to shake the rafters." Longley was a paragon of Yankee yeomanry, in Benton's admiring eyes, a man able to quote Shakespeare or Longfellow just as easily as he could the current price for milk or apples, who honored his cows first with the speeches he would deliver the next day in the Massachusetts legislature. Benton, who for a time during his youth entertained the idea of becoming a farmer, relished the chance to drive Longley's manure cart or to sit for long, hot hours astride Sally, Longley's horse, as the corn crop was planted and cultivated. And the boy was never prouder than on the day of the big picnic to celebrate the end of haying season, when "Melvin fitted up his big blue wagon, filled it with every kid in town, and handed me the reins: No Roman hero ever reflected more glamor from his chariot." For Benton, as the years went on, Longley became the "symbol of a whole way of life—scenic, cultural and political" that was represented by Shirley Center, "the seat of a rural civilization in marked contrast to any other within America, and perhaps the world."[45]

❖

By the end of 1891, Steele had embarked on the climactic endeavor of his career, his ill-fated Spectatorium, which he predicted would be the most popular feature of the upcoming World's Columbian Exposition planned for Chicago in 1893. Until his death in February 1894, Steele spent all his time promoting and building his stupendous theater. Equipped with his remarkable collection of mechanical and lighting devices, the pleasure palace was designed to stage a giant six-act pantomime depicting Columbus's discovery of the Americas. The family waited hopefully for the promised fortune the Spectatorium would beget.[46]

Benton saw little of his father during these years. "You have been gone so long I would hardly know you from Adam," Benton wrote his father plaintively from Shirley in July of 1892. The melancholy boy also expressed to Steele a desire to "get a job in some telegraph office . . . till I got old and large enough to be a brakerman and thence conductor."[47] Benton grew up traveling on the railroads as the family shuttled about the northeastern states. When members of the family planned their journeys and rendezvous, they often turned to Benton to plan the least expensive and most expeditious itineraries. He romanticized the life of the railroadmen, like those on the Fitchburg Railroad. Engineers Jerry Cushing and Frank Wheeler sometimes let Benton ride in the locomotive on local freights in the years when, as he later recalled, "I thought of nothing else but 'railroad,' and the sun rose and set by the engine whistles (two long and two short) heard from my home in Shirley Center 2 1/2 miles north of the track."[48]

One day in the autumn of 1892, Benton and Percy rode the trunk line west from Shirley Village, beyond Fitchburg and Gardner and Greenfield to the Hoosac Tunnel, one of the dramatic engineering feats of the age. Five miles long, the tunnel through Hoosac Mountain had been completed in 1875 at the expense of 195 lives and $20 million. Yet it had made possible a direct route between Boston and the rest of the continent, and it dramatized to a student of the landscape how men's efforts could affect not just the flow of commodities but also the progress of human ambitions. "It was," Benton would write, "the product of a dream to connect, by the easiest grades, Boston Harbor with the Great Lakes and the 'Marvels of the New West' (meaning the rich Ohio farmland then being cleared of its thick hardwood forest)."[49]

Benton came of age in an era of transportation by railroad, horse-and-buggy, and—the craze of the 1890s—the bicycle. He would come to measure by his own experience the dramatic manner in which the automobile began to transform society and the American landscape in the early years of the new century. "I myself have seen the change from horse and mud to gas and cement," he later wrote.[50]

❧

During the winters of 1891/92 and 1892/93, Benton lived in New York with his mother at 55 West 19th Street and attended a small private school. Feeling restless and confined by his urban surroundings, he yearned to escape "that horrible place New York."[51] With some of his schoolmates he explored the city streets, and with Hazel, Jamie, and Percy he walked across the frozen Hudson River on a cold January Sunday. He also undertook a rigorous program of self-discipline and self-education. He described for his father, in Chicago, the daily regimen revolving around his conception of the "Cosmopolitan Organization," which Percy described as "a world-scheme for the abolition of all national barriers and prejudices."[52] The serious-minded thirteen-year-old was already laying out the program and the habits of his solitary, dogged, fiercely personal lifetime studies.

Benton reported to Steele:

> I told you last time of my plans for my daily life this winter and this time I can tell you that I finished this morning a book which I am going to study the four studies of Zoology, Drawing, History and Geography in.
>
> I first divided the earth's surface into twelve parts, which I call sections.
>
> I am going to put down in that book every Lowland Highland and division and everything I can find out about each.
>
> I shall also have a map of each section and after I have learnt what animals, plants, etc that live in each section I shall draw them from stuffed animals at the American Museum of Natural History or from the Menagerie at Central Park.[53]

He combined his love of railroads, his enthusiasm for geography and natural history, and the example of his Washington scientific mentors to conceive the "Cosmopolitan Survey." His detailed sketches depicted an elaborate one hundred–car train that would serve as a rolling university, laboratory, museum, and natural history excursion, traveling wherever the continent's railroad network reached. Benton's whimsical Cosmopolitan Survey was designed to take the nomadic scholar directly to the sources and scenes of natural phenomena.[54]

The spring of 1893, after his industrious winter in New York City, he returned to Shirley Center to apply his geographical methods to the "expeditions" that took him to Hunting Hill. His trenchant hilltop reflections may have been influenced by the grim news emanating from Chicago, where his father's Spectatorium had been declared bankrupt. Steele, whose health was suffering from a more serious ill than mere overwork, sent for his three youngest sons. For two September days, Benton, Percy, and James explored the Columbian Exposition's dazzling "White City," a carefully planned assemblage of neoclassical architecture emblematic of the era's City Beautiful movement of urban design. From a wheelchair pushed along by his sons, Steele showed the boys some of the more exotic attractions along the Midway Plaisance. He also led them out to the massive skeleton of his uncompleted Spectatorium on the edge of the exposition grounds. Silently, they ascended "an iron stairway that climbed steeply upward to nowhere." When they could go no higher, Steele, his voice quivering, said, "Boys, this is where it was to have been." The next day, the three returned east, never again to see their father.[55]

"To me, one's greatest object in life is to get interested in something and to learn all he can about it," Benton wrote his father upon returning to Shirley. "He can never exhaust knowledge on any subject, as the more he learns the more he sees there is to learn. The study of subjects which interest me, under such a light, is the greatest pleasure of my life to me, and the more I study them the more I enjoy it. I would rather work in this way than do anything else, so don't worry about me. . . ."

In his educational declaration of independence, Benton explained that "so far I like Physical Geography better than anything else I have tackeld yet."[56] And indeed, Benton's teacher in Shirley remembered sixty years later, writing to her one-time student, "Your pet interest of that time was geography and travel."[57] The first line of the first geography lesson in Miss Nutting's schoolhouse was etched permanently in MacKaye's memory: "The earth is the planet on which we live."[58] Benton, following his own interests in his own manner, had already arrived at the subject matter that he would pursue under various names—geography, forestry, regional planning, geotechnics—for the rest of his life.

In one of his last letters to his youngest son, Steele wrote, "The great thing at your time of life, is to realise that the life which you must eventually face among your fellow men is a battle, and the victory depends upon certain qualities of heart, mind and will, which have always been successful."[59] A few weeks later, Steele MacKaye was dead. Benton was not quite fifteen years old.

⚜

At the Grand Masquerade Ball at the Shirley town hall a few weeks before his father's death, Benton "graduated to long pants" and "a high stove pipe collar."[60] His more mature wardrobe reflected the new masculine role thrust upon him. His other brothers had either begun their careers or were enrolled in college. Now Benton was the man of the household, which usually included Aunt Sadie, his mother, and his sister.

The family gathered in New York for Steele's funeral services, but in the late winter of 1894 Benton returned to Shirley, along with Aunt Sadie. "Emile is so alone there, poor boy," his brother Hal wrote to his mother, who had remained in New York, "and he will be very likely to feel neglected unless much is done to make him feel that we love him."[61] Benton harbored his share of confused sorrow and anger at the loss of the father whom he had seen so little. His walks in Shirley were more somber and introspective than his geographic "expeditions" of the year before. Now his thoughts turned to "pondering on the Universe, so to speak, or internaly discussing life." He had momentarily "longed for" suicide, he confessed to his mother, but instead "made up my mind that the world will kill me, before it will forse me to do it myself."[62] As Benton turned fifteen, he considered how to prepare for the entrance examinations at Harvard, where his brothers James and Percy were both studying.

In September, Benton enrolled at the Cambridge Latin School, not far from the Harvard campus. Jamie and Percy lived at 35 Divinity Hall, on campus. Benton moved in with family friends, the Davenports, whose Watertown home stood just over the Cambridge city line, near Mount Auburn Cemetery. He estimated that it would take "two or three years" to complete his course of studies at Cambridge Latin; and he was able to secure a modest scholarship from Harvard. The Davenport home provided a convenient headquarters for Benton's program of self-education. He explored Cambridge and parts of Boston on foot, on his bicycle, and by trolley. He regularly visited his brothers at Harvard, sometimes accompanying Jamie to his chemistry laboratory, or Percy to hear the legendary Charles Eliot Norton lecture on art.

His diary from that school year depicts a somewhat lonely and desultory time, however. The Davenport children, who had been frequent visitors at Shirley, were older than Benton. He was already experiencing the stomach and intestinal troubles that would plague him for the rest of his life; he took the period of recuperation from one such attack to read Thoreau's Walden. His interest in school flagged. "I was late this morning," he noted in his diary, "as I intend to be hereafter as 'Brad' [the principal] let me off."[63]

However tedious Benton found his formal studies, his eyes were opened to other stimulating influences and ideas. Charles Davenport, the patriarch of the family with whom he lived, could still recall Lafayette's 1824 visit to Harvard. A successful businessman, Davenport had long dreamed of a plan for the residential and commercial development of the banks of the Charles River. The elaborate plans for what Davenport called his "povements" hung on the walls of an upstairs room. His scheme was a precursor of the movement that resulted in Frederick Law Olmsted's design for the "emerald necklace" of parks along Boston's creeks and rivers. For Benton, Davenport's plan suggested how a map could depict an idealized, but realizable, natural and communal landscape.[64]

Jamie graduated from Harvard in June of 1895 and began a career in Boston as a research engineer. As for his own studies, Benton learned that he had passed all of his courses and "was no. 5 in exams for the year."[65] The next school year was a happier one for Benton. With his mother and Hazel, he moved into 1 Berkeley Street, just outside Harvard Square, sharing a large house with two other friendly and interesting families, the Shermans and Whittemores. Some evenings the family would attend lectures by their neighbor, historian John Fiske. Other nights, gathering with the Shermans and Whittemores in what they called the "Tactless Club," the families played charades and improvised other entertainments.[66]

Benton's situation at Cambridge Latin was less satisfactory, however. He joined the football team ("I made a damn fool of myself and that is about all"), and he found some of his teachers "very rank." He talked to the principal about reducing his course load, but his request was denied. By December, he had dropped out of school, having decided that he could prepare for the Harvard entrance examinations on his own.[67] At the end of that month, he stopped keeping the diary he had carefully maintained since 1890. He may have remained in Cambridge with his mother and sister during the winter and spring, but the family returned to Shirley for the summer. In an arboreal study he constructed in a maple tree overlooking the Shirley Center grammar school, Benton set to work reading from Harvard's list of specified books, which included *A Midsummer Night's Dream*, Defoe's *History of the Plague in London*, and Longfellow's *Evangeline*. He also looked to Percy for help in literature and languages and to Jamie, during his visits from Boston, for assistance with algebra, geometry, and physics. And his brother Hal, at a distance, worried about the ultimate consequences of Benton's approach to his studies. "Poor E.B.!" he wrote to his mother. "I am afraid he may hurt himself studying too hard. It is as bad to be too ambitious as not to be ambitious enough, isn't it? It would be a joke wouldn't it, if Emile could overlap into Percy's stay at Harvard as Percy did onto Jack's, and the same room 35 remain in the family for four years more."[68]

Benton barely squeaked into Harvard; it would be another two years before he made up "deficiencies" in German, French, algebra, and physics. "I am greatly in fear that [Benton's] success in entering school after a course of study avowedly for the sole purpose of cramming and not for information, may give him the habit of shirking until the last moment—a habit that will be ruinous," Hal wrote to Jamie two days after Benton registered for Harvard. "I am disappointed thus far in observing in [him] no ambition higher than 'getting through' certain tasks. For Heaven's sake let us see something better there, if possible. Let his ambition be what it will—it would be better than none."[69] Benton's approach to his work and his studies was fixed by the time he entered Harvard, and it would never really change. He became, in his own word, a "humper," intensely pursuing his own unorthodox interests, while others sometimes wondered just what it was he was doing or talking about. But his brother need not have worried: he was a MacKaye, after all. Benton, it would become clear, was never to be without ambition.

From Harvard Yard to the
"Primaevial Forest"

1896–1903

Every year, Harvard geologist William Morris Davis began his first lecture in Geography A, "Elementary Physiography," in the same dramatic manner. Standing before the students assembled in his Agassiz Museum classroom, Davis held in his hand a six-inch globe.

"Gentlemen," the eminent earth scientist intoned, "here is the subject of our study—the planet, its lands, waters, atmosphere, and life; the abode of plant, animal, and man—*the earth as a habitable globe.*"

"That was my guidepost No. 1," MacKaye wrote more than a half-century after hearing Davis's lecture in 1898, "pointing me toward no mean field of work." From that later vantage point, MacKaye recalled Davis's exhortation as the clue to a vocation, but during his college years, the earnest young scholar never arrived at a firm idea of just what his life's work would be.[1]

When seventeen-year-old Benton registered as a Harvard freshman on October 1, 1896, he was no stranger to the venerable environs of Harvard Yard. Not only had he lived and studied nearby for the previous two years, but he already belonged to a Cambridge network of family and friends. His brother James, who had graduated from Harvard in 1895, still lived nearby, while working in a series of minor and frustrating jobs at the university. And Percy, having already made a mark for himself as a poet and playwright, was beginning his senior year.

Turn-of-the-century Harvard was in its "golden age," as the era came to be called with the university's customary air of self-satisfaction. That period, historian Samuel Eliot Morison noted, "marked the zenith of undergraduate liberty at Harvard."[2] Under the long presidency of chemist Charles W. Eliot, the college had grown steadily, from 570 students when Eliot took over in 1869 to 1,754 in Benton's first year. The institution had gradually thrown off many of the moral and disciplinary shackles that were the heritage of its Puritan origins. Chapel attendance was no longer mandatory; class attendance requirements had been loosened. More important, Eliot had pioneered establishment of the elective system, which would become a standard of American higher education. Eliot's Harvard College offered students a broad, liberal education rather than significant professional training.[3] As it happened, the Harvard of the late 1890s was the ideal place at the ideal time for a student of Benton MacKaye's interests, aptitudes, and peculiar brand of self-discipline.

As the youngest of the MacKaye boys, Benton perhaps felt less keenly than James and Percy the loss of wealth and status which their prodigal father's undisciplined habits and finances had brought upon the family. Harvard's laissez-faire approach to the lodging of its young scholars only exacerbated the social and economic differences between students. Tuition was a $150 for all students, but the living quarters offered to Harvard gentlemen varied in spaciousness and splendor depending upon ability to pay. Along the plummy "Gold Coast" of Mount Auburn Street, developers had recently built plush "private dormitories" for wealthier undergraduates. Across Harvard Yard, on the other hand, at the northern fringe of the campus, stood more spartan digs, where Percy and James had resided, among the least expensive offered to undergraduates. As his brother Hal had speculated, Benton followed his brothers by taking the very same room, at 35 Divinity Hall, during his four college years.[4]

The social chasm that yawned across the Yard was a sometimes harsh fact of Harvard life. Benton could not expect an invitation to join the college's social clubs, such as the Institute of 1770, the Hasty Pudding, or one of the exclusive "final" clubs, like Porcellian, A.D., or Fly. His athletic ambitions were thwarted when he failed to make the freshman football team. And he lamented his inability to scrape together the few dollars necessary to join the Weld Boat Club. Hence, much of Benton's social life revolved around family friends who lived in the Cambridge area, such as the Hildreths and the Davenports. In fact, he defrayed some of the expense of his first three college years—and contributed to his mother's support—by tutoring and reading for his friend Horace Hildreth, who needed the help, due to weak eyesight.[5]

Harvard Yard was a quiet sanctuary at the heart of the growing, industrial city of Cambridge. This evocative setting was "distinctly Bohemian," according to philosopher George Santayana, then on the faculty, and stimulated those most talented of Harvard students on their passages to accomplishment and fame.[6] Gertrude Stein, Wallace Stevens, Robert Frost, Frank Norris, and Herbert Croly, for instance, were among the undergraduates destined for literary renown who were on campus during Benton's undergraduate years. However bohemian the campus atmosphere, though, Benton shared with a majority of his 477 freshman classmates a Protestant New England upbringing. Jews, Catholics, and blacks were sparsely represented among the student body.[7]

The Harvard faculty then comprised an exceptionally self-confident group of scholars during a confident era in American history. Some were already legendary, others just beginning to carve out their niches in university folklore. Such figures as philosophers Santayana, William James, and Josiah Royce, art historian Charles Eliot Norton, and English professors Barrett Wendell, George Lyman Kittredge, and Charles Townsend Copeland ("Copey") "were trying, more or less blatantly, to make converts to their own view of the world," according to a later Harvard historian.[8]

Benton's self-designed crash program for passing the Harvard entrance examination had not entirely prepared him for the scholastic rigors of his freshman year under the tutelage of such powerful personalities. As might have been expected from a student whose family had attained a measure of literary distinction, Benton fared reasonably well in his freshman English composition course, under A. S. Hill, Le Baron Russell Briggs, and the colorful Copeland. That year he also studied English literature and constitutional government. He scraped by in French, failed Spanish, and ended his freshman year on probation because of his lackluster academic performance. Dean Briggs, reporting to the struggling student's mother, attributed Benton's poor grades to "his hasty, short-cut preparation for College and the natural reaction after a great effort to get in."[9]

One of his freshman courses, Geology 4, had opened an intellectual door for Benton, however. Nathaniel Southgate Shaler's "Elementary Geology" was the most popular course in the college, "the first great gut," providing "all the geology necessary to a gentleman." The imposing, white-bearded professor was an eloquent, colorful, and entertaining lecturer, who commanded the strict attention of the 300 students with whom Benton gathered at the Fogg Museum lecture hall. An influential and representative figure of late-nineteenth-century American science, the venerated geologist-geographer was cut from the same intellectual cloth as John Wesley Powell and the Smithsonian researchers MacKaye had met in Washington a few years earlier. A native of Kentucky who had fought with the Union Army in the Civil War, Shaler had been a student or teacher at Harvard since the late 1850s. He had studied and worked with the Swiss-born zoologist and glaciologist Louis Agassiz; so he had been on hand in the 1860s during the great battle over Darwin's theory of evolution, which had roiled Harvard's academic waters, pitting Agassiz, a leading opponent of Darwin, against the equally respected botanist Asa Gray. Shaler remained on good terms with his mentor Agassiz, but was soon won over to the evolutionary doctrine, even as he attempted in his own fashion, as one biographer observed, "to integrate nature, humanity, and God."[10]

Determined to bridge the gap between the scientific community and the public, Shaler was an effective popularizer, as well as an energetic advocate of social reforms based on applied science. Books and articles on a sweeping variety of subjects, including poetry and historical fiction, flowed from his pen at awesome speed and length. An outspoken advocate of the ideas of the Vermont polymath George Perkins Marsh, a personal acquaintance and professional colleague of John Wesley Powell and forester Gifford Pinchot, Shaler shared the prevailing utilitarian aims and perspective of the nascent American conservation movement. He strove to shape a society that was at once efficient, democratic, and in balance with natural forces. Shaler's concerns about the social implications of transportation networks and technology would be echoed in Benton MacKaye's career. On the threshold of the automobile age, Shaler advocated, in his book *American Highways* (1896) and in articles, the construction of an extensive, well-engineered road network, to foster a "real communal life" and a more prosperous economy.[11]

A "poet in the world of science," Shaler also addressed issues such as leisure and solitude which would later preoccupy MacKaye. The art of landscape study, the geographer asserted, was an antidote to the speeded-up pace and constricted purpose of life in an ever more industrial society, enabling an individual "to set himself against the spirit of the age." MacKaye, in his own preservationist polemics, would one day bring his own interpretations to Shaler's claim that the "contemplative attitude demands solitude, or at least a mental isolation from our fellow men."[12]

If Shaler provided the model of the scientist as generalist and reformer, William Morris Davis was the theorizer and technician. Davis, who earned bachelor and masters degrees at Harvard, had joined the faculty under Shaler in 1877. The ancient, eroded, knotty structure of the Appalachian mountains, as interpreted by Davis, provided a central paradigm of American geology and physical geography for decades. Davis's grand theory of the "geographical cycle" had already made a profound impact on the field of geology by the time Benton, as a sophomore, entered his classroom in 1898. Using three variables—"structure, process, and time"—Davis, a dedicated evolutionist, set out to develop what he called a "genetic classification of land forms," a theoretical framework that relied on organic metaphor to explain the creation and condition of the natural landscape.[13]

Davis's theory was that, like biological species, landforms such as mountains and rivers each had a "life history," passing through stages, which he described as "young," "mature," and "old." The cycle of erosion began with the uplifting of land masses and the creation of mountain ranges. Mountains were gradually eroded, over the eons, by the forces of water, gravity, and weather, until reduced to a "peneplain." Then the cycle began anew, as the peneplain was lifted up and dissected by rivers. Davis pointed his students to the nearby landscape of central Massachusetts and southern New Hampshire, Benton's home terrain, for a classic example of the late stages of the cycle. From any high point in the region, an observer could notice that the gentle granite and schist hills in fact melded into an almost level and uniform topography. The vista was occasionally broken, however, by taller projections of harder, less erodible rock. Davis borrowed the name of such an eminence, the solitary 3,165-foot Mount Monadnock, visible over the New Hampshire border to the northwest of Shirley Center, as the generic term for such a geologic remnant—a *monadnock*. Although Davis had probably coined the term some years earlier, Benton liked to recall that his professor had consulted him about its use and appropriateness. In any case, the original Monadnock became a key vantage point for MacKaye's own dynamic perspective of the regional landscape.[14]

Benton absorbed Davis's theories, techniques, and his grand perspective of "the earth as a habitable globe." He earned his best grades, B's and his single A, in earth science courses. For the "forensic" in his junior English composition class, MacKaye drew from his work under Davis to write a long thesis on the glacial origins of the sand plains of eastern Massachusetts. Studying the physiography of Europe under Davis during his senior year, MacKaye produced a tightly argued explanation of the evolution of river valleys in England; the paper was profusely illustrated with his detailed and colorful geological maps. Throughout his life, MacKaye's principal source of reference on matters geological and geographical was a densely annotated copy of Davis's elementary 1899 text, *Physical Geography*. Even as twentieth-century geological science itself evolved, Benton remained satisfied with the metaphors, images, and techniques he had been taught by Davis, who had described "a historically dynamic landscape, in which landforms were continually dissolving before time."[15]

Benton looked forward to the weekly evening "geological conference" held in the basement of the Agassiz Museum, where a handful of students could mingle with the professors in Harvard's well-regarded geology department and listen to debates about current theories of geology, evolution, and other scientific fields. (Among those who attended the conferences with Benton was the occupant of the room next to his in Divinity Hall, Alfred Brooks, the eponymous future geological explorer of the great Alaskan mountain range.) "The big show of the evening was the ever-ready bout of Shaler versus Davis," Benton later reminisced. "Either the two profs did not agree on a certain theory or else Shaler saw to it that there be enough difference to stir mirth among the gang. It was a great bearded grizzly bear versus a sharp-bearded, bead-eyed, well-perched catamount." Under Shaler and Davis, MacKaye was indelibly imprinted with the gospel of evolution, a sense for the depths of geological and cosmological time, and an appreciation for the vast scope of geographical studies. "From these men I imbibed the notion of the earth as a habitation in astronomic space, and the life thereon as a family dating from evolutionary time."[16]

✧

Inspired by such magisterial earth scientists, Benton was ready for more ambitious explorations of his own. At eighteen, after his freshman year, he set out on the first, and perhaps the most memorable, of the hikes he would make in the mountains of northern New England. On the afternoon of August 14, 1897, he waded across the Swift River just north of James Shackford's Passaconaway House, a modest New Hampshire hostelry that was a landmark for hikers in the remote Albany Intervale, fifteen rugged miles west of the town of Conway. With two Harvard companions, Draper Maury and Sturgis Pray, he began a hike that he would recall as the time he "first saw the true wilderness."

"It was a journey of Ulysses," he remembered thirty-five years later. "I graduated from dung to spruce—from the tang of the barnyard to the aroma of the virgin smoke amid the far night roar of Bolles Brook tumbling from Passaconaway."[17] MacKaye, Maury, and Pray had set out by bicycle from Shirley ten days earlier, riding their "wheels" north over rugged backroads through the picturesque hill towns of southern New Hampshire. "The moonlight on the wild and weird landscape of the Swift River valley was something which I had never seen before and was as beautiful as anything I ever expect to see," Benton recorded in his diary upon their nighttime arrival at Shackford's.[18]

The White Mountains were a revelation to the wiry, observant, energetic young man, who characteristically kept a vivid and meticulous record of his trip—in a journal, letters to his family and friends, and even a six-chapter semifictionalized account that he later wrote for his Harvard composition class. "The country we are about to traverse is one, I am told, undisturbed by civilization in any form," MacKaye noted on the morning he began his hike.

"Besides us, only the bear, deer, catamount, porcupine and coon disturb the solitude of the primaevial forest. No longer will women folk look shocked or embarassed as they did when they met me clothed in a sleaveless undershirt or stripped to the waist. We have said 'good-bye' to the bicycles and civilization and will now pursue our way on foot through the White Mountains."[19]

Benton's introduction to the White Mountains came at a climactic moment in the region's history. While the awestruck student, taking in this northern terrain for the first time, saw undefiled wilderness, other observers, distraught, bemoaned the rapid destruction of the forest under the onslaught of voracious logging companies. Until the late 1860s, most of the land making up the White Mountain range had been owned by the state, but the reviving economy of the post–Civil War years and the emergence of the railroad made the previously unattractive and inaccessible timber resources of the mountains more alluring. The state sold its vast holdings for a pittance. By the 1870s, private logging rail lines had begun snaking through almost every notch and along every valley in the mountains, carrying out logs for lumber and pulp at an accelerating rate. During the decade of the 1890s, in response to the demand created by northern New England's rapidly growing paper industry, the value of pulpwood harvested in the state multiplied sixfold. A 1904 report prepared by federal forester Alfred Chittenden rated New Hampshire "the most intensively lumbered state" in the country.[20]

Mountain slopes denuded by clearcutting, uncontrolled forest fires set by cinder-spewing locomotives, unsightly piles of slash and rubble left by loggers, sudden and devastating downstream floods—such were the disturbing effects a growing number and variety of critics laid at the feet of the "wood butchers" who were transforming the face of the mountains. An unlikely coalition began to agitate for greater public control of the region's forest resources: Influential textile-mill owners in Manchester complained about the increasingly unpredictable flow of the Merrimack River. The entrepreneurs of New Hampshire's growing tourist business worried that the logging companies were ruining the vistas and pastoral atmosphere that attracted vacationers to their hotels, inns, and boarding houses. Artists and academics who summered in the mountains lent their voices and talents to the growing clamor for public protection of the region's mountains and forests. In the 1880s, historian Francis Parkman, who had explored the mountains in the early 1840s, and Harvard botanist Charles Sprague Sargent were among the early voices calling for stricter regulation of forestry practices or public ownership of forest lands. By the 1890s, crusading journalists and local pamphleteers were attacking the timber kings who by then owned most of the White Mountain forests. "The glory of the mountains is departing, and the mountain-lover mourns," lamented Benton's future professor A. B. Hart in *The Nation* in 1896. Investment in roads, paths, and inns for the tourist trade would provide a sounder, more stable basis for the region's economy, and for its natural and scenic resources, Hart argued, than the cut-and-run practices of the loggers.[21]

Not the least active and influential among those seeking to salvage the mountain landscape were the growing number of hikers heading into the hills for recreation and adventure. Trampers and climbers had been traversing the White Mountains and blazing paths there since before the Civil War. The Appalachian Mountain Club, organized in 1876 by a group M.I.T. and Harvard professors, was one significant indication of the increasing popularity of the rigorous pastime MacKaye was enjoying for the first time. The club's charter membership of thirty-four had grown to almost one thousand by the late 1890s.[22]

So the "primaevial forest" Benton entered on the north bank of the Swift River was not so undisturbed as he imagined. Nonetheless, the profiles of 3,384-foot Mount Tremont and its companion, Owl Cliff, presented a rugged prospect to MacKaye and his fellow refugees from civilization. Rising to the north of the Swift River across the broad, flat Albany Intervale, these mountains were distant enough, wild enough, and solitary enough to quicken the imagination of a young man bent on adventure.

Benton and his companions found a route north to Mount Tremont, then over Mount Lowell and Mount Anderson to Crawford Notch. At the Crawford House, a popular resort hotel, they were joined by another friend, Rob Mitchell. An attempt to climb to Mount Washington by way of the popular Crawford Path was aborted by torrential rains. So they retraced their steps to Crawford Notch, rode the train to Gorham, and ascended Mount Washington from the Glen Road. After a night at the Summit House, they hiked south along the Presidential Range to Crawford Notch, caught another train to Franconia Notch, made their way south to North Woodstock, and returned east along the East Branch of the Pemigewasset and the Hancock Branch, the domain of the J. E. Henry and Sons lumber operation. The "happy but fagged" young men completed their challenging circuit of the mountains by climbing over Osceola and Tripyramid mountains on their way back to Shackford's. They had covered a large portion of the terrain that would later, with MacKaye's modest assistance, be incorporated into the White Mountain National Forest.[23]

But it was MacKaye's experience atop that first modest mountain—Tremont—that forever changed the way he responded to the world around him.

"A second world—and promise!" he wrote his close friend and fellow Wilderness Society founder Harvey Broome in 1932, recalling the occasion. During their approach to Tremont, the hikers spent a full day traversing a blowdown of trees, "the most horrible thing on earth," "crawling like snakes" through "the trees which lay like jack straws interwoven with bushes and briers."[24]

"We found ourselves in a cloud with the rain coming down like 'pitchforks,'" he wrote of their progress up the mountain. "The scene was beyond description," as "the thunder kept up an incessant roar." As the three climbed toward the summit, "the lightening struck within a few rods of us, almost giving us a shock."[25] In darkness, during a lull in the storm, the hikers built a rough shelter on what they believed to be Tremont's summit. Then they were enveloped in another thunderstorm. "We were soaked, all our things were soaked, the shelter inside and out was soaked and still the rain poured." After midnight, when the storm abated, they tried to sleep, but Benton awoke "cold and wet" well before dawn. "I started a little fire, got my shoes on and was up on our summit just as the sun was rising," he recorded. "The first thing I discovered was that the true summit was about 1/4 miles beyond ours. It was about 4:30, the clouds were in the valleys below and a brisk cool breeze was blowing from the southwest. The grandest sight I ever saw was now before me, nothing but a sea of mountains and clouds."[26]

It was that moment, as the sun rose and the weather cleared, that MacKaye would always remember as one of revelation, perhaps the closest thing to a religious experience that he ever mentioned or recorded. The vista from Mount Tremont that misty, mystical August dawn in 1897 made "the cold chills go up your back," MacKaye reported to a Shirley friend:

I saw the sun rise over the mountains making one side of them day and the other, night. Several hundred feet below in the valleys there were white clouds and occasionally parts of them would rise and chase themselves by me.

On one side Mt. Washington would emerge now and then from a cloud, on another, to the northwest, the Franconia range was seen through Carrigain Notch; to the northeast I could see over into Maine, and on the south away in the distance I could make out the hills of old Massachusetts. I felt then how much I resembled in size one of the hairs on the eye tooth of a flea, to use a vulgar expression.[27]

Later that morning, when the three climbed to the "true summit," they were surprised by a black bear "about 10 rods down on the rocks," then awed as they "saw two eagles soar through the air and over the valleys." The events of that day, he observed in his diary, "have been great experiences for us." He climbed grander peaks and gazed over broader vistas during the next two weeks; but the rest of his trip was something of an anticlimax after this initiation into the thrilling and humbling power of the wilderness.[28]

By the time he graduated, MacKaye had augmented his studies under Shaler and Davis with courses in other physical sciences, including astronomy, meteorology, oceanography, and glacial geology. Under Albert Bushnell Hart, he studied the constitutional and political history of the United States from 1781 to 1868, an era he would return to repeatedly in developing a rationale for reorganizing the nation's political framework along what he saw as more rational biogeographical grounds. For a course on U.S. economic history, he prepared a lengthy paper on the history of the nation's transportation system.[29]

One practical element of his undergraduate training that served Benton well was Harvard's rigorous training in literary composition. He later recalled English 22, sophomore daily themes, as "the best downright *practice* in writing that I had."[30] Encouraged to experiment by taskmaster Lewis Gates, Benton gained confidence in his writing style and voice. In his themes he inveighed against the horrors of suburban life, the desecration of forests, the poor condition of country roads, and other subjects that would remain lifelong vocational interests. For one paper, he described with chagrin the changes taking place on the cross-section of landscape, from bustling city to peaceful countryside, that he traversed on his trips by train and bicycle between Cambridge and Shirley Center:

> Although the growth of cities can be accomplished perhaps in no other way than by the institution of suburbs, the last should be regulated, it seems, with more care than is now taken. Instead of pushing further into the country and turning farms, not into parts of real cities, but into a deplorable "half-way" condition, why not spend the time and energy in building up what is already in a "half-way" state? Give me noisy streets, heaving crowds, and egotistical policemen, rather than a country of smoky sky, with oozy ground and manure flavored atmosphere known so terseley [sic] as "Lonesomehurst"![31]

As the end of his senior year approached, Benton was indecisive about his career plans. During the winter, he had been confined to his chilly Divinity Hall room for weeks with an attack of sciatica, followed by the flu. But he arranged to take his mid-year exams in his room and to have his meals and coal for his fireplace brought to him. By spring, despite his poor health and uncertain future, he was feeling sanguine and content about his experience at Har-

vard, especially his studies in geology. Benton's immediate problem as com-
mencement approached was his unpaid tuition bill—a difficulty overcome
with the grudging last-minute assistance of his aunt Emmie von Hesse.[32]

Although never an exceptional student, Benton had fashioned a Harvard
education that allowed him to pursue his own interests, while providing a
sound and diverse basis for his later work in forestry, regional planning, and
"geotechnics." His course of study did display gaps and prejudices that were
the product of his autodidactic inclinations. Since his solitary studies at the
Smithsonian and his Shirley "expeditions," he had, almost intuitively, envisioned his calling and his method, but it would take some time for him to
fashion a career. The family's finances did not permit him the indulgence of
leisurely graduate study or a grand tour: Benton would have to get a job. Percy,
who had spent several years teaching in private academies in New York City
while trying to establish himself as a playwright, helped him locate a teaching
position in New York. Out of necessity, but without much enthusiasm or
sense of purpose, the young Harvard graduate prepared to return in the fall of
1900 to the city he had only recently described as "a joyous one to leave."[33]

Before buckling down to the responsibilities of adulthood, Benton that
summer of 1900 embarked on a footloose northcountry adventure. In mid-
July, with his brother Percy and his friend Horace Hildreth, he boarded a train
destined "upcountry" from the Shirley Village depot. Headed for Vermont,
their goal was to climb many of the state's highest mountains, beginning in
the south and working their way north. "By back road, cart path, and sheer
bushwhacking," MacKaye and Hildreth (Percy departed halfway through their
trip) followed the Green Mountains north, over the summits of Stratton,
Haystack, Bromley, Killington, Camel's Hump, and Mount Mansfield, Vermont's highest peak. Such a trek was no mean feat in 1900. Although paths
led up many of these mountains, the Long Trail, the 265-mile footpath along
the main Green Mountain ridge from Massachusetts to Quebec did not yet
exist; it was not proposed until 1910 and was not completed until 1930. No
hikers' shelters existed along their sparsely settled route, so the trampers
camped along the road and in the mountains, or relied upon the courtesy of
none-too-prosperous farm families willing to provide food and a place to
sleep.[34]

MacKaye's 1900 Vermont excursion—"our Classic of the Green Mountains," as Hildreth later described it—was only the latest chapter in a variety of
educational experience that contrasted dramatically, in both environment

and method, with serene and historic Harvard.[35] Beginning with his exhilarating White Mountain adventure in 1897, and for every summer thereafter until he entered the Forest Service in 1905, Benton spent his vacations exploring New England's mountains, forests, and rural byways. He returned to the Swift River valley again for several weeks in 1898. Accompanied by his brothers James and Hal, he rendezvoused at an abandoned logging hut on the slopes of Mount Passaconaway with Pray and Mitchell, who had previously discovered and renovated what they dubbed "the camp of the 'Bloody Bikers,' or B.B. Camp." The next summer, 1899, Benton explored other corners of New England. He spent part of the season on a farm near Brattleboro, Vermont, investigating the terrain of the West River valley while riding a flatbed car on the narrow-gauge railroad that wended its way into the hills. Toward the end of that summer, he traveled alone by bicycle along the unpaved roads from Springfield, Massachusetts to New York City.[36]

His spartan northcountry expeditions were partly an enjoyable means to avoid draining the family's meager financial resources. More important, though, his travels opened up new physical, emotional, and vocational possibilities. Rambling across New England, Benton absorbed the terrain, creating mental maps of a democratic landscape, open to rovers and walkers, unsullied by industry and commerce. Benton's boyhood expeditions in the countryside around Shirley Center and his enthusiasm for the stories of explorers like Powell, Peary, and Humboldt had whetted his appetite for experiencing wild geography. At Harvard, he became acquainted with other enthusiastic and experienced young outdoorsmen, who introduced him to New England's mountain ranges, and to the skills of camping, climbing, and wilderness routefinding. As much as his formal academic studies, MacKaye's backcountry excursions during these years sowed in his fertile imagination the seeds of ideas about long-distance trails and wilderness preservation.

The two leading figures of the young, growing, and already divergent American conservation movement—John Muir and Gifford Pinchot—had visited Harvard briefly at either end of Benton's undergraduate years. Muir, the writer, mountaineer, and Sierra Club founder, was awarded an honorary Harvard degree in 1896, in recognition of his literary and political achievements in opening the eyes of Americans to the precious beauty of their wild lands. And in the spring of 1900, Yale man Pinchot, the wealthy and charismatic chief of what was then the U.S. Division of Forestry, delivered a talk titled "Forestry as a Profession." Benton attended the lecture presented by his future boss, but he did not immediately hear the vocational call to forestry. Nevertheless, the

concerns that Muir and Pinchot so passionately—and sometimes conten-
tiously—expressed about the fate of America's forests and wild lands struck a
chord with Benton, harmonizing with his own observations of changes
throughout New England.[37]

"New York is as certainly the first place to get 'Mon,' as it is the last place to
live," Benton had once written his brother Jamie, "and my scheme now is to
use it for the first as much as possible, and for the second just as little as can
possibly be."[38] After his tramp through Vermont during the summer of 1900,
MacKaye took up his duties as a tutor of schoolboys from several well-heeled
Manhattan families. During the winter and spring, however, he fell ill. His
"rheumatism" brought on a case of iritis, an eye inflammation that kept him
confined to darkened rooms at his mother's apartment on West 101st Street
for several weeks. His convalescence provided an opportunity to think more
carefully about his future and his unsettled vocational status.[39]

Always attuned to his youngest brother's interests and ambitions, James
had sent Benton a set of elaborate geological maps for his twenty-second
birthday. In a birthday letter offering career advice, Jamie also urged Benton
not to be one of those who "lose sight of the end in their effort to attain the
means."[40] Benton's lengthy response sketched out his vocational future with
prophetic precision. "As you know, I have always been a crank on the subject
of maps, and always expect to be," he wrote. "They mean more to me than verbal descriptions by far, and much more than pictures." Tutoring, he confessed,
was a temporary and confining means of making a living.

> But once more, what is the career? . . . What is the problem of the day which I
> want to tackle? It is the problem of distribution, (both of men and his neces-
> sities), and embraces the questions of transportation, of natural resources, of
> "Lonesomehurst" [his derogatory codeword for suburban life], of the sacrifice of
> the country's good soils to the ignorant and the covetous. To the mechanism of
> that industry which connects all others, and includes all the above problems and
> subjects, to the industry of transportation (railroading), I mean to give my atten-
> tion. But only the whole problem in its greatest breadth do I mean to pursue, and
> as you say about your Thermodynamics, I want to learn the whole thing in all its
> branches "from A to Z." . . . A career which takes me over the country, out of
> doors, and into all communities and places, is the one for me; it is a scheme to be
> developed and difficult to be attained, but one toward which I shall "hammer,
> hammer, hammer." Such are my vague ideas for the future.[41]

Only gradually did Benton's vague vocational ideas take concrete form. He worked two more years in New York as a tutor, but he grew restless and plaintive about his prospects. He contemplated writing an elementary geography text, toyed with his own design for railroad timetables, and, his mother observed, "was in a chronic state of towering passion over the New York prices for bed and board."[42] "We are in another generation from Poog [Percy's nickname]," he wrote to his sister Hazel as he looked ahead to another year of tutoring, "and I do wish that we could do something to declare our independence."[43]

His customary northcountry explorations brought relief from such frustrations. With a friend, he canoed a 50-mile stretch of the Connecticut River from Fairlee to Bellows Falls in early June of 1901, paddling downriver with the log drive. "Big John," boss of the rivermen, was "the 'he-ist' man I ever met," MacKaye remembered. According to river lore, Big John had once ridden a log over Bellows Falls and was said by his men to "be scared not of 'God, man or devil.'"[44]

Benton spent the summer of 1902 as a counselor at Camp Penacook, a boys' camp on a small lake at North Sutton, New Hampshire. His brother Hal had introduced him to the camp director, a legal client with the evocative name of Virgil Prettyman, who was also principal of New York's Horace Mann School. Benton pursued with zest his duties as a leader of youth, in an era when northcountry summer camps for "gentlemanly, well-bred boys," as Camp Penacook's brochure read, were becoming increasingly popular. Besides leading hikes up nearby Mount Kearsage, MacKaye sought to teach his young charges the rudiments of surveying.[45]

That fall, Benton rendezvoused with Sturgis Pray, Pray's new wife Florence, and a female companion, possibly a family friend named Mabel Marks, at Shackford's in Albany Intervale.[46] Pray was pursuing his duties as the Appalachian Mountain Club's Councillor of Improvements, and Benton assisted him with surveying and maintaining nearby trails. His 1902–1903 school year in New York was enlivened by the opportunity to study law at Columbia University. As a tutor for a blind student at the law school, Benton accompanied the student to classes and read the course texts.[47] Benton briefly considered completing legal studies; indeed, when he left New York in the spring of 1903 he assumed that he would be returning to Columbia in the fall.

His summer plans were up in the air. His job as a counselor was unavailable. Virgil Prettyman had moved farther north to build a new camp that summer, at a site on Upper Baker Pond, just south of Mount Moosilauke, which Benton had helped him choose the year before during a horse-and-buggy reconnaissance trip.[48] Nonetheless, the summer proved eventful in several ways. He spent part of the season working again with Sturgis Pray, "the Moses who was to lead me *into* the wilderness."[49] Pray was laying out a trail for the Appalachian Mountain Club (AMC) along the Swift River from Sabbaday Brook to Kancamagus Pass—roughly the route of the present-day Kancamagus Highway. Pray's Swift River Trail was a landmark in the history of the White Mountains recreational landscape: it linked for the first time the trails of the Sandwich Range with those of the East Branch of the Pemigewasset River and the Franconia Range. The trail was a harbinger of a new era of hiking activity, which emphasized connecting local trail networks—previously radiating from resort hotels and railroad stops—into one grand network. Pray, after his graduation from Harvard in 1898, had gone to work with the Olmsted Brothers landscape architecture firm, quickly establishing a reputation in the popular new profession. In 1902 he started teaching at Harvard, where he would eventually head the university's landscape architecture program; and he is credited with being the first to offer a course in city planning in an American university.[50]

Pray was also a vocal and influential figure in the New England hiking community. He urged his fellow AMC members to take a more systematic and comprehensive approach to the development and maintenance of trails. Specifically he proposed that the AMC construct and maintain major "trunk" trails, like his Swift River Trail, "connecting one centre of interest with another. . . . Such paths should always lead from one settlement or quasi-highway to another, and if properly chosen these lines will effectively cover the whole White Mountain region." In Pray's scheme, individual trail clubs would maintain "local sets of paths, sometimes leading to features of particular local interest, and always to one side from the most convenient through routes." Pray worried, however, that the AMC's efforts to build and maintain trails were often being undone by the activities of the logging industry—a problem, he accurately predicted, that would only be solved by "the adoption of more far-seeing forestry methods" or "the inauguration of national control."[51]

Sturgis Pray's personal and practical influence on MacKaye's ideas about hiking trails cannot be overstated. Not only did Pray introduce MacKaye to the techniques, rigors, and outright fun of mountain tramping. He also enunciated the standards and principles of trail design that Benton unequivocally adopted. Pray brought a sensitive but disciplined eye to the art of trail design, a matter that had customarily been left to the whims and skills of individual trail builders. "If ever there was a 'pathfinder' it was Sturgis," MacKaye said of his friend's trailmaking style. "His ideal was simple: a footpath is a trail made with the foot. The hand is mere assistant; its main use is to 'spot'—to clear enough for all to see and to widen enough for a fat man to comfortably barge through."[52] It was from Pray—"my old woods teacher," "a pioneer in keeping improvements *out* of the wilderness,"—Benton gladly and frequently admitted, that he borrowed the idea of "a path through a pathless wood."[53]

❖

That summer Benton also joined his brother James for an excursion into Canada. The two disciples of Thoreau explored the same region of Quebec which the Concord saunterer had described in *A Yankee in Canada,* the account of his 1850 journey with Ellery Channing. Benton and James traveled by train to the city of Quebec. By steamboat, they voyaged up the St. Lawrence River in the moonlight, and "viewed the aurora borealis on the banks of the Saguenay." These novel sights and landscapes impressed Benton, who shared with James his vocational daydreams, such as an oceangoing summer camp aboard a schooner. He was equally moved, however, by his brother's struggle to wrestle all of humanity's endeavors into an orderly and progressive philosophy. Thirty-one years old, Jamie now worked as an industrial chemist for a Boston engineering firm. But as a self-styled philosopher, he had substantial intellectual ambitions that reached far beyond the narrow limits of his job. "We sat on the upper deck all by ourselves," Benton later recalled. "He talked of science and its powers for achieving a greater humanity and set forth his idea for a book thereon."[54]

His brother's inspirational musings along the St. Lawrence were fresh in MacKaye's mind a few weeks later, when he survived a harrowing skirmish with death. At Shirley in early September, he suffered an attack of acute appendicitis. One night a doctor was summoned to the Cottage, where he performed emergency surgery beneath kerosene lanterns as Benton lay on the kitchen table. This frightening brush with fate shocked MacKaye out of his vocational indecision and vacillation. While recovering, he swiftly began making specific plans—to become a forester.[55]

"I seem to wake up after this sickness under entirely different conditions from those before the sickness," he wrote to his brother James in a long letter describing the evolution of his career plans.

As you know for several years now I have been trying to work out some method wherby I could make a living in those lines which interested me. Finally I decided that the way to deal most nearly with such subjects—the geography etc.—was to write on them and get up my name as a writer. As a regular job, this at best is uncertain. I am also very interested in the Summer Camp work—the schooner scheme etc.—which is along the same lines and wherein money could be made. The necessity of some steady work—other than tutoring—work which would count toward a future, and which would be in the same line with the writing and Camps etc.—this necessity is making itself felt more and more as time goes on. And so Forestry has been suggested to me.[56]

During his tramps and excursions through the mountains and forests of northern New England, Benton had observed firsthand the extensive and destructive character of the logging industry's operations. He had also experienced the incipient battle between competing interests—lumbermen versus hikers and tourists—over the use and control of the northcountry terrain. According to the tenets of his brother's evolving philosophy, science could reconcile such a conflict. And forestry, as Benton now saw it, was the science that held the greatest promise of productively managing and protecting the landscape he had come to cherish.

He received encouragement to pursue forestry from his closest trail companions, already well launched on their own careers. Draper Maury, now a physician overseeing his friend's recuperation, advised Benton "that a course in Forestry would be the very best thing" for him. And Benton credited Sturgis Pray as his "mainspring inspiration as a young forester."[57]

Benton made inquiries about Yale's forestry school, whose program appealed to him because "a little of the whole range of natural science is taught but each little counts toward producing a practical Forester."[58] But Yale's director, Henry S. Graves, discouraged his notion of undertaking a crash program at Yale. Instead, Graves referred Benton to Harvard, where, he was surprised to discover, his alma mater had just begun to offer training in forestry. Within a matter of weeks, he had registered for a program of courses at Harvard that would earn him the first graduate degree in forestry awarded by the university.

❖

His abrupt confrontation with the stark fact of his mortality had brought twenty-four-year-old Benton MacKaye's experiences and ambitions into sharp focus. A Harvard education and his northcountry adventures had combined to suggest a career. But beyond that, a bold idea had begun to take shape in his mind, inspired by the vistas from those New England mountain summits—a hiking trail along the Appalachian range. Already he had broached the thought to his closest trail companions. By one account, he had first conceived of the idea during his 1900 Vermont hike with Horace Hildreth, when he had climbed a tree atop Stratton Mountain to take in the view. "It may have been this summer with Horace; or perhaps back in '97 on that trip with Sturgis when we used to talk about Mount Mitchell [North Carolina] and the Ranges which lay between," he remembered on another occasion. In the summer of 1902 he first suggested to Virgil Prettyman, his camp boss, the idea "that there should be a trail from Mount Washington to Mount Mitchell. He was sanguine for all my projects but this was a trifle beyond the imagination of that day. 'A damn fool scheme, Mac,' was his blunt reply."[59]

Almost two decades would pass before MacKaye resurrected his "damn fool scheme." In the meantime, he helped to shape the fledgling profession of American forestry.

The Education of a
Progressive Forester

1903–1911

The fervor of the dashing, patrician Gifford Pinchot, whom historian Richard Lillard called "a most respectable radical," inspired many young men of privileged backgrounds to pursue careers in forestry. MacKaye later told Pinchot, recalling the visit the nation's chief forester had made to Harvard in 1900, that he "regretted not signing up with him that night as one of his 'student assistants.'" Pinchot's avid promotion of forestry, combined with the fact that the most prestigious universities provided some of the first educational opportunities in professional American forestry, attracted an educated, upper-crust cohort to the profession in its formative years. And when Theodore Roosevelt assumed the presidency in September 1901, a career embodying what Roosevelt called the "strenuous life" became all the more socially acceptable. Vice-president Roosevelt had been climbing Mount Marcy, the highest of New York's Adirondacks, when he received the news of William McKinley's death.[1]

If Benton had wanted to study forestry immediately after his graduation in 1900, his opportunities would have been limited. In fact, his New York teaching interlude had been providential in one respect: during that brief period the forestry profession had begun to establish itself firmly in American soil. Pinchot, after graduating from Yale in 1889, had traveled to France and Germany to learn the most up-to-date techniques of intensive, sustained-yield forestry. Not until 1898 was formal forestry training offered in the United States; it was taught by two German-trained pioneers of scientific forestry. Bernhard Fernow, Pinchot's predecessor as head of the federal government's Division of Forestry, that year started a four-year program at Cornell. On the vast forest tracts of the Vanderbilt estate in North Carolina, which Pinchot had managed in the early 1890s, Dr. Carl A. Schenck initiated a one-year course of instruction at the Biltmore Forest School. And in 1900, with the financial support of Pinchot's family, Yale started a program leading to a masters degree. As Harvard instituted its forestry curriculum a few years later, other schools in the East and Midwest also began training students in the discipline. Moreover, the

U.S. Division of Forestry, under Pinchot's supervision, in 1899 initiated a popular student assistant program that paid college men $25 a month to study and practice forestry during the summer. By 1902, 300 young men were enrolled in the program.[2]

For MacKaye, the decision to study forestry coincided with the most practical and economical way for him to do so, in Harvard's familiar surroundings. When Benton returned to Harvard in the fall of 1903, he was the first student to pursue a graduate forestry degree from the university.[3] Harvard's Lawrence Scientific School had announced a modest forestry curriculum a year earlier, at the instigation of Benton's former teacher Nathaniel S. Shaler. Organized instruction under a trained forester did not begin, however, until the semester Benton enrolled, when he was one of seven students, including undergraduates, taking courses in the increasingly popular field. He would remain involved with the Harvard forestry program, as either student or instructor, for the next six and a half years.[4]

The Harvard forestry program was at first a patchwork affair. Only three technical forestry courses—silviculture, forest measurements, and dendrology—were taught that first year; Benton supplemented these with basic courses in zoology and botany. The university as yet had no forest field station, so there were limited opportunities for on-the-ground training. The program's faculty comprised two instructors, John G. Jack and Richard T. Fisher. Jack was a botanist affiliated with Harvard's Arnold Arboretum. Fisher, the program's newly appointed director, only three years older than MacKaye, had earned his forestry degree at Yale after graduating from Harvard in 1898.[5]

Fisher was keenly aware of the deficiencies of Harvard's forestry program, but he was certain of its objective: "to train foresters in the practice as well as the theory of the economical management of woodland, so as to equip them as scientific advisers and managers to forest owners." Fisher asserted that a distinctly "American forestry" demanded not only technical training, in the European tradition, but a thorough familiarity "with the business and practical problems which confront the timber owner at home."[6]

Establishment of a forestry curriculum at Harvard paralleled the development of the university's landscape architecture program. During the 1900–1901 school year, Harvard had offered its first course in that discipline, taught by Frederick Law Olmsted Jr. and Arthur A. Shurtleff. President Eliot took a particular interest in the field, as his son, Charles Eliot, was one of its leading practitioners, trained in the Brookline, Massachusetts, office of the eminent Frederick Law Olmsted Sr. MacKaye's career would frequently intersect with the practice of landscape architecture. Indeed, his friend and trail companion

Sturgis Pray was appointed to teach landscape architecture at the university in 1902. Foresters worked in the nation's forests and countryside, but landscape architects usually found their clients in cities and suburbs. The distinction between forestry and landscape architecture as taught at Harvard reflected the divergent utilitarian, aesthetic, and preservationist impulses that were emerging within the American conservation movement.[7]

During his second year, as the forestry curriculum expanded, Benton was introduced to the history of forestry and to such practical matters as forest management and lumbering practices. To fulfill his fieldwork requirement, he arranged an extended trip to Pennsylvania, where he studied the operations of the Goodyear Lumber Company. "The attitude toward forestry both by the owners and the employees seems to be one of ignorance rather than of prejudice," he wrote in his detailed report, adopting the rhetoric of the appealing, but virtually untested, gospel of "scientific" forestry. "Neither pretends to know much about it. They see only one way to get timber and that is to 'slaughter' it. They regard it as pityable but unavoidable."[8]

Benton financed his studies partly with loans from the father of his college and tramping friend Horace Hildreth. During his second year, 1904–1905, he received the $250 Francis Hathaway Cummings Scholarship, awarded annually to graduates of Harvard College who wished to pursue graduate work "as will prepare them for the profession of Landscape Gardener, or for the efficient practice of Horticulture, Arboriculture or Forestry."[9] He also controlled his expenses by living with his brother James, only a few blocks from the Harvard campus. Money, as always, preoccupied the MacKayes, principally because lack of it hampered the "larger work" which they assumed to be their destiny. Jamie, still employed as an industrial chemist, worked late into the night to set down on paper the philosophy of reform that Benton had found so stirring during their Canadian excursion. Benton's education in forestry paralleled his indoctrination in *The Economy of Happiness* (1906), as Jamie titled his thick treatise. Benton later claimed that his two years with Jamie "constituted my 'third' education" (Melvin Longley's barnyard being the first). "I learned *forestry* in the Forest School, but I learned *geotechnics* from James at home."[10]

James MacKaye attempted to provide a scientific foundation for the solution of the "problem of happiness." His "philosophy of common sense" was rooted in the utilitarian principles of Jeremy Bentham—principles which, in MacKaye's view, had been corrupted in the nineteenth century by John Stuart Mill, who "followed Adam Smith into commercialism and perpetuated the separation of politics and morality."[11] He also borrowed from the population theories of Thomas Malthus, the ideas of Charles Darwin and Herbert

Spencer on evolution and heredity, and the political economics of Karl Marx. Moreover, his book had roots in an American tradition of utopian collectivism most widely popularized in Edward Bellamy's *Looking Backward* (1888).

By James MacKaye's rigid yardstick of social utility, the overarching purpose of human life was to maximize the "output of happiness." He coined a unit of measurement, the "pathedon," to quantify the balance between pleasure and pain.[12] The equations and "happiness curves" scattered throughout his book were designed "to ascertain the conditions under which happiness, regarded as a commodity," might be "produced with the greatest efficiency."[13] The product of his strenuous logical exercise—possibly the most humorless and austere tract ever devoted to the subject of happiness—was a socialist program he termed "pantocracy" ("all to rule"). James's eight-point pantocratic plan called for, among other things: "public ownership and control of the means of production"; centralized, government-operated bureaucracies of distribution, marketing, and labor, with powers to adjust supply and demand of products and labor; a national inspection system, with national standards of product quality; old-age insurance; and national research laboratories.[14] In fact, the sprawling five-hundred-plus-page *Economy of Happiness* contained not only a radical critique of capitalism but an attack on Theodore Roosevelt's brand of progressivism, which he described as "pseudo-socialism," or "the acquiescence in, and national regulation of, private monopoly."[15]

The Economy of Happiness addressed several issues that would become Benton's primary concerns, such as the use and conservation of natural resources and the social role of recreation and leisure. James MacKaye assailed as "practomaniacs" those who promoted production and profits as supreme goals but who did not take into account the "vast and increasing surplus of misery" that he foresaw as the likely consequence of failing to emphasize efficient consumption:

> The resources of the country should be conserved and husbanded with the greatest care; their development should be delayed as long as possible; the activities of men should be turned to developing individuals who are fitted to convert the potentiality of happiness involved in natural resources into actual happiness; the efficiency of consumption should be stimulated and then, when the husbanded resources are at last developed, the nation will have something to show for their dissipation.[16]

The federal forest reserves provided a model of the sort of "public monopoly" he envisioned to replace the wasteful, inefficient private ownership and exploitation of productive resources. MacKaye attacked as "a political myth" claims for the "benificence of competition." Competition, he argued, failed every test of justice, and was not genuinely preferred or practiced in any case. "If the nation does not own the monopolies, the monopolies will own the nation," he emphasized. "Socialism is but consistent democracy."[17]

James MacKaye's complex and ostensibly objective theoretical apparatus did not conceal his disdain for conventional politics, which he hoped to supplant with a form of political engineering. "When the time comes, politics will take its place with applied mechanics, electricity, and chemistry, as a branch of technology," he predicted, "and will become as thoroughly revolutionized as were alchemy, astrology, and the other varieties of mysticism from which the sciences of to-day have been evolved."[18]

A disciple of Thoreau, James argued that "simple tastes are better than any others. A nation whose tastes are of the simplest can, other things being equal, dispense in greatest proportion with the most universal of needs—the need of labor. It is for this reason that luxurious tastes are so uneconomic."[19] His plea for simplicity was an argument for more efficient use of human labor, which by his calculation would produce a greater "output" of human happiness.

"Simple tastes are better than any others" was the axiom by which his brother Benton would consciously live his life. "Jamie has certainly written an epoch-making book," he reported to his mother. "It may sound fanciful to say it, but it requires only reason to show that he stands today at the head of philosophical thinkers. . . . And his is the philosophy which will bring emancipation."[20] In later years, Benton credited his brother's thinking as "the rock-bottom foundation," the *chief source* of inspiration," for his own writing and work. But never in his own published writings did he acknowledge his intellectual debt to James. In fact, on one occasion he squelched a scholar's attempt to state the connection, commenting that "folks would just laugh"— whether at his brother's obscurity or at the very details of his philosophy he did not make clear.[21] James, in his softspoken but relentlessly logical manner, made a profound impression on those he met personally, some of whom became devoted disciples—victims of "MacKayomania," as one of his students later put it—even if they sometimes found his writing and his ideas bewilderingly abstruse.[22]

Benton nonetheless remained his brother's most faithful disciple. His life's work embodied Jamie's "pantocratic" ideas for a socialistic utopia, a cooperative commonwealth in which the skills and knowledge of scientists and technicians would serve the end of happiness, not simply material accumulation. For Benton, forestry provided the skills to pursue a more ambitious and altruistic end than the mere cultivation of trees.

When he completed his forestry studies in the spring of 1905, Benton could at last look ahead with some confidence to his career prospects. Preparing for the Civil Service exams that would qualify him for government employment, he felt keenly the pressure to secure a respectable job, as the expense of his education had limited his ability to support his mother and sister. When Benton received his masters degree in June, his Aunt Emmie von Hesse, to whom the family had often turned for financial assistance, pointedly urged him to "succeed in some business enterprise *very soon.*"23

As he embarked on his new career, Benton used the occasion to change his name. Thereafter, he informed his family and friends, he would no longer use Emile, "my transcendental first name." His A.M. degree was awarded in the name of Benton MacKaye. "When I register in the Forest Service," the young forester informed his mother, "it will be as 'Benton MacKaye' as my *full* name."24

During the summer of 1904, between his two years of forestry study, he had returned to New Hampshire to spend the summer as a counselor at Virgil Prettyman's Camp Moosilauke, newly opened on the western fringe of the White Mountains. He and another counselor, Knowlton "Cub" Durham, led eight boys on a strenuous ten-day hike over 4,810-foot Mount Moosilauke, into the heart of the White Mountain range. They referred to themselves as the "Tattered Ten." Benton, whom the campers nicknamed "Farmer," entertained them around the campfire with his harmonica. They traipsed several established paths that would ultimately be incorporated into the Appalachian Trail, wending their way to the summit of Mount Washington, along the Presidential Range, and over Mount Madison—an ambitious trip for a group of inexperienced young hikers. All of this terrain was now familiar to Benton, but when the campers were caught in a storm on the notoriously dangerous slopes of Mount Washington, he led them "crawling yard by yard, from one cairn to the next," and was relieved when they came upon the way to safety. "I have never forgotten the sight of the cog railway, looming suddenly out of the fog, that would lead us safely to the Summit House."25

That summer, Benton saw the White Mountain landscape through a *forester's* eyes. He recorded this fresh perspective in his first published article, "Our White Mountain Trip: Its Organization and Methods," which appeared in *The Log of Camp Moosilauke.* MacKaye recounted in careful detail the outfitting and the camping practices of the ten-day hike, considering such practical matters as food, equipment, and campsite selection.26 In the article's conclud-

ing passage, under the heading "Camp Ethics," he summarized his own growing sense of responsibility for the environment he had come to enjoy and to
know so well since his first mountain adventure in 1897. "The simplest rule of
conduct for a camper is to leave a place as he would like to find it," he urged.
"Such conduct is becoming more and more the tradition, and hence camping
as an institution is giving greater enjoyment every year." But he scorned the
behavior of the "picknicker," who leaves his litter everywhere and "improves
the rocky summit of every mountain with his initials."27

"Camping," he concluded, was a pastime that embodied good citizenship,
common sense, and self-respect.

> There is no other sport or mode of living which so clearly exemplifies the need of
> each to do his share and the dependence of all upon the resources of nature. If we
> are to have these resources, whether lumber or other; if things are to be used and
> not dissipated; if we are to have a camping ground and not a desert, we must
> work and fight for these ends. The duty of the camper, as one with greater oppor-
> tunities in this respect than the average citizen, is to preserve the resources which
> nature has bestowed and to cherish the land as he would his home.28

Together, his technical training in forestry, his firsthand experience roving
the New England northcountry, and his brother's singular philosophy inspired MacKaye's gradually coalescing vision of a wild, mountain landscape,
protected from exploitation by private commercial interests and preserved for
public uses.

The completion of Benton's graduate work in the spring of 1905 coincided
with one of the most important events in the annals of American public
lands. When the Transfer Act was passed on February 1, Gifford Pinchot realized his ambition to control the federal forest reserves. The law provided that
administrative jurisdiction over the forest reserves, first established in 1891,
would pass from the Interior Department's Land Office to Pinchot's bureau in
the Agriculture Department on July 1. A "study agency . . . became an action
agency," as MacKaye put it, when Pinchot's bureau—renamed the United
States Forest Service—took over the 86-million acres of federally owned forests. Pinchot assumed the new title of Chief Forester. In 1907, the forest reserves would be retitled "national forests."29

A peerless bureaucratic empire builder, Pinchot was employing foresters almost as fast as they came out of school. With the creation of the Forest Service in 1905, 821 employees came under his command, 153 of whom were forest-

ers. Benton was hired as a "forest assistant," at $1,000 a year. Unlike "forest rangers," who were chosen primarily for their capabilities as riders, woods-men, and packers, the forest assistants were typically college men trained to develop working plans for forest maintenance and cultivation. MacKaye ex-pected an assignment somewhere in the vast western forest reserves—the only lands, in fact, which the Forest Service then administered. But the direction of his career changed, literally as he rode the train to Washington to report for work on September 1. En route, he received a telegram informing him to re-turn to New England. His instructions were to travel to Keene, New Hamp-shire, to assist in a study of forests in the southern part of that state.[30]

When MacKaye checked into Keene's Cheshire House he was in familiar surroundings. In fact, his very first task for the Forest Service was a timber sur-vey of the forests owned by the Faulkner and Colony Company, managed by the MacKaye family's close friend, Robert Faulkner. The New Hampshire as-signment proved to be fortunate and fateful for Benton's career. In his very first days with the Forest Service, he met one of Gifford Pinchot's closest and most influential colleagues, Raphael Zon. Five years older than Benton, the Russian-born Zon, a high school classmate of Lenin, had studied biology at university. He had fled his native country in the mid-1890s after receiving an eleven-year prison sentence for his political and union-organizing activities. After brief sojourns in Germany, Belgium, and England, where he attended Fabian Society meetings, Zon made his way to the United States and earned a forestry degree at Cornell. Since joining Pinchot's agency in 1901, he had quickly earned respect for both his technical expertise and his sweeping perspective of forest conditions. During the course of his long Forest Service career, Zon would be instrumental in establishing the agency's research activ-ities; in later years he headed the Lake States Forest Experiment Station. He also brought to the Forest Service a concern about "the social goals of forestry" that sometimes placed him at odds with agency colleagues and priorities.[31]

"Mr. Zon is at the head of the new section of Silviculture," Benton noted in his diary, "and is looking around."[32] In fact, the Forest Service's plans for MacKaye and his equally unseasoned colleague, Anton Boisen, were uncer-tain. Soon after Pinchot assumed leadership of the Division of Forestry in 1898, his modest agency had begun offering free technical assistance to lumber companies and private landowners. "Circular 21 work"—so called af-ter the publication describing the program—sent government foresters into private woodlands across the country, spreading the gospel of sustained-yield forest management. By the year MacKaye joined the Forest Service, some eleven million acres of private forests were enrolled in the program. In the East, where there were as yet no federally owned forests, Circular 21 work provided virtually the only role and presence for the Forest Service.[33]

The agency was stepping gingerly into areas and responsibilities beyond the borders of the forest reserves under its jurisdiction. Efforts to create new federal forests in the South and East were making only halting progress in Congress. Zon was prospecting for a project that could increase the Forest Service's influence in the region, without raising political hackles. From their outpost in Keene, New Hampshire, the small band of government foresters spent two weeks making forays into the surrounding countryside, sometimes by bicycle, to study forest tracts. During the evening, they engaged in long conversations that strayed far from forestry concerns. "Zon is a very interesting man and seems brighter than the average," Benton wrote in his diary after one such "ethical discussion." In Zon, Benton had found another of the many influential mentors and patrons who appreciated his unique talents and insights—and understood his shortcomings and outright eccentricities. Zon was intellectually cosmopolitan, politically liberal, and scientifically astute—the "savant" of the Forest Service, as MacKaye later put it. The friendship between the two socially conscious foresters would last until Zon's death in 1956.[34]

Zon left his young crew with vague instructions concerning their next duties. Benton returned briefly to Cambridge for a family gathering, a bittersweet occasion, because Aunt Sadie was terminally ill. Presently he received news that he was being sent to Wyoming. He appealed to Washington to remain in New Hampshire, and headquarters capitulated to his wishes. For the rest of the autumn and on into the winter he traveled southern and central New Hampshire surveying and inventorying sample forest plots of white pine. It was a challenging assignment that demanded a wide array of skills and ample physical stamina. Tramping the woods in raw weather, marking and measuring trees, mapping and photographing plots, compiling statistics, calculating and plotting "yield curves," then writing up reports, all with minimal supervision—the tasks provided ideal training for a young, ambitious forester learning his trade. "This is mighty interesting and healthy work and I think the future prospects are good," he reported to a friend not long after he began his Forest Service duties. "I guess I'm started now for fair, if it is toward the end of my life."[35]

In the late spring of 1906, Benton received a proposition from Richard T. Fisher at Harvard. His former professor, upon Pinchot's personal recommendation, had offered him a one-semester appointment as a forestry instructor, beginning that autumn. The Forest Service had planned to assign MacKaye to the West in July. The agency assured Benton that summer and part-time work would still be available, so he accepted the job at his alma mater.[36]

Forty students, divided equally between graduate students and undergraduates, were registered in Harvard's expanding forestry program that semester, double the previous year's enrollment. In August and September, Benton had been engaged in Circular 21 work for the notorious J. E. Henry and Sons lumbering operation in the heart of the White Mountains, whose ill-treated lands along the East Branch of the Pemigewasset he had been traversing for a decade. When he reported to Harvard to teach courses in forest history and forest measurements, he had only one year of practical experience under his belt. But given the young profession's thin ranks and its relatively untested techniques, MacKaye was already something of a veteran forester.[37]

That year Harvard required his services for only the fall semester, so he became available to the Forest Service for Circular 21 work again the next spring. During the spring and summer of 1907, his independent duties for the Forest Service, "marking logs with lumberjacks and marking trees for country gentleman," took him to private forests in eastern Massachusetts, southern New Hampshire, the eastern Adirondacks of New York, and western Maine.[38] That spring he also undertook a Forest Service task that had fateful and auspicious consequences for his role in shaping the public landscape of the Northeast. The Appalachian Mountain Club had recently received a gift of some 300 acres in Fitzwilliam, New Hampshire. The land was distinctive because of its 12-acre stand of the magnificently blossoming *Rhododendron maximum,* a variety rarely found so far north. Allen Chamberlain, a Boston journalist and activist prominent in the AMC and several other regional conservation groups, was the club's Councillor of Exploration and Forestry. When he asked the Forest Service for assistance in preparing a working plan for the reservation, he was referred to MacKaye.[39]

In April 1907, MacKaye and Chamberlain met at the old tavern on Fitzwilliam's common. The two outdoorsmen quickly found that they shared many interests and enthusiasms. After their investigations on the AMC property, they hiked up nearby Mount Monadnock. During their brief sojourn in the pastoral but rugged landscape of southwestern New Hampshire, they cemented a friendship that carried them through many conservation crusades over the next four decades. At the time the two men met, in fact, Chamberlain ("the forester's big brother," Benton later wrote) was mobilizing New England's conservationists on a variety of issues, including the creation of national forests in the White Mountains and southern Appalachians and opposition to a proposed dam in Yosemite National Park's Hetch Hetchy Valley. In years ahead, especially during the initial efforts to promote the Appalachian Trail, the well-regarded Chamberlain would serve as MacKaye's principal contact in the New England conservation and hiking community. With the help of four Harvard forestry students, Benton prepared a detailed working plan for the Rhododendron Reservation. His program, by Chamberlain's account, called for the removal of the existing timber, which was "generally of an unsatisfactory nature," and its replacement by "progressive annual planting." The AMC's hope, and Benton's optimistic claim, was that, after fifty years, the periodic harvest of the managed stand of trees would provide income sufficient to support the tract.[40]

In the fall of 1907, MacKaye was appointed to a proctorship at Harvard, a position which provided him a room in Harvard Yard, at 21 Stoughton Hall. He remained on the teaching staff throughout that school year, but Fisher looked to his conscientious colleague for help in other matters as well. Benton prepared a statement in Fisher's name for delivery at a January 1908 hearing of the House Committee on Agriculture, arguing for passage of legislation to create eastern national forests. Recreationists like his new friend Chamberlain were spearheading the movement for national forests in New England. Consistent with the prevailing utilitarian philosophy of the Forest Service and the forestry profession, however, MacKaye, as his boss's amanuensis, did not espouse recreation or scenic preservation as reasons for creating new national forests. The statement instead emphasized timber production, flood control, and fire prevention as justifications for the proposed federal forests.[41]

After almost three years of intermittent government service, Benton had not yet visited the Forest Service's headquarters. When Pinchot came to Harvard in the spring of 1908, Benton met with his Forest Service chief ("a splendid man," he assured his mother), who promised that a position would always be available for him in the agency. MacKaye wasted little time securing a post for the summer. Instead of the customary New England assignment, however, he had an opportunity to investigate new terrain, in the mountains of eastern Kentucky. Benton stopped for instructions en route at the Forest Service's modest but hectic headquarters on F Street in Washington. There he found Pinchot, his chief aide Overton Price, and other Forest Service leaders "planning the management of a forest empire covering half of the U.S.A." In fact, Pinchot and Price had just been appointed by President Roosevelt to a new National Conservation Commission, chaired by Pinchot and charged with inventorying the nation's natural resources. The commission had been proposed at a climactic event in the conservation crusade, the Governor's Conference, which had convened in Washington only a few weeks before Benton passed through the city.[42]

MacKaye carried the missionary spirit of that crusade into a region that was entirely new to him. A photograph from his Kentucky excursion depicts the lean, smiling young forester on horseback, with the mountains stretching off behind. He later recounted in humorous terms some of his encounters with the friendly mountainfolk. In any event, Benton's brief Kentucky sojourn extended southward the horizon of his firsthand familiarity with the physical and human character of the Appalachian terrain.[43]

When MacKaye returned from the Kentucky hills that fall, Richard Fisher had an urgent task for him. A wealthy young Harvard graduate and aspiring forester, John Ames, had purchased a large forest tract for Harvard in the central Massachusetts town of Petersham. At last the university's forestry program would have its own field station, for research and for the demonstration of forestry and logging practices. Fisher assigned MacKaye the task of organizing the facilities at the 2,000-acre property.[44] Benton relished this opportunity, the first time he had been handed any significant administrative responsibility. The gift property included a farm, as well as some ramshackle buildings where MacKaye stayed. In this rustic domain, he felt like the lord of the manor. "As soon as you can you want to see my farm here," he wrote his sister. "We have all kinds of cattle, fruit, vegetable and cream, besides hot and cold water in the bath room. . . . Got a pair of horses the other day—they weigh 30 hundred. I also have some serfs whom I give orders through my supt. like an earl."[45]

That fall, in the Harvard Forest's first year, Benton taught his courses at Petersham. During the winter months, he returned to Cambridge to conduct additional courses in forest history, forest law, and forest products. His room at 21 Stoughton Hall soon became a focal point for a campus in political ferment. The "Harvard Renaissance," as journalist John Reed later described it, reflected both the progressive mood of the times and the remarkable group of undergraduates who were then at Harvard. Among at least some students and faculty, there was a new spirit, an acute consciousness of events and ideas beyond the fenced-in borders of Harvard Yard. And Benton, along with his brothers James and Percy, was at the center of the circle of students, professors, visitors, and hangers-on who instigated this brief, exhilarating moment in the university's history.[46]

Perhaps the central spirit of campus political activism among the students was the brilliant and ambitious Walter Lippmann, like John Reed a member of

the class of 1910, who was destined to be the most influential American journalist of his era. A contributor to an array of campus publications and a
protégé of William James and George Santayana, Lippmann was the most precocious member of a class that his biographer called "probably the most illustrious Harvard ever produced." It included poets T. S. Eliot and Alan Seeger,
journalists Reed and Heywood Broun, and stage designer Robert Edmond
Jones. Lippmann could not wait to make his mark on the world. In early 1908,
he helped organize relief efforts when a huge fire in Chelsea, a working-class
community just north of Boston, left thousands homeless. That spring, Lippmann collaborated in drafting the constitution and bylaws of the Harvard Socialist Club, the purpose of which would be "the study of Socialism and all
other radical programs of reform which aim at a better organic development
of society."[47]

Forty-one students signed on as founding members. During the 1908–1909
school year the Socialist Club maintained a busy schedule of meetings and
lectures. The young radicals heard from eminent reformers about subjects
ranging from child labor laws and woman suffrage to the Single Tax and European social movements. The Harvard socialists were admirers of Britain's Fabian Society, which included such well-known figures as George Bernard
Shaw, H. G. Wells, and Beatrice and Sidney Webb. The Fabians advocated a
moderate, incremental, nonviolent brand of socialist reform, under the leadership of enlightened intellectuals and technicians.[48]

Two of the leading local influences on the Harvard Socialist Club were
James MacKaye and the renowned muckraking journalist Lincoln Steffens.
Benton's brother James was something of an oracle to undergraduates like
Lippmann, his *Economy of Happiness* "a sort of bible." On a hike in the Boston
suburbs, Benton, Lippmann, and Carl Binger, later a distinguished psychia-
trist, resolved to study Jamie's lengthy book chapter by chapter. Upon returning to Stoughton 21 to begin their reading, tired from their hike, they instantly
dozed off. Waking three hours later, the trio shared a laugh at the gap between
their earnest intentions and their limited energies.[49]

Steffens, in Boston during 1908 and 1909 to assist the reform-minded merchant Edward A. Filene in exposing municipal corruption, was an occasional
visitor at the "pow wows" Benton sponsored in his rooms. The righteous and
acerbic crusader took Lippmann under his journalistic wing when the two
met at a dinner, attended by Benton, at James MacKaye's apartment. Steffens
recruited them all as self-styled "Conspirators" for an educational experiment, designed to stir up political thinking and activity at complacent Harvard. Working with a small group of Harvard faculty members and a handful

of students organized by Lippmann, Steffens plotted a series of lectures, including one by Jamie, intended "to leave burning questions instead of convictions in the minds of the student audiences." Steffens declared the experiment a success.[50]

For a time, the MacKaye name and influence seemed pervasive on the Harvard campus. Benton's thirtieth birthday coincided with the Socialist Club's first-anniversary dinner. Holding a seat of honor beside club president Lippmann, he was called upon to give an impromptu speech to the thirty student socialists in attendance. With Lippmann and his brother Percy, Benton arranged and publicized Jamie's well-attended six-lecture series on "political engineering." And Percy, earning a reputation as a playwright and poet, was on campus to produce two of his plays, *The Canterbury Pilgrims* and *The Scarecrow*. His Harvard lecture urging "the emancipation of the drama and the dramatist from the chains of commercial bondage" was included in his 1909 book, *The Playhouse and the Play*. In a subsequent volume of lectures and essays, *The Civic Theatre* (1912), Percy further elaborated his notions about drama's potential to promote "the conscious awakening of a people to self-government in the activities of its leisure."[51]

The MacKaye brothers were flattered by the attention of so many bright, eager undergraduates. John Reed, a consummate campus joiner, was on the business staff for production of *The Scarecrow*. He and other members of the production who would go on to make names for themselves, such as Kenneth MacGowan, Lee Simonson, and Robert Edmond Jones, visited with the MacKayes at Percy's apartment, "where the discussions were as likely to be political as literary," acccording to Reed's biographer Granville Hicks. "Indeed, these gatherings at MacKaye's were evidence of the alliance between politics and literature that was the foundation of the renaissance. The same men were active in the *Monthly*, the Dramatic Club, and the Socialist Club, and the same impulses dominated the three organizations."[52]

Benton reveled in this most sociable brand of socialism, which was motivated as much by moral righteousness as by any sense of solidarity with the working class. Its aim was "to change a system and not human nature." He lectured his sister, "One of the ways to call attention to the harm of the system is to point out the 'bad men.'"[53]

After a summer stint for the Forest Service in the eastern Adirondacks and a vacation in Shirley, MacKaye's prospects seemed as bright as his mood in the fall of 1909 as he began his fourth year teaching forestry. Taking up his duties at Petersham, he planned the forestry textbook he had been contemplating and helped Fisher draft the forestry division's new plan of instruction. And, not least, he looked ahead to a lively season of campus political activity. But there was also a surprise in store for his family and friends: he had become engaged. Mabel Foster Abbott, his fiancée, was a recent Radcliffe graduate and a friend of Hazel's. The two young women had become acquainted when Hazel was enrolled as a special student at Radcliffe, where they both took part in theatrical activities. Hazel had introduced her new friend, nicknamed "Lucy," to Benton in 1907.[54]

While the MacKayes gradually came to accept Lucy as a prospective member of their tight-knit family, some of the clan were surprised to learn of Benton's matrimonial intentions. "What astonishment!" the ever-skeptical Hal wrote to his mother. "Why last summer when I last saw Ben—such mysogynistic language—such contempt for anyone who could think of marrying. Isn't it a curious—a mysterious thing! . . . You may have heard it stated that 'wonders never cease'—I now believe it."[55]

Benton was downright secretive about his betrothal, boasting that during a number of public appearances the two "sat together on all occasions but (except to those taken in) I continue to 'deny it.'"[56] From the start of the relationship the couple channeled much of their passion and energy into their respective careers. Lucy, who had been president of Radcliffe's prestigious English Club, hoped to become a journalist. Benton's forestry background and his connections with the suddenly beleaguered Forest Service provided her with a ready subject as the politics of conservation grew in national popularity and controversy. "I wanted to know Benton's subject for my own good," Lucy wrote Hazel at the beginning of 1910, as her first articles about conservation appeared in print.[57]

When William Howard Taft succeeded Theodore Roosevelt in March 1909, the political momentum of the conservation movement, which was already meeting congressional resistance from westerners and Democrats, abruptly slowed. During the Roosevelt years, Pinchot, with the president's unflinching support, had shaped the government's natural resource policies. Pinchot credited W. J. McGee, of the U.S. Geological Survey, with the definition of conservation that had motivated him and his like-minded disciples: "the use of natural resources for the greatest good of the greatest number for the longest time."[58] Rhetorically appealing, this overarching doctrine generated controversy, however, when applied to particular cases and resources. Nonetheless, Roosevelt's conservation legacy was formidable. When he left office, 195 million acres were set aside as national forests, as opposed to the 46 million acres reserved at the beginning of his administration. Roosevelt won congressional support for five new national parks; he created fifty-one federal wildlife refuges by executive order; after passage of the 1906 Antiquities Act he created sixteen national monuments. He supported the 1902 Newlands irrigation and reclamation act, which altered the face of the arid West, as well as the proposed legislation for eastern national forests. His creation of the Inland Waterways Commission (1907), the Governors Conference (1908), and the National Conservation Commission (1909) defined a bold and broad—if not always popular—agenda for the nation's natural resource policies.[59]

In the Taft administration, however, Pinchot soon found his powers and ambitions severely constrained. The cautious, legalistic Taft felt that Roosevelt and his subordinates, like Pinchot and Secretary of the Interior James Garfield, had often exceeded the intent of Congress in interpreting laws dealing with irrigation, grazing rights, water-power sites, minerals, and other natural resources.[60] In January 1910, after months of tension and dissension within the government over natural resource policy, Taft fired the increasingly recalcitrant Pinchot, who did not leave the public scene quietly.

Throughout 1910 and 1911, Benton and Lucy closely followed the dramatic and oft-recounted Ballinger-Pinchot affair, a defining landmark in American conservation history. The duel of egos and reputations between Pinchot and Taft's appointee as Interior Secretary, Richard A. Ballinger, was played out to a muddled conclusion during protracted congressional investigatory hearings on conservation issues in early 1910. The hearings and public controversy were precipitated in part by the allegations of Louis R. Glavis, a righteous young investigator in the Land Office. Glavis, on Pinchot's urging, had taken directly to the president his charge that Ballinger, during a previous stint as commissioner of the Land Office, had improperly intervened on behalf of former legal clients, who had ties to the notorious Morgan-Guggenheim syndicate, to secure coal fields on federal lands in southeastern Alaska. Taft quickly fired Glavis and exonerated Ballinger. Glavis then aired his evidence, much of it circumstantial, in the crusading *Collier's Weekly*. For Pinchot and other conservationists, the charges against Ballinger and the special interests he allegedly served provided a personal and moralistic focus for the arcane, often ambiguous debate about government natural resource policy.[61]

"Lucy has a job 'muck raking' on 'conservation,'" Benton reported to Hazel as the congressional hearings began. "I am helping her. We are thick in 'Ballinger-Pinchot' and the latest pow wow in Washington. I suppose you know that Pinchot was dismissed. If you want some fun, follow the 'B-P' inquiry forming in Congress and the 'insurgent' movement there."[62] For the *Twentieth-Century Magazine,* a reform-minded (and short-lived) Boston journal, Lucy had already begun writing a hard-hitting pro-Pinchot column titled "Conservation News." For the same magazine, Benton helped Lucy work up a lengthy five-article series, "A Brief History of the Conservation Movement," which paralleled in substantial measure the material he presented to his Harvard forestry students. When the first installment, illustrated with Benton's maps and charts, appeared in March, Lucy had begun to carve out a journalistic niche; her letterhead read "Conservation Writer and Editor."[63]

The apparently auspicious state of Benton's affairs suddenly unraveled in the spring of 1910. In mid-March, he received a brief but civil note from Fisher containing the unexpected news that he would not be invited back to teach for the next academic year. Benton, shocked by Fisher's action, pleaded for an explanation for his dismissal. He had no previous indication, he claimed, that after almost seven years as both student and colleague his relations with his superior had been anything but satisfactory. Benton, after talking with Fisher, claimed that "he practically admits that he has nothing against me except that our ideas are different."[64] Nonetheless, his termination raises a question: Did Benton's political opinions and affiliations cost him his job at Harvard? Fisher did not raise any such issues in his correspondence with MacKaye, nor, apparently, did Benton mention such a motive in explaining his firing to his family. The circumstances surrounding Benton's dismissal suggest at least the possibility, however, that his left-wing campus politics and his behind-the-scenes involvement in national conservation affairs may have provoked Harvard's decision not to rehire him.

Fisher gave Benton the opportunity to save face by resigning, and offered to recommend him for another position; but the incident unsettled MacKaye's personal and professional plans. He and Lucy bravely carried on with their April engagement party in Shirley—but without announcing a wedding date. Mary MacKaye, ever sensitive to any blot on the family reputation, told acquaintances that her youngest son had resigned "of his own free will" because of the limited prospects for promotion at Harvard and his desire to pursue "a wider field." Benton, confronted with new leadership at the Forest Service, decided that his future in the profession now depended on the completion and publication of the forestry textbook he had been contemplating for several years. His dilemma created a crisis for the entire family. Still in debt for his forestry studies, having given as much as half of his income to his mother in "all but two in the last ten years," Benton pleaded with his brothers to share in her support. "My status in the Forest Service is continuous," he gamely assured Hal, "but my chances of advancement there or elsewhere depend upon what I have to contribute as a result of my teaching lure."[65]

While Benton made arrangements for Lucy to live in New York with one of his aunts, his sister helped him find an opportunity to work on his forestry text. Hazel, after her brief stint at Radcliffe, was attempting to establish her own acting career. She met Mary MacDowell, widow of the respected composer Edward MacDowell, who was seeking someone to plan the grounds of the Peterborough, New Hampshire, estate she was transforming into an artists' colony. Benton applied for the job and undertook the work at Peterborough in the fall of 1910. For Mrs. MacDowell, he prepared the sort of forest working plan at which he had become proficient through his Circular 21 work. But he exploited the work in another way, by using the MacDowell property as the key case study for his forestry textbook.[66]

After Benton finished this consulting job early that winter, he worked intensely on his textbook, at Percy's home in Cornish, New Hampshire, making occasional trips back to Shirley Center. The family still fretted about his situation and, in some measure, about his state of mind. "I hear he is overworking in Cornish," a concerned Jamie wrote his mother. "I wish he would read Thoreau, and get his equilibrium."[67] Benton was secretive and defensive about his writing, declaring to family members that he still had much work to do on the arcane statistical "curves" that were critical to the text. Finally, he unveiled his ideas, charts, and diagrams to Jamie and the ever-faithful, if a bit bewildered, Hazel. "We sat around the dining room table, and the dear boy began to unfold his scheme and show us his work," she wrote her mother. "I have never seen anything that impressed me more. Instead of thinking him slow, I think he has done it in an incredibly short time."[68]

Jamie, meanwhile, had been following the progress of the Weeks Act, the proposed federal law that would authorize the establishment of new national forests in the East. "If [Benton] could be in Washington, on deck, for a job by the time this bill is passed," he explained to Mary MacKaye, "he might secure a good post in the White Mts. He could show the people in Washington that he was peculiarly fitted for such a place because of his familiarity with the region."[69]

The Weeks Act, enacted on March 1, 1911, did not immediately open the door to a new Forest Service job, but Benton did attempt to cultivate his tenuous employment prospects with the agency. In January, he had received a letter from the new chief forester, Henry S. Graves, who requested MacKaye's formal resignation as a part-time, per diem employee, while holding out the prospect of a "special appointment as expert" sometime in the future. It is not clear whether MacKaye in fact submitted his resignation, but he did plead with the agency for "more time" to complete his writing work, which he intended to present to Graves. So sensitive was he about his status that he asked his sister to accompany him when he visited Harvard Square, to help fend off the inquiries from acquaintances about what his mother called "this crisis in Benton's life."[70]

The strain of his unsettled circumstances and his intense work on his book had begun to take a toll on Benton, however. He decided to go to New York, to live for a time with Hal and to be closer to his fiancée and to job prospects; but he suffered a collapse, including an intense recurrence of his digestive troubles. "Benton seems to be having a kind of 'let down' after his strenuous work," Hazel reported. "He isn't ill, so far as one can tell, only sort of 'all in.'" After a brief rest, he departed for New York, and his spirits and energies were refreshed when he arrived there. "Benton and Lucy came to our house Saturday and have been sparking before the hall fireplace by the romantic process of discussing forestry charts and 'curves,'" Hal teased. "It gives Lucy a great opportunity to tell Benton what a wonder he is."[71]

Conflicting emotions churned within the unemployed thirty-two-year-old forester, as he attempted to sort out his past disappointments and his future ambitions. He described them in a letter to him mother:

> I have been chewing on these propositions six years beginning next month, and for the first time I see them in their entirety. . . . I am tired of meditating, of walking the floor and of "sweating out." I have meditated so much that it may account for some of my curious ways. A poor excuse is worse that none perhaps, but we'll put some portion of the blame on the meditative mind. A rough calculation indicates that during the past six years I have walked back and forth, cogitating, a distance about that between Harlem N.Y. and Kalamazoo, Mich., and I mean to stop before reaching Omaha. No telling what the journey's end will find, but I hope at least a change in my working mind.[72]

Benton had reason to hope that his lonely work, whether published or not, would at least demonstrate his capabilities and convictions to the Forest Service. "A Theory of Forest Management," as he titled the product of his meditations, was intended to be a state-of-the-art treatise on scientific forest management, with particular emphasis on the "timber problem of the Northeast." His "curves" illustrated the timber yields over time of different regimens of forestry, from the carefree "laissez-faire" practices prevailing over most of the region's woodlands, to a system of intensive management on an expanding area of public forests established by the federal and state governments. He concluded that technical forestry was most efficient when applied over larger areas rather than on small, separate parcels of land. "Scientific forest organizing" would most likely succeed, he proposed, under government direction on "organized localities" consisting of publicly owned forests.[73]

By conviction and by design, MacKaye was stating ideas that were consistent with those then prevalent in the Forest Service. With passage of the Weeks Act, the agency could look forward to the rapid and substantial expansion of public forests in the East. As a young progressive forester, MacKaye espoused a single factor—timber production—as the index of successful forestry. Over the years, he would gravitate to a far more preservationist view, advocating the protection of wild forests rather than their exploitation for timber production. But by whatever measure or for whatever purpose—economic, biological, aesthetic, recreational, spiritual—he consistently supported public ownership of forests.

✤

Benton and Lucy spent an April evening as guests of Judge George Woodruff, a long-time confidant and legal adviser of Pinchot. Woodruff praised Mac-Kaye's manuscript as a "splendid piece of work." Woodruff also wrote to Henry Graves at the Forest Service on the young forester's behalf, but no job offer was yet forthcoming.[74] Lucy's career, by contrast, was suddenly flourishing. In the early months of 1911, she had assiduously cultivated her journalistic prospects in New York; and in March, she achieved a breakthrough: a by-lined article in *Collier's* about a corporate and legal battle for control of Maine's abundant hydroelectric resources.[75] *Collier's* editor, Norman Hapgood, was a friend and ally of Pinchot and other progressives. Now, Abbott's name was identified with the leading journalistic tribune of the conservation crusade. Just as swiftly as her career took off, though, her work, and perhaps her marriage plans, were thwarted by her involvement in the acrimonious politics of conservation.

Her article about Maine caught Pinchot's eye. Benton wrote his mother, "It seems he wrote Hapgood that her facts . . . were wrong, and then she showed him they were correct [and] he admitted it." By May, Lucy was in regular contact with Pinchot and some of his colleagues, exchanging information and documents.[76] In early March 1911, Richard Ballinger resigned as Interior secretary. As his replacement, Taft appointed Walter L. Fisher, a progressive who was considered sympathetic to conservationist concerns. Lucy's muckraking soon stirred another round of controversy about the control of Alaska's coal resources, the issue that had been at the heart of the Ballinger-Pinchot affair. For *Collier's* she wrote a hard-hitting article (which included a map prepared by Benton), raising new questions about coal leases at Controller Bay in Alaska's Chugach National Forest. In particular, Abbott recounted the circumstances surrounding Taft's October 1910 executive order to restore for public entry, under the homestead laws, 12,800 acres at Controller Bay that had been withdrawn during Roosevelt's administration. Controller Bay, she asserted, was a strategic and commercially valuable location because it provided the only access by sea to the Bering River coalfields inland. Finally, she questioned the role of Richard S. Ryan, president of the Controller Bay Railway and Navigation Company, whom she called "a well-known factor in Morgan-Guggenheim development."[77]

Abbott's article raised intriguing questions, but her evidence was mostly circumstantial and suggestive. In the weeks after publication, she attempted to put more flesh on the bones of her story. She met several times with Pinchot and showed him various documents she had turned up during her research. Soon she had prepared a provocative sequel to her initial story about Controller Bay. Hapgood refused to publish the story in *Collier's*, however. Instead, Lucy placed her story with two newspaper syndicates. When the article first appeared in the *Philadelphia North American* on July 7, the young muckraker immediately found herself at the center of a national political firestorm. Her most sensational charge seemed to implicate President Taft directly in a conspiracy with Ballinger, to secure the Controller Bay leases for the Morgan-Guggenheim interests. As evidence, she cited a postscript in a letter from Ryan to Interior Secretary Ballinger—what quickly came to be called the "Dick to Dick" letter—which pointed to the president's brother, Charles P. Taft, as go-between in the alleged scheme.[78]

During the summer and early fall of 1911, Lucy's name was prominent in the national news, as accusations flew. She claimed that she had come across the letter while studying files relating to the Alaskan situation in the office of the new Interior Secretary, Walter Fisher. But the respected Fisher said he could not find the document containing the damning quote. President Taft's personal credibility had been undermined during the previous year's Ballinger-Pinchot hearings, by damaging testimony involving a key document which, with his knowledge, had been predated, then suppressed. That testimony had been elicited by the brilliant investigation and cross-examination conducted by the young Boston attorney, Louis Brandeis, who represented the Land Office whistle blower Glavis and *Collier's* during the lengthy hearings. The chairman of the House Investigations Committee, Illinois Democrat

James M. Graham, now secured Brandeis's services to get to the bottom of the Controller Bay matter.[79]

Taft, for his part, took the unusual step of conducting a personal investigation. In late July, he produced a detailed, closely argued statement, which vigorously refuted the charges raised in Lucy's article. The president called the Dick-to-Dick letter "a wicked fabrication" and the accusation involving his brother "utterly unfounded." But he was careful not to charge Mabel Abbott directly as the perpetrator of a hoax. Like his defenders in the press and in Congress, who accused the "ultra-conservationist element" and "Insurgent Republicans" of exploiting the controversy, the fuming Taft retorted that the incident's "only significance is the light it throws on the bitterness and venom of some of those who take part in every discussion of Alaskan issues."[80]

Neither Taft nor anyone else provided specific evidence that the young reporter had concocted the cryptic passage on which she based her most damning charges. But neither did any prominent figure spring to her defense.[81] Lucy's difficult situation also created a painful dilemma for Benton and the other MacKayes. Benton, who had begun negotiating for a full-time Forest Service position in the midst of all the turmoil, worried that if his relationship with Lucy were publicly revealed, his hope for a government job might be shattered. Mary MacKaye saw the ordeal as another chapter in her perennial battle to preserve the MacKayes' reputation. "The one great thing now is for us to keep our family name out of the papers," she fretted in the days after Lucy's article first appeared. But Benton, arriving in Washington from New York shortly after Taft issued his report, assured his mother that the MacKayes would not be associated with the controversy. He stood by his fiancée's account of the affair. "I am way on the inside as regards information," he confided to his mother. Lucy, he said, "has the protection and the council of powerful men—newspaper men, Congressmen, and such men as Senator La-Follette and Gifford Pinchot. She has been right-hand man to Brandeis—the attorney in the case. She has the esteem of the whole crowd, men and wives. . . . She is at present having the time of her life helping to work up the case."[82]

In the end, it was the word of the President of the United States against that of an inexperienced, unknown woman journalist. By autumn, none of the parties in the dispute was eager to proceed with the congressional investigation or to compel Lucy's testimony. The immediate controversy petered out when Fisher, after an investigatory trip to Alaska, promised in October to reform Alaskan development policies and pressured Ryan to withdraw his claims at Controller Bay. Lucy, and the Pinchot forces through her, were partly vindicated on the larger issue, but her own rapidly ascendant journalistic career was suddenly brought up short.

The events of the year, whether because of the public turmoil or a simple cooling of affection, had taken a toll on Benton's attenuated marriage plans. Nor did he have any success finding a publisher for his forestry textbook. But he had shown the manuscript to Forest Service chief Henry Graves, who offered him a job as a "forest examiner." In December 1911, when he assumed his official duties in Washington, Lucy was still living in the city. But the two soon went their separate ways. The next year, Lucy returned to New York. By then, MacKaye's circle of personal and professional relationships no longer included Mabel Abbott. In later years, Benton—otherwise voluble and precise in recollecting his own eventful past—did not discuss or write about this intense personal episode, which had combined a fragile romance with high political drama. Almost fifty years later, though, he noted in his diary that he was "burning old letters vintage 1907–1910"—the period of his relationship with Mabel Abbott.[83]

✢ 4 ✢

Raising Hell

1911–1915

W hen MacKaye took up his new Forest Service duties in December 1911, the exhilarating and righteous spirit of the Progressive Era was near its highwater mark. Before long, Benton was moving in a social and political circle that included numerous leading lights of the conservation movement, as well as many of Washington's prominent liberal and left-wing politicians, government officials, journalists, and activists. His professional duties, political activities, and personal life, as they had been during the recent years at Harvard, soon became inextricably intertwined. From his modest new position as a forest examiner in the Forest Service's Silvics section, Benton optimistically joined the crusade for the "wise use" of natural resources, a fight that embodied the Progressive tenets of antimonopolism, centralized government planning and control, and dispassionate technical expertise. Some historians, notably Samuel P. Hays, have asserted that a development-oriented "gospel of efficiency" characterized the era's conservation movement, especially as promoted by the federal authorities with whom MacKaye was now allied. But MacKaye's efforts and evolving ideas increasingly encompassed a social dimension that challenged the prevailing, production-oriented precepts of the conservationist faith.[1]

Benton's hiring coincided with Chief Forester Henry S. Graves's effort to expand and systematize the Forest Service's research capabilities. In January 1912, Graves named MacKaye's mentor and immediate superior, Raphael Zon, to a new Central Investigative Committee, charged with coordinating the agency's research in silviculture, grazing, and forest-product utilization.[2] Graves may have hoped to depoliticize the Forest Service's scientific activities, but simmering controversies over western grazing and mining rights, the withdrawal and development of federal water-power sites, and the disposition of Alaska's natural resources nevertheless kept the politics of conservation prominent in the public eye. The battle involved more than the alleged misdeeds of a few venal political figures and businessmen. Conservationists faced resistance, especially in the West, from ranchers, miners, congressmen, editorialists, utility companies, and other foes of "Pinchotism," who sought freer access to federal lands and resources. From political exile, Pinchot continued to wield substantial influence, through his sympathetic accomplices in the press and the federal government, as well as through organizations that he significantly controlled and financed, such as the National Conservation Association and the Progressive Party.[3]

During the first decade of the century, multiple-purpose development of river basins had emerged as a central principle of conservationist doctrine. The concept had been most thoroughly articulated in the 1908 report of the Inland Waterways Commission. "Every river system, from its headwaters in the forest to its mouth on the coast, is a single unit and should be treated as such," the commission declared.[4] Throughout the nineteenth century, federal policy toward the nation's water resources emphasized the development of navigation and the control of floods. The Progressive conservationists believed that such concerns should be integrated with other uses of American waterways, including irrigation, municipal water supplies, sanitary disposal, control of soil erosion, preservation of forests, and, not least important, the development of hydroelectric power.

The American roots of river basin planning reached at least as far back as George Perkins Marsh's 1864 classic, *Man and Nature*. Planning and development according to river basin boundaries had been promoted as a government policy since the 1870s by the U.S. Geological Survey's John Wesley Powell. His *Report on the Lands of the Arid Region of the United States* (1878) took direct aim at the Homestead Act. Powell urged that the traditional 160-acre land grant be scrapped in the West. Instead, he envisioned the establishment of cooperative organizations, arranged according to watershed boundaries, to manage common pastures and develop irrigation works. But western boomers and politicians successfully resisted his "blueprint for a dryland democracy."[5]

Nonetheless, policies designed to heed the organic integrity of watershed boundaries continued to appeal to the orderly sensibilities of scientists and reformers. A close circle of hydrographers, geologists, and engineers associated with the Geological Survey and the Reclamation Service, including such figures such as W. J. McGee, Frederick H. Newell, and Marshall O. Leighton, refined the ecological and technological rationale for comprehensive river basin development. Their ideas meshed neatly with the key axiom of scientific forestry, which held that forest cover restrained the run-off of water, thereby controlling floods, reducing soil erosion, and moderating the seasonal flow of rivers.[6] Other federal scientists and agencies, however, notably the Army Corps of Engineers, challenged the conservationists' ideas about multiple-purpose river basin development and the effects of headwater forests on downstream river flow.[7]

Some of the Americans who worried about the exploitation and despoliation of the nation's natural resources found themselves at odds with policies pursued in the name of conservation. The most notorious such battle of those years raged over the fate of the scenic Hetch Hetchy Valley in California's Yosemite National Park. Pinchot, invoking "the greatest good for the greatest number" doctrine, was a stalwart supporter of efforts by the city of San Francisco to enhance its municipal water supply by damming the Tuolumne River as it flowed through the park. Preservationists across the country vigorously opposed the plan. Led by John Muir and including such figures as MacKaye's friend Allen Chamberlain, the project's enemies viewed it as a dangerous precedent for future assaults on national parks. In 1913, however, Congress finally approved the Hetch Hetchy dam and reservoir. Hetch Hetchy marked, on the landscape and in the public consciousness, the boundaries between use and preservation. MacKaye, from his modest federal position, took no part in the Hetch Hetchy dispute. Nor had he yet begun to cultivate the rhetoric of wilderness preservation on which his later public reputation was substantially based. As a disciple of Pinchot and a friend of Chamberlain, though, he could perhaps recognize that distinctions between conservation and preservation were not always precise.[8]

MacKaye subscribed enthusiastically to the prevailing Forest Service creed that policies for managing national forests and the nation's rivers should be seamlessly integrated. Pinchot "worked from both ends of the watershed: in the forests downward from the headwaters; on the waterways upward from the main stream," as Benton later explained. "Hence, conservation and multiple use went hand in hand on American watersheds."[9] Imbued with this conviction, he was soon handed an official task that hinged on the question of federal authority over the nation's water resources. The Forest Service and the Geological Survey were then grappling with the delicate politics of implementing the Weeks Act, which was enacted in March 1911 after a decade-long battle over the creation of new national forests in the East. The law's linchpin was the Constitution's "commerce clause," under which the Supreme Court had sanctioned broad federal authority to protect and develop America's navigable waterways. Opponents of the Weeks Act had resisted the idea that the federal government could condemn and buy private lands to establish new national forests; previously, most national forests had simply been reserved from western lands already in federal ownership. The legislation's proponents, on the other hand, maintained that preservation of headwater forests, by outright federal purchase if necessary, would thereby protect downstream navigability. The theory met stiff resistance from influential scientists, like Army engineer Hiram Chittenden; nonetheless, the law won approval when Representative Weeks amended his bill to create a National Forest Reservation Commission charged with validating that purchased lands directly influenced the navigability of specific rivers.[10]

The Forest Service and the Geological Survey immediately turned their attentions to the White Mountains of New Hampshire. The Forest Service, still suspected of empire building despite Pinchot's departure, was charged under the Weeks Act only with identifying desirable forest tracts; the Geological Survey was responsible for demonstrating whether such lands did indeed affect the navigability of the region's rivers. Marshall O. Leighton, Chief Hydrographer of the Survey's Water Resources Branch, designed the methodology for coordinating measurements of downriver streamflow with headwater forest cover in ten river basins in the White Mountains. When Leighton asked Forest Service chief Graves to supply a man qualified to make the necessary forestry measurements, the job was offered to MacKaye, on the recommendation of Karl Woodward of the Division of Acquisitions. Benton possessed a sure and current knowledge of the White Mountain terrain, from his experiences as both a hiker and a working forester. He was also a competent forestry technician, as Zon recognized from past experience and as MacKaye's unpublished textbook had demonstrated to his superiors. The agencies knew exactly the results they wanted from their investigation. MacKaye's task was to secure the needed data and to present the results in a scientifically and politically credible form. To meet the strict terms of the Weeks Act, Benton was temporarily reassigned to the Geological Survey. Before heading to New Hampshire that summer of 1912, he received his instructions in a series of talks with Leighton. He briefly joined another Survey employee, John Hoyt, in the heart of the White Mountains. Their makeshift headquarters was a shack beside the East Branch of the Pemigewasset, along the logging railroad running through the timber domain of J. E. Henry and Sons.[11]

Throughout the late summer and fall, "armed with plane table" and specially prepared Geological Survey topographic maps, MacKaye surveyed the forest cover of eight small river basins, encompassing headwater tributaries of two larger rivers, the Pemigewasset and the Ammonoosuc, which rose in the White Mountains. "I sketched the overall cover from the high points, and made cross-sections along the profiles of the critical slopes," he wrote many years later, when Forest Service officials resurrected his study as a model of technical analysis—and were surprised to learn that its still well-informed author lived nearby at the Cosmos Club. "I recall especially the section along the cog railroad from the base to the summit of Mount Washington. This span contained every one of C. Hart Merriam's 'life zones' between the latitudes of N.H. and Labrador, from 85' maple and 40' spruce and fir to 5' jungle of everything to the subarctic tundra." After the last tourists of the season departed, Benton bunked with the keeper of the railway base station. He rode in the engine as the railway made its last trip of the year to Mount Washington's snow-covered summit, where the wind spun him around "like a drunken man." His superiors, Woodward and Zon, joined him briefly in the mountains; at the end of Zon's visit Benton "had the joy of giving him a fifteen mile race with daylight through Zealand Notch" to catch a train. But he carried on most of the work alone that autumn in the White Mountains, "an audience of one in the kingpin scene of northeastern U.S.A."[12]

That winter, back in Washington, MacKaye prepared an elaborate series of charts and tables compiling his measurements of tree species, crown density, slope, altitude, and other elements of forest cover. His findings were correlated with the Survey's waterflow measurements in an unpublished 1913 report, "The Relation of Forests to Stream Flow," which corroborated a direct relationship between the density of mountain forest cover and the moderation of river flood waves downstream. "All in all, so far as my own part was concerned, it was a thumb-nail sketch, an attempt during a few man-weeks to bridge the techniques of a forthcoming science still in the making," MacKaye later recollected of his first significant exercise in "geotechnics." He had used his technical skills to achieve a specific social and political end. He had helped, in a modest way, to refine the discipline of scientific forestry in America, which had proceeded as much on faith and uplifting rhetoric as on empirically documented fact. His efforts had also buttressed the legal foundation of the Weeks Act, which withstood subsequent court challenges. And, not least important, the eastern national forests that soon began to take shape (including the White Mountain National Forest) would later provide a protected environment for significant stretches of the Appalachian Trail.[13]

❖

Benton was among the nearly one million Americans who cast their 1912 presidential votes for Eugene V. Debs, the Socialist Party candidate.[14] Debs's backers represented only six percent of the voters, but never before—or since—had a socialist candidate for president won so much support among the American electorate. In fact, Debs could claim a moral victory by comparison with the humiliating showing of President Taft. The incumbent ran third behind his predecessor Teddy Roosevelt, who campaigned under the Bull Moose banner of the Progressive Party, and the Democratic victor, former New Jersey governor and Princeton University president Woodrow Wilson.

For Washington activists of a liberal bent, the advent of a new administration offered fresh opportunities to advance both their ideas and their personal prospects. After returning from the White Mountains in the fall of 1912, Benton had moved from the boarding house where he had previously resided to an apartment at 2829 27th Street, off Connecticut Avenue near the National Zoo. His new roommate was William Leavitt Stoddard, a correspondent for the *Boston Evening Transcript,* whom he had met shortly after his arrival in Washington the year before.[15] The two young men shared more than bachelor status and New England backgrounds; they also both took a keen interest in the whole range of issues and events reverberating through the nation's capital at the time. Soon, at their apartment and at Stoddard's office, they organized after-hour "pow-wows," consisting of a "bakers dozen" of reform-minded bureaucrats, journalists, and activists.

"Stod and I have plans for running salons next year. (We call them saloons)," Benton reported to his mother. MacKaye later said of the "Hell Raisers," as the comrades-in-political-arms dubbed themselves, "There were three brands: newspaper men, . . . government men, . . . institutions on two legs." Besides Stoddard, the journalists belonging to this informal fellowship included Fred Kerby, correspondent for the Newspaper Enterprise Association and a hero of the Ballinger-Pinchot affair; Laurence Todd, who would go on to become the long-time U.S. correspondent for the Soviet Union's Tass news agency; Charles Ervin and Paul Hannah, both writers for the socialist *New York Call;* and Art Young, political cartoonist for *The Masses.* One of the "government men," other than MacKaye, was Hugh Hanna, an economist with the Bureau of Labor Statistics. By "institutions on two legs," Benton referred to some of Washington's persistent freelance reformers, lobbyists, and gadflies, such as Judson King, director of the National Popular Government League; Pinchot's faithful operative Harry Slattery, secretary of the National Conservation Association; and Basil Manly, an economist who shuttled between newspaper jobs and staff positions with such government agencies as the Commission on Industrial Relations and the Federal Trade Commission.[16]

Benton and his colleagues worked behind the scenes, writing legislation and speeches for congressmen sympathetic with their aims, orchestrating publicity campaigns through their press connections, and promoting their causes within government agencies. In Washington's tight-knit political community, the Hell Raisers were representative of what one historian has called the era's leading "independent progressives," who attempted to influence political debate on issues ranging from the support of Mexican revolutionaries to the cause of woman suffrage.[17]

One of the group's efforts involved the promotion of a 1913 bill, drafted by Benton on the basis of the ideas of his brother James, for the development and administration of the natural resources of Alaska. The Ballinger-Pinchot affair and Mabel Abbott's subsequent allegations had highlighted, but not resolved, persistent issues concerning the territory's prospects. Since the days of the Harvard Socialist Club, Walter Lippmann had urged James to develop practical applications for his abstruse theories of "pantocracy" and "political engineering." Now Benton could help his brother find a political foothold. The Alaska bill, filed in July 1913 by two progressives from the state of Washington, Senator Miles Poindexter and Representative James W. Bryan, was an even more ambitious version of proposals and bills already advanced by Pinchot, La Follette, Brandeis, and other reformers.[18] The various progressive schemes to date had all included measures to thwart private monopolization of coal and other minerals, to maintain significant federal control over resources on public lands by a system of leasing, and to construct a government-supported railroad in the resource-rich territory. Benton and James, by contrast, saw Alaska as the laboratory for a large-scale political and social experiment in the public development of land and natural resources.

The MacKaye plan provided for creation of the Alaska Transportation Service, a federal agency to build and operate not only a railway but also an interconnected coastal and river navigation service. The Alaska Mining Service, the other key provision of the bill, smacked of outright socialism, however. The MacKayes proposed that the government lease half of Alaska's coal reserves to private developers. But the other half, in their plan, would be retained and developed by the government itself. Under this system of "regulated competition," private leaseholders and the government mining service would meet comparable standards for minimum wages, maximum working hours and prices, worker safety, life and health insurance, and housing. In Progressive parlance, public development would provide a yardstick by which to measure private enterprises's true efficiency and profitability. Perhaps the most radical feature of the legislation, though—dooming any hope of enactment—was a proposed dividend fund, which would split the government mining service's annual surpluses between employees and customers.[19]

Benton and his fellow Hell Raisers William Leavitt Stoddard and Fred Kerby organized a press campaign to promote the initiative.[20] The MacKaye-inspired bill did not win passage, but compromise legislation for Alaskan development was enacted in 1914. A law initiating a modest leasing system for Alaskan coal set a precedent for leasing of mining rights rather than outright sale of mineral lands in the public domain. In addition, federal funds were approved that year for a railroad from Seward to Fairbanks. Benton MacKaye's initial legislative effort did not achieve its grand aims. But he maintained a keen interest in Alaska, and though he never traveled to the territory he described as "the last American frontier," he long held in mind the image of Alaska as "one of the most promising areas now left on the globe for those seeking a new start in life."[21]

MacKaye's work on the Alaska bill was not a formal Forest Service assignment, although he enjoyed wide latitude and a relatively free rein within the agency, under the general supervision—and with the apparent approval—of Raphael Zon. Throughout the next thirty-five years of MacKaye's intermittent federal employment, he would often promote policies and legislation that stretched the limits of debate within the agencies for which he worked.

♣

On March 3, 1913, the day before Woodrow Wilson's inauguration, MacKaye looked on as 8,000 marchers, parading down Pennsylvania Avenue in support of woman suffrage, were met with taunts and jibes by heckling onlookers. The riotous scene on the capital's streets marked the advent of a more combative phase in the battle for women's right to vote. The women of the MacKaye clan were well represented among the suffrage demonstrators. Mary MacKaye had marched the first steps with a delegation of New York women as they set out to walk from their home state to Washington for the inaugural display of female solidarity. Hazel had recently been enlisted by the charismatic, iron-willed Alice Paul, head of the Congressional Union (CU), a militant suffrage lobby then splintering from the more conservative National American Woman Suffrage Association (NAWSA). Hazel's job was to write and produce pageants for the cause; her well-received suffrage pageant, *Allegory,* was staged on the steps of the Treasury Building as part of the unofficial inaugural activities.[22]

A high point of Benton's Washington social life occurred that October, when he accompanied Percy to the White House. The playwright had been summoned for an afternoon of tea and dancing in the East Room with the President's daughter, Eleanor, who had performed in the playwright's *Bird Masque* that summer in New Hampshire. "We had a pretty good time, chewing the rag," Benton reported to his sister, "mostly about play acting."[23] It was among Washington's sisterhood of suffrage activists, however, that MacKaye and his hell-raising friends found politically like-minded female comrades, as well as social companions. Before long, some of the intelligent, ambitious women affiliated with the Congressional Union were regular companions on the weekend tramps that Benton had begun leading in the countryside surrounding Washington—"suffragette hikes," he called them. The sociable bachelors and suffragists often got about town in Stoddard's Model T, which they dubbed the "Jumart" (a mythical cross between horse and cow). Benton later remarked that it was "a rolling cradle of reform; without it, Amendment [Nineteen] might still be in committee."[24]

Washington's customarily sultry summer atmosphere seemed all the more oppressive in 1914, beause of the ominous events unfolding in Europe. At this emotionally charged political moment, he first met Jessie Hardy Stubbs, a Congressional Union activist who shared his own antiwar sentiments. "It was in 1914, in the opening days of the European War, that she and I first found each other," he later remembered. "Our lives were bound together and enfolded by sympathetic forces—beyond our comprehension—which were let loose at that hour on the world."[25]

Indeed, in early July, as he prepared to head to the upper Midwest on Forest Service duties, Benton penned an exuberant, emotional letter to his new acquaintance, recently departed for New York. His effervescent rhetoric somehow obscured his true feelings, however. He spun out a parable about the evolution of the sexes, describing himself in the third person as one of the "Jumart twins" (the other was his roommate Stoddard). The "'three great S's'—Suffrage, Socialism and Sex"—were conjoined, Benton claimed, in Stubbs and her fellow suffrage crusaders, who "had the real salt of life." His thoughts, he confessed, had "left the realm of language" after he had seen her off for New York on the train, and he had broken down in tears of joy. He related in comic tones his failed efforts to make spiritual contact with her at an appointed hour, an experiment they had previously arranged. And he concluded by describing how his own emotional revelation merely served to illustrate his theory of humanity's place in the universe's progressive evolutionary scheme:

Homo sapiens has not yet begun to live within the universe; he simply grunts upon the earth. He will not begin till both his halves find themselves and one another. When this happens the harmony, the waves to which we respond will make present mankind look like 29 cents. All friends will then have more than lovers have now. After we solve the crude problems of grub and mechanical welfare the problem of human harmony will really start. But even now its first faint breathings are heard. They have been heard by the Jumart twin.[26]

In other words, even though he couldn't quite bring himself to say so straight out, Benton MacKaye was in love. "Betty" Stubbs (a nickname she took after performing the part of "Betty Sinclair" in an amateur play), born in Chicago in 1875, was almost four years older than Benton.[27] By the time she first became acquainted with the enthusiastic, socially conscious government forester, she had spent nearly two decades working for the political and social liberation of American womanhood. Her personal experiences had sometimes been harsh. As a nineteen-year-old art student, her vocational plans shifted course abruptly when her mother died. "It happened so suddenly that it made a woman of me immediately," she told a reporter years later. She enrolled as a nursing student at Chicago's St. Luke's Hospital, where she soon became head surgical nurse. In 1896 she married a doctor, F. Gurney Stubbs, who encouraged her independence and idealism as she enrolled for courses at the University of Chicago. But he died only three years later, leaving his childless young widow with sufficient financial means to indulge her hunger for education, personal fulfillment, and social justice.[28]

Betty lectured before the genteel women's clubs of Chicago society on subjects ranging from the lives of Henry D. Thoreau and Booker T. Washington to the art of Michelangelo and the French Impressionists. But her fundamental mission, she claimed, was "to devote her life to the solution of sexual problems" and to "renovating the morals of the race."[29] She worked during the early years of the century in a series of Illinois political campaigns involving women's issues, including raising the age of sexual consent from fourteen to sixteen and tightening the state's prostitution laws. Finding herself "handicapped because I was a woman . . . I decided then to go after the ballot," first as an activist and officer of the Illinois Equal Suffrage Association, which helped secure limited voting rights for the state's women in 1913.[30]

Betty's involvement with the Illinois campaign introduced her to the national network of suffrage activists. She came to New York to work for Harriet Stanton Blatch's Women's Political Union, quickly winning a reputation as an organizer and public speaker. She organized and walked the full length of a suffrage march from New York to Albany, and spent much of the summer of 1913 in Dutchess County, New York, during the campaign for a state suffrage amendment. By August of that year she was in Washington to take part in a major suffrage demonstration.[31]

Betty's talents and commitment soon came to the attention of the Congressional Union's Alice Paul, who was assembling a team of determined, capable activists to wage her aggressive campaign for a federal amendment ensuring woman suffrage. (The more cautious NAWSA still advocated the adoption of state suffrage amendments.) Betty joined a tight-knit cadre of suffrage shock-troopers, such as Lucy Burns, Doris Stevens, Crystal Eastman, and Inez Milholland, who carried out Paul's toughest political and organizing assignments. Paul recruited Betty to organize the CU's press department. By early 1914 Stubbs was traveling throughout the country to organize nationwide demonstrations, set for May 2, demanding passage of the "Anthony amendment" (named for Susan B. Anthony). She was back in Washington by May 9, for a 15,000-woman march through the city, carrying the CU banner as head marshal.[32]

Hazel MacKaye had been involved with the CU virtually since its inception. It was apparently through his sister and her suffragist friends that Benton first met Betty Stubbs. In the frenzied political atmosphere of Washington, debate over war, suffrage, conservation, and a myriad of public issues often replaced conventional social discourse and colored domestic relationships. This was the stimulating, if sometimes enervating, background against which was played out the couple's brief courtship and ill-fated marriage.

In the summer or 1914, shortly after his euphoric parting with Betty, MacKaye was dispatched by the Forest Service to the Great Lakes states—Wisconsin, Michigan, and Minnesota—to study conditions in the region's "cutover" district. This northern swath of the three states had been largely denuded by the forest industry and abandoned by many settlers. "I'm glad to be getting away," he admitted to Percy. "I need the change, and hope to return with a less rusty mind."[33] Working out of the agency's recently established Forest Products Laboratory in Madison, Wisconsin, often traveling with forester E. A. Frothingham, Benton crisscrossed the northern counties of the three states during the late summer and fall, investigating ways to help farmers and landowners market their lumber products. The trip, followed by another extended stint in the region a year later, opened his eyes to new physical terrain. He also encountered troubling social and political conditions that overshadowed the mundane details of wood-products marketing.

En route to Madison, MacKaye stopped in Chicago, where he visited the weed-covered, trash-strewn site of his father's failed Spectatorium on the shore of Lake Michigan. He described the scene in heated rhetoric to his brother Percy:

The water remains a big blue highway skirting to the Gods. But the land!—In the foreground, on the beach, occurred a layer of sardined humanity in bathing-suits, having as high an "output of happiness" as probably the average American ever gets. Here were "the players." Over their heads, and miles beyond, stood out the huge steel plants of South Chicago, with their twenty smoke-stacks, each issuing a grim black cloud, that streamed indefinitely across the prairie and closed it from the sun. Here were "the workers."—The picture made an exact diagram of play and work and commercialism in America. Here on the beach was our feeble attempt at attaining Heaven; back in the phalanx of smoke-stacks was our titanic triumph in attaining Hell.[34]

For the rest of his trip—indeed, for the rest of his life—MacKaye perceived an America laid out according to that "exact diagram of play and work and commercialism." As he traveled the upper Midwest, he sent a steady stream of reports to his family back east, describing both the evidence of capitalist depredations and the prospects for cooperative alternatives. "It's a big experience to wander around the country and meet big people," he wrote his mother upon arriving in Madison, "the most advanced town in the United States." While living at the University Club between his forays into the countryside, he met other Progressive activists, who exemplified what came to be called the "Wisconsin Idea," including labor economist John Commons and members of the staff of *La Follette's Weekly*. In Milwaukee, he discussed the war with Victor Berger, editor of the socialist *Milwaukee Leader* and a prominent leader of the Socialist Party.[35]

In the "cutover"—the counties of the northern Lakes states that lumber companies had stripped of white pine during the last decades of the nineteenth century—MacKaye witnessed a social and environmental tragedy in the making. As lumber barons like Frederick Weyerhaeuser deployed their capital in the still-abundant forests of the South, West, and Pacific Northwest, they left behind an economically and environmentally devastated landscape. Until the turn of the century, the region's settlers had briefly survived, even thrived, on a mixed economy of agriculture, logging, and mining. But they were soon faced with a shorn forest, abandoned sawmills, and competition from western iron and copper mines. In response, lumbermen, land development companies, and government agencies conceived a cunning plan to salvage the regional economy, by preying on the hopes of Americans and would-be Americans who still yearned for farms of their own. The best farmlands elsewhere in the Middle West had already been homesteaded. Poor farmers unable to afford land prices and rents elsewhere, city dwellers eager to build a better life, and European immigrants, mostly from Germany and Scandinavia—all were lured to the cutover by extravagant claims about its prospects, fertility, low land prices, and deceptively easy credit.[36]

Cash-strapped settlers were quickly overwhelmed by the impossible challenge of blasting, burning, or pulling stumps from those marginally fertile farmlands, however. "At no time in American history was the prospective settler at so great a disadvantage," observes one historian of the region.[37] Many years later, MacKaye remembered the desperate situation faced by settlers. "The region had been the scene of giant first-growth pine trees whose stumps were equally gigantic," he wrote. "The game was to sell these lands to the prospective settler, omitting to advertise the stumps. I reported to one of the outfits the large number I counted on one of his acres. His response was swift and pungent, 'For God's sake, don't tell anybody!'"[38]

In the plight of one such Wisconsin settler, Vincent Was, MacKaye witnessed the grim situation in dramatic and personal terms. He compiled a detailed biography of Was, who was eager to recount his baleful, archetypal story to a sympathetic listener. Was had acquired his eighty-acre farm in Conrath, in northwestern Wisconsin, in 1909. The Polish immigrant, a veteran of the Prussian army, had arrived in America only two years earlier. In Milwaukee, he had worked in a tannery, lumber yard, and a brewery. A land company advertisement in a Polish-language paper had enticed him to make a downpayment towards the farm and to assume an $800 mortgage. He had paid the company another $200 for the house on the property. He had brought his wife and two children to the farm only to find it covered with pine stumps and piles of old logs. Then his situation had quickly worsened. He had run through his savings while clearing and improving the farm. Although he had hired himself out as a day laborer, he had slid still deeper into debt acquiring livestock and equipment.[39]

MacKaye viewed in sardonic but sympathetic terms the fate of Was and countless settlers in similar circumstances. "With good luck, perfect health, a strong wife, not too many children, a 15 to 18 hour working day, good soil, no accidents, merchantable timber and a market for it," Benton wrote, "he may within a decade be able to clear his farm sufficiently to pay his interest, taxes, and other carrying charges,—and also make a living. Such is the theoretic possibility and some actually attain it."[40]

For more than a year after MacKaye's visit to Wisconsin, the desperate farmer kept up his correspondence with the "government man" who had at least taken the trouble to listen; but Benton was exasperated and helpless himself, uneasily explaining to Was that there was little he could do to allay the farmer's troubles.[41] As MacKaye witnessed the plight of solitary, hardworking Americans like Was struggling for a livelihood, he began to conceive a broader, alternative program for the development of the nation's public lands—a program that would strive to fulfill an American tradition of community rather than unbridled individualism. He began to visualize a salvaged, recreated American landscape, sustaining permanently both the people who lived on it and the natural resources it encompassed.

MacKaye's 1914 field trip through the Lakes states coincided with the outbreak of war in Europe. Wherever he went in his travels, he reported to his sister, people were eager to talk about the growing conflict:

> It is easy to get their ear. They will listen and devour anything about the war; they will not look at anything else. I won't myself. I find myself reading everything about the war and nothing about any other thing. They will listen as to the cause of war—the guarantee of bona fide peace—to putting down militarism. The bottom cause is capitalism and here is the chance to introduce to many minds the idea of socialism. It is the chance to introduce a bigger idea than socialism and one which starts not counterprejudice: I mean the idea of political engineering.[42]

In his travels, he observed on the American landscape the symptoms of a social disease more malignant than the war itself. MacKaye, like many of his Hell-Raiser friends, was a determined neutralist, opposed to U.S. involvement in the growing war. He hewed to the view held by many on the left that the conflict was largely instigated and promoted by profiteering munitions manufacturers and other capitalists. After visiting the iron ore mines in Hibbing, Minnesota, he reported, "Here's where the steel trust spoons out the stuff with which to make the dreadnaughts—the ultimate source."[43] At copper mines in Calumet, Michigan, where he campaigned with the Progressive congressman William J. MacDonald, MacKaye got his "first real smell of the revolution," he told Hazel. "This is one of the vent holes of the volcano of the U.S. . . . A cloud of sinister suspicion hangs over all."[44]

Disturbed by his observations but encouraged by the example of the likeminded men he met during his travels, MacKaye turned his energies to the battle against militarism and U.S. participation in the war. The situation looked to him "like America's great chance—the chance perhaps of a nation's lifetime—to stand off quietly with three thousand safe miles of salt water between us and hell, and contemplate the latter," he wrote Betty, who was in Atlantic City that summer managing a suffrage campaign for the Congressional Union. Benton's attitude combined a strict isolationism with a righteous vision of America as leader of a new world order. The conflict's moral, he wrote, "is for us to *learn* and then *teach* the story of this war."[45]

Upon his return to Washington in November, MacKaye, joined by William Stoddard and by Harry Slattery, Pinchot's principal operative, moved to an apartment house closer to the city center. They took over rooms previously occupied by Betty Stubbs and other CU colleagues at The Milton, which served as an informal headquarters of reform, at 1729 H Street. (She was campaigning that fall in Oregon during the first test of Alice Paul's electoral strategy.) MacKaye and his roommates, joined by another Hell-Raiser, Julian Leavitt, plotted a "war plan" to attack the causes of war on several fronts. A self-styled "Provisional Committee," they drafted a statement calling for a "War Council of Peace," to take place at year's end. Their hope was to mobilize "a string of hell-raising groups" in cities throughout the country to support a program of legislation and propaganda intended to quash the growing sentiment for expanding the nation's military capabilities. Benton sent the manifesto off to his brothers Percy and James, to Walter Lippmann in New York, and to acquaintances in other cities. He also met with the prominent pacifist David Starr Jordan, retired president of Stanford College and director of the World Peace Foundation.[46]

Congressman William J. MacDonald, with whom MacKaye had campaigned in Michigan, filed a "war preparation bill"; Benton seems to have been little help on the hustings, for MacDonald was now a lame duck.[47] MacKaye drafted a "Government Munitions Bill," filed in late December by Representative Robert Crosser, an Ohio Democrat, designed to eliminate the profit motive for instigating and pursuing war. The proposed legislation called for the complete government takeover of the manufacturing of all military and naval equipment, the prohibition of the export of privately made military munitions, and government reservation for military use of coal and oil resources on public lands. The bill reflected the belief of many progressives that the federal government alone could present an effective countervailing force to avaricious corporate interests. For some on the left, the most direct means of preventing war was to socialize the machinery of warmaking. This episode illustrated a dilemma with which MacKaye struggled throughout his career: could the alluring, large-scale social-engineering capabilities of federal authority be truly reconciled with the grassroots powers, needs, and desires of American citizens?[48]

Antiwar activity was on the rise throughout the country. The Women's Peace Society, led by Jane Addams, and groups like the League to Limit Armaments and the American Union Against Militarism locked horns with those arguing for an increased American commitment to the war effort. Woodrow Wilson himself, in December 1914, seemed to heed the call of the antiwar activists when he told Congress that strengthening the National Guard, rather than building armaments, was a sufficient national response to the European war.[49] MacKaye's idea for a War Council of Peace and the Crosser munitions bill won little immediate support. As a low-level government employee with no public reputation, MacKaye kept his name off the documents that emanated from The Milton. But during the next two years of intensifying debate and controversy that led up to America's entrance into the war in April 1917, he and his friends remained close to the center of Washington's network of persistent, if ultimately defeated, pacifists and antiwar activists.

MacKaye's unofficial political activity was becoming more public, but its efficacy was doubtful. In early May, 1915, he gave a talk titled "A Substitute for the Economy of Wealth" to a meeting of the District of Columbia's Socialist Party. "'Red' Agitator Stirs Hearers to Yawns," read the headline on a newspaper story describing the meeting. MacKaye drew a version of his brother James's daily "mood diagram," which purported to depict "the joys and sorrows that come to the average man in a single day." But some in the audience of eighteen, wrote the acerbic reporter, "enjoyed almost uninterrupted slumber."[50]

Benton was persuasive in a more intimate forum, however. With little warning to their acquaintances, Benton and Betty Stubbs were married on June 1, 1915, in a small ceremony at the Chevy Chase, Maryland, home of a friend of Betty's.[51] A photograph depicts the small wedding party posed in front of the handsome, tree-shrouded country residence of their hostess. Benton is flanked on one side by his new wife, on the other by his sister and mother. Hazel and Mary MacKaye were forced to adjust to a sudden change in the family's center of gravity. For twenty years, except during the period of his engagement to Lucy Abbott, Benton had been theirs alone. Now they were competing with another woman for his attention, his affection, and, not least important, his financial support.

When Benton MacKaye and Jessie Hardy Stubbs married, they had been acquainted for not much more than a year. During that interval they had been separated for extended periods while pursuing their respective callings and causes. For his part, Benton was serene about this dramatic new chapter in his life. His wife, however, may not have sensed the difficulty of penetrating the MacKaye family circle—or of escaping its orbit. "Betty is a far more wonderful character herself than even I had realized—this comes home to me more and more as I see her in a somewhat daily life here," he wrote Percy. "I wrote Jake that she reminds me in some ways of Aunt Sadie—her ability to do things about the house and her interest in it. I don't think I am exagerrating [*sic*] when I say that she strikes me as a young Aunt Sadie. She gives me a sense of security that I didn't know existed."[52]

Betty's sudden matrimonial plans "surprised the suffrage world," according to one bemused account in a Washington newspaper.[53] But she had a ready answer when asked how she could reconcile the duties of marriage with her commitment to the suffrage cause. "Of course I want to be a good home maker," Betty told one reporter. "And there are people I might mention who say I am a good cook. After all, suffrage for women simply means a larger home-making, a larger home-building. Woman suffrage means that women shall take the home influence out into the world. . . . The new idea of home making is that women should go out with her home making influence and drive the evil out of the world—in other words, to make the world a better home for the human race." Modern women, Betty told another reporter only a few weeks before the wedding, "are increasingly unwilling to enter the bonds of matrimony on any other ground than that of mutual love and mutual responsibility. They do not intend to descend to man's level of morality, but demand that he scramble up to theirs."[54]

The events of Benton's married life would provide ample evidence that he shared many of Betty's high-flown conjugal aims. Their relationship was based on genuine affection, tenderness, and mutual respect. Although they were idealists, their personal experiences had not left them starry-eyed about life's exigencies and unpredictability. Nonetheless, Benton and Betty remained devoted to the larger national issues and causes that occupied their daily energies. The strains and stresses created by those sometimes futile public efforts could not help but spill over into their personal lives.

⁕ 5 ⁕

Reclaiming America's Wild Lands
for Work and Play

1915–1916

When Benton returned to Minnesota and Wisconsin in the fall of 1915 to continue his Forest Service fieldwork among farmers in the cutover districts, Betty traveled throughout the two states as well, speaking and organizing for the suffrage cause. The midwestern working excursion provided the newlyweds a practical opportunity to share their fervor for common political and social causes. "Now Benton, bless him, has heaps inside," Betty wrote to her sister-in-law Hazel, "but it has not come out. He has withal such a *passion* to serve the world that it is touching. Everything must be done to make it possible for him to fulfill his true destiny."[1]

Benton foresaw a possible change in his current status as a government employee. "Of course I may go rgt. on with the F.S.," he reported separately to his sister, "but unless some big changes in attitude are shown, I want to get out—sooner or later—or that office will make a cave man of me." Betty and he were contemplating the Thoreauvian alternative "of lowering the standard of living and thus gaining independence," Benton added.[2]

As he surveyed the bleak state of the cutover lands that fall, Benton pondered the fundamental relationship between human labor and natural resources—a relationship, he now believed, that had broken down on the American landscape. Back in Washington at the end of the year, settling into his H Street apartment with Betty, he mounted a personal campaign to radically transform federal policies for the development of land and natural resources.

MacKaye worried that the Forest Service was becoming increasingly oriented toward the economic priorities of the lumber industry, at the expense of other social interests and objectives. Changes in the direction of Forest Service policy in such areas as research, timber management, and recreation influenced both his hopes for and his prospects with the agency. For years Raphael Zon had promoted expansion of the Forest Service's commitment to scientific forest research. In 1908, he had proposed establishing experiment stations in national forests throughout the country; the first was opened that year in Arizona. He had also urged the Forest Service to create an independent forest research division, insulated from day-to-day administrative pressures and influences. But in June of 1915, Henry Graves replaced the system of central and district investigating committees, which had operated since 1912, with a centralized Branch of Research, headed by Earle H. Clapp. The centralization of control of research in Washington shifted bureaucratic power from district foresters to Forest Service headquarters. Under Clapp's leadership, MacKaye's freedom of operation within the agency may have been constrained, as the scope of Zon's influence over research activities narrowed.[3]

The Forest Service was also struggling to rationalize—in every sense of the term—its timber-management policies. Timber prices had already been softening for several years before the national economy entered a severe and prolonged economic slump in 1913. It was not the slower economy alone that accounted for the weak lumber market, however. Economically competitive alternative fuels and building materials, such as oil, coal, iron, and steel, replaced wood products for many purposes. Furthermore, the lumber industry's capacity had expanded substantially during the early years of the century, especially in the Southeast and the Northwest.

During his tenure as chief forester, Gifford Pinchot had regularly predicted that the nation was on the verge of a "timber famine." The weak timber market now seemed to belie Pinchot's dramatic claim; but it did not prevent reformers from worrying about the apparent trend toward monopolization and concentrated ownership of private forestlands, especially in the Northwest. A three-volume report by the Bureau of Corporations issued in 1913 and 1914 seemed to buttress trustbuster claims that a handful of companies controlled lumber supplies and prices. Almost 80 percent of the nation's standing timber was privately owned, according to the report. Three companies alone—Weyerhaeuser, the Northern Pacific Railway, and the Southern Pacific Railway—held 11 percent of America's timber. The Bureau of Corporations charged that such companies could and did hold timber off the market, thereby maintaining artificially high prices.[4]

The Forest Service itself controlled a relatively small portion of the nation's marketable timber; but the agency's timber-selling policies could still influence lumber prices, particularly in localities near national forests. Moreover, the Forest Service was under persistent pressure from Congress to demonstrate the agency's claim, made under both Pinchot and Graves but not convincingly substantiated in practice, that the revenue generated from national forest timber sales would make the reserves economically self-sustaining.

Graves's chief of timber management was Yale-trained William B. Greeley, an exact contemporary of MacKaye's. Greeley's sympathy with lumber industry priorities placed him at odds with agency liberals like MacKaye, who favored increased public ownership and regulation of the nation's forests. Greeley prepared a 1917 Forest Service study, *Some Public and Economic Aspects of the Lumber Industry,* to counter the anti-industry thrust of the Bureau of Corporations report. MacKaye was able to examine an early draft of Greeley's work, which only confirmed his fears about the trend of Forest Service policy. Greeley supported lumbermen's claims that the industry was burdened with excess mill capacity, unreasonable taxes, volatile markets, high carrying costs for "cheap timber acquired from the public domain," inefficiencies in manufacturing and distribution, the risk of loss from forest fires, and competition that was "not only keen but often destructive." To help the overcapitalized industry cope with such economic conditions, Greeley maintained, the Forest Service should set price stability as its principal timber-selling objective. Cooperation—both among lumber producers and between the government and the industry—became the Forest Service's guiding principle under Greeley's influence, and later direction.[5]

Pinchot, president of the National Conservation Association during these years, called Greeley's report "a whitewash of destructive lumbering." MacKaye, sharing his former boss's skepticism about the lumber industry's influence on forest management, wrote his own lengthy assessment, "The National Forest Timber Policy," which challenged many of Greeley's conclusions. He maintained that more aggressive national-forest management would provide the surest antidote to timber monopolization. His plan to preserve the nation's timber supply and to thwart what he called the "Law of Concentration" then at work in the American forest called for market-rate sales of national-forest timber to commercial operators; strict enforcement of timber-sale contracts, to ensure that commercial producers did not withhold timber from the market; and public regulation of private logging railroads.[6]

The growing influence within the Forest Service of the industry-friendly policies articulated by Greeley was reflected in his appointment as chief forester in 1920. MacKaye's report was not published by the agency; it indicates the trend of his own thinking about the optimal use of natural resources. He was seeking means and measurements of forest management that emphasized social values other than just the production of more lumber, wages, or profits. Benton's ideas about the uses of national forests were diverging sharply from the Forest Service priorities as expressed by Greeley.

❧

While reading the 1915 annual report of the Department of Labor, MacKaye had taken notice of Labor Secretary William B. Wilson's appeal for a more active federal role in "making new opportunities for employment." Wilson proposed that the government use portions of the public domain, as well as certain private lands, "for promoting labor opportunities as advantageously as other areas have been acquired or retained by it for the creation of public parks."[7]

"It will not be enough to hunt 'manless jobs' for 'jobless men,'" Wilson wrote. Instead, he sketched out a program for workers based on the effective repeal of the nation's hallowed homestead laws. "Those laws relieved the industrial congestions of their day by opening the West to workers of pioneering spirit who set up individual homes and created independent farms in waste places," Wilson concluded. "But the day of the individual pioneer is over."[8]

Wilson proposed the cooperation of the Departments of Labor, Agriculture, and Interior "to bring the right men to the right places on the soil and settle them there under favorable circumstances." A government-subsidized "rotary fund" would help settlers finance equipment and improvements; and the agencies would educate farmers in up-to-date agricultural techniques. The program's most important and controversial element, though, was its proposal that the federal government retain title to remaining public lands. Government title was essential, Wilson maintained, to prevent "the inflation of land values" that undermined the prospects of stable settlement and secure employment.[9]

The Labor Department report had been prepared principally by Wilson's assistant secretary, Louis F. Post. Hence, its assault on conventional notions of private land tenure reflected the single-tax rhetoric and reasoning of Henry George. As editor of the Chicago *Public* and as "Single Taxer No. 2," Post, "a philosopher with an intellectual curiosity that never rested," had long been George's most ardent and persuasive disciple.[10] George had laid out his controversial philosophy in a wildly popular 1880 book, *Progress and Poverty,* which attracted the support of many Americans who felt they had been deprived of the bounty of the Gilded Age. A key tenet of George's doctrine was that the public was entitled to receive the "unearned increment" of land value created by the community itself. A steep annual "single tax" on land, George contended, would eliminate land speculation, monopoly ownership, and tenancy, while providing incentives for more democratic and productive access to land and other resources.

Benton may already have known of "land man" Post through his Hell-Raiser acquaintances in the Department of Labor. In any event, when he wrote to the assistant labor secretary in late December, proposing a joint study of colonization possibilities on national forests by the Labor Department and the Forest Service, MacKaye could anticipate a favorable response. "The question of developing wild lands upon modern principles is a national and not a local question," MacKaye asserted. Such a "new homestead principle," he concluded, could not be viewed from the narrow perspective of one profession or government agency. "The forester's task, then—that of marketing the wood-lot and small timber products—is found to be but one link in a chain of difficulties which, taken as a whole, constitute the one big problem of reclaiming and colonizing wild lands," he wrote. "Thus the Department of Agriculture in this work has before it the problem of bringing together the man and the land; our Department approaches the question from the standpoint of the *land* while your Department approaches it from the standpoint of the *man*." In "A Government Policy for Reclaiming and Colonizing Wild Lands," one of the several memorandums MacKaye prepared to elaborate his ideas for Post and other officials, he insisted that the federal government "humanize" its conservation policy.[11]

"The kingpin question with 'Uncle Louis,'" MacKaye later wrote, "was this one of land and labor. How to make them meet?"[12] The assistant labor secretary had fashioned a policy that "emphasized an aspect of the conservation of natural resources somewhat overlooked by the conservationists, namely the steady supply of the earth's material as the basis of 'profitable employment.'"[13] Post, in MacKaye's view, had focused on human needs in a way that the commodity-oriented rhetoric and policies of the Forest Service did not.

The Labor Department at first did not act formally on MacKaye's proposal for a joint colonization study by the two departments. Nonetheless, with Raphael Zon's encouragement, MacKaye began collaborating intensely with Post. Soon, he had fleshed out a legislative program embodying his ideas for colonization and public employment on public lands. At the instigation of journalist friend Fred Kerby, MacKaye sought the legislative sponsorship of Congressman Robert Crosser. The Ohio Democrat, who had previously sponsored MacKaye's 1914 Government Munitions Bill, introduced the "National Colonization Bill" in February 1916.[14]

MacKaye's scheme was as ambitious, utopian, radical—and, it turned out, as politically unachievable—as any public lands program offered since Powell's 1878 *Arid Regions* report. He proposed to end the American tradition of dispersing and selling the public domain by offering each individual or family the opportunity to develop an independent farm or ranch. As an alternative, MacKaye outlined a sweeping program for federal development of entire communities, supported by farming, grazing, mining, and lumbering on public lands.

The bill presumed a degree of cooperation among federal agencies the likelihood of which was unsupported by history or precedent. MacKaye called for the creation of a National Colonization Board that would include the secretaries of labor (as chair), agriculture, and the interior. The board would have broad powers, with presidential approval, to identify, classify, and withdraw sites on public lands (as well to purchase adjoining private lands) suitable for development of "colonizing communities" or "farm-colony reserves." The bill also allowed for the development of irrigation works, mines, and water-power sites to support the new settlements and provided specifically for colonies devoted to lumbering and forestry operations. Comprehensive community plans would provide for advance development of roads, drainage, and utilities, and cooperative purchasing and marketing facilities—but only with "a reasonable presumption that the soil and other physical conditions and the markets and other economic conditions involved in such a project" would "permit of immediate, continuous, permanent employment for the settlers." The whole program would be initiated by a $50 million "colonization fund," financed by a bond issue.[15]

The most provocative element of the ambitious program was its provision that the government retain permanent title to land in the reserves. Settlers would secure tenure by long-term permits or leases. An "improvement charge," payable over fifty years, would reimburse the government for its original development costs. A "tax charge"—the rough equivalent of George's single tax—based on a percentage of the land's assessed value, would also be paid annually into the colonization fund, partly to support state, county, and local services, partly to replenish the revolving fund for new colonization projects. Finally, the Colonization Board would be granted sweeping powers to administer working conditions and standards. Regulations mandating an eight-hour work day, minimum wages based on local averages, job safety, worker's compensation, health and life insurance, and prohibition of child labor would all come under the board's purview.[16]

During hearings before the House Committee on Labor that year, testimony was offered solely by the bill's supporters. The severe economic downturn still gripped the nation; the proponents exploited fears that the nation's Jeffersonian agrarian ideal was threatened by immigration, unemployment, industrialization, urban congestion, migration from rural districts, and high costs of land and credit. The reformers depicted a new communal Arcadia, under the federal government's benevolent guidance.

Benjamin Marsh, an early advocate of city planning who was then serving as executive secretary of the New York Congestion Committee, told the congressmen that the bill would simultaneously assist would-be farmers and urban laborers. "Peasants who flock to-day to the great industrial centers," he observed, "constitute a menace to the trained factory operators, and reduce wages below the subsistence point."[17] Frederic C. Howe, commissioner of immigration for the Port of New York and a veteran of many reform battles, cited in almost lyrical terms the successes of farm colonization plans in Denmark, France, and Germany. "We must treat agriculture with as much intelligence as we treat the science of city government and town planning," he testified.[18]

Secretary of Labor Wilson supported the bill's "general principles" and its provisions "to establish community life" by creating new rural colonies, but he was not ready to abandon traditional principles of American agricultural settlement. "I would not stop solely with the colony," Wilson continued. "If the individual wants to find expression for himself, separate and apart from the colony, I would give him the opportunity." Wilson's concluding comment to the committee was perfectly equivocal: "I could not approve the bill in its present state, nor am I prepared to disapprove it."[19] Louis Post invoked Henry George's ideas about the "unearned increment" on land in his staunch defense of the bill's provision for government retention of title to colonized land.[20]

The committee heard at length from Elwood Mead, a leading figure in the settlement and development of irrigable lands. A water engineer who had held important government positions in Wyoming, California, and Australia, Mead was advising the agency he would eventually head—the Reclamation Service of the Department of the Interior. Potential agricultural settlers, Mead explained, faced challenges many nineteenth-century homesteaders had avoided: "the disappearance of free fertile public land"; higher costs of land, irrigation, and credit; less access to free range and timber. "The individual working alone is not efficient," Mead asserted.[21] "The great merit of this bill is that it provides organization, practical experience and the use of adequate capital in carrying out preliminary work necessary to successful settlement. It substitutes community and cooperative action for that of the individual."[22]

Despite the support of the respected Mead and others, the Labor Department's bold alternative program for the public lands won little immediate support; in fact, it never came up for a vote. Employment and economic conditions improved dramatically during 1916, in response to wartime demand, thus damping enthusiasm for new public programs. Nonetheless, MacKaye, Post, Zon, and their sympathetic colleagues in the Labor Department and the

Forest Service continued to pursue and revamp their colonization program. By June, MacKaye had prodded Post to request formally that the Department of Agriculture join his department in a study of employment possibilities on public lands. He had also attempted "to set a new standard in the building of a new community" in a document titled "Settling Timber Lands," an early draft of the colonization study that would occupy him for the next three years.[23]

✢

MacKaye's efforts at drafting legislation that spring of 1916 resulted in bills on two other, interrelated issues of intense interest to progressive conservationists: river regulation and public water power. For almost a decade, Senator Francis Newlands of Nevada, author of the 1902 Reclamation Act, had been promoting measures to implement multiple-purpose river basin planning and development. Destructive floods in the Mississippi and Ohio River valleys in 1912 and 1913 renewed the urgency of overhauling national policies concerning flood control and river development. But legislation with that intention remained bogged down by disputes between upstream and downstream political and financial interests—and by controversy among government scientists and officials about the validity of different theories of flood control.

In early 1916, Newlands filed a new river regulation bill designed to accommodate the recommendations of the Interdepartmental Committee of Service Chiefs, which included the secretaries of interior, agriculture, war, and commerce. President Woodrow Wilson had appointed the committee in the hope of resolving the legislative impasse. Robert Crosser was a principal sponsor and spokesman for the Newlands program in the House. MacKaye, having collaborated with Crosser in authoring the colonization bill, helped the congressman draft his own version of the Newlands measure. Crosser's bill requested $60 million for a nationwide system of flood storage reservoirs at river headwaters, planned by a National Waterway Council. Despite efforts to work out an alliance with Mississippi Valley congressmen who sought a similar amount, predominantly for levee building, legislation reflecting the Newlands-Crosser approach was thwarted. In 1917, Congress approved creation of a toothless waterway commission, but Woodrow Wilson, distracted by the war, never appointed its members.[24]

Also in the spring of 1916, a key year in the debate over federal water policy, MacKaye helped Representative Clyde H. Tavenner, an Illinois Democrat, draft a bill providing for government development and operation of water-power sites on the public domain and on navigable streams. Like Newlands's crusade for multiple-purpose river development, the campaign for public water power had preoccupied Progressives and conservationists for more than a decade, especially as private utilities and "power trusts" sought to profit from the rapidly growing demand for electricity. Not until passage of the Water Power Act of 1920, however, would federal authority over hydropower sites on navigable streams and public lands be established in law. And not until the days of the New Deal would the federal government pursue the extensive implementation of the principles of multiple-purpose river-basin development and the public development of hydroelectric power.[25]

∻

Two laws that were passed during the summer of 1916—the Federal Aid Road Act and the National Park Service Act—would prove to have special significance for MacKaye. The Federal Aid Road Act matched state road construction appropriations dollar-for-dollar and provided uniform national standards for certain roads. A major step toward the creation of a national highway system, the law demonstrated the automobile's revolutionary impact on the nation's economic, political, and cultural affairs. In 1912, just under one million motor vehicles of all types were registered in the United States. By 1920, eight million passenger cars alone would be registered. During the years from 1909 to 1921, Henry Ford increased his company's annual sales of Model T's from 6,181 to 1,250,000—while dropping the price of his popular car from $950 to as little as $355.[26] As MacKaye, during the 1920s and 1930s, would come to understand and articulate, the Federal Aid Road Act and its funding mechanism would transform not only the nation's transportation system but the overall pattern of American land use and development—indeed, the fundamental relationship between Americans and their environment.

The Forest Service was one beneficiary of the 1916 Federal Aid Road Act. The law authorized $10 million for the construction of roads and trails in national forests. The agency emphasized resource development and protection—fire prevention, lumber production, and economic development of local communities—as fundamental justifications for additional roads within its forest domain. Chief Forester Graves predicted that roads would soon reach "regions hitherto inaccessible."[27] But the impending prospect of a larger road network in the national forests served to dramatize the possibility—indeed, the likelihood—that some of the most spectacular and remote of federal lands could be defiled by logging, mining, and the onslaught of automobiles. The National Park Service Act, enacted August 25, 1916, at first glance seemed to respond to the concerns of worried preservationists, creating a single agency to protect national parks. In fact, however, the National Park Service itself, by promoting "recreational tourism," would soon become an active advocate of roadbuilding on the lands composing its official domain.[28]

Legislation to create a separate federal agency to manage America's existing scenic federal parklands had first been filed in 1900. (Yellowstone, the first national park, had been designated in 1872.) "Nowhere in official Washington can an inquirer find an office of national parks or a single desk devoted solely to their management," complained national parks advocate J. Horace McFarland in 1911.[29] A Pennsylvania businessman who spearheaded the campaign to save Niagara Falls and became the first president of the American Civic Association, McFarland exemplified the growth of preservationist sentiment and activism during the early years of the century. Amateur enthusiasts in groups like the Sierra Club, the Appalachian Mountain Club, and the Boone and Crockett Club fought to protect the Hetch Hetchy Valley, Niagara Falls, the White Mountains, and other scenic treasures. John Muir was the patron saint of the preservationist cause, but other devotees gradually knit together a nationwide movement in support of a national park system, which would establish the priority of scenic, recreational, and wilderness uses of certain federal lands. Others fighting for the cause were Sierra Club secretary William Colby; MacKaye's friend Allen Chamberlain, who was influential in New England conservation circles and as a journalist; George Bird Grinnell, sportsman, author, and a founder of the Boone and Crockett Club; Robert Underwood Johnson, editor of *Century* magazine; and the Rocky Mountain author and guide Enos Mills. Some of these men were among the organizers in 1910 of the Society for the Preservation of National Parks. In addition, western railroad companies, alert to the prospect of increased tourist business, were influential and enthusiastic supporters of the national parks.[30]

The National Park Service's fundamental purpose, described in a passage of the law attributed to landscape architect Frederick Law Olmsted Jr., was "to conserve the scenery and the natural and historic objects and the wild life therein and to provide for the enjoyment of the same in such manner and by such means as will leave them unimpaired for the enjoyment of future generations."[31] The energetic, wealthy, and publicity-shrewd head of the new National Park Service, Stephen T. Mather, along with his deputies Horace Albright and Robert Sterling Yard, succeeded brilliantly in generating political and public support for national parks—and for the aesthetic, scenic, and recreational values they represented. Mather and his allies, eager to expand their national park empire, worried little about the paradox presented by their legal mandate: Was it, in fact, possible "to provide for the enjoyment" of the parks by one generation while leaving them "unimpaired for the enjoyment of future generations"?

The Forest Service, under both Pinchot and Graves, had in fact resisted creation of a separate national parks agency. Some of the parks had been cleaved out of national forests, and the Forest Service feared further encroachments on its domain. Philosophically, foresters believed that recreation could be accommodated within the profession's credo of multiple-purpose and "wise" use. Indeed, in 1904 Pinchot had sought unsuccessfully to transfer control and administration of all the national parks to the Department of Agriculture. But preservationists remained skeptical that Pinchot and his successors were genuinely sensitive to and committed to their cause. The Hetch Hetchy battle had galvanized the will of those who were determined to prevent the Forest Service from controlling the national parks. "There seems to be continuous trouble over the National Parks," Henry S. Graves lamented in April 1916, just four months before passage of the National Park Service Act.[32]

As debate over the Park Service bill reached its climax, MacKaye prepared his own manifesto supporting a shift toward recreation as a high-priority use of national forests. Shortly after its passage, he spelled out his bold ideas in the article "Recreational Possibilities of Public Forests," which appeared in the *Journal of the New York State Forestry Association* that October. As a forester, MacKaye had played a role in the establishment of the White Mountain National Forest, but he had first experienced those New Hampshire hills and woodlands as an adventurous young hiker, not as a working woodsman. The formulation of recreation policy had not, apparently, been part of his official Forest Service duties, nor had he been directly involved with the activities of any amateur hiking clubs or conservation groups since he had come to Washington. But his 1916 article, adapted from a speech he had made that summer to a group of foresters gathered in the Adirondacks, outlines the fundamentals of MacKaye's later work and thinking on outdoor recreation and wilderness preservation. The appeal to his professional colleagues contained the seed of his proposal to create the Appalachian Trail, published five years later. It also foreshadowed the important role he would play (along with fellow foresters Aldo Leopold and Robert Marshall) in creating the Wilderness Society during the mid-1930s.[33]

His article was inspired in part by another 1916 article, "Recreational Use of Public Forests in New England," written by Allen Chamberlain for presentation to the Society of American Foresters.[34] Indeed, MacKaye himself read a version of his friend's paper to a gathering in Washington of the Baked Apple Club, as the foresters' group was nicknamed during the years it met in Gifford

Pinchot's home. Chamberlain urged his forester friends to commit their energies and imaginations to the promotion of recreational activities in national
forests. He described the progress being made by such organizations as the
AMC, the Green Mountain Club, and various college outing clubs in building
and expanding a network of footpaths and overnight shelters throughout the
mountains of New England. And he depicted the possibility of "a linked-up
system" of trails that might stretch from Quebec, across northern Vermont
and New Hampshire, down through Massachusetts, Connecticut, New York,
and into northern New Jersey, following the Berkshires, the Taconic Range,
and the Hudson Highlands.[35] "The present primitiveness [of the national for-
ests] will fade," Chamberlain predicted. "It is unlikely that man's fondness for
wilderness will likewise pass. It seems more probable that it will increase as
opportunities for indulging it narrow."[36]

MacKaye applied some of Chamberlain's ideas on a nationwide scale. While
confessing the difficulty of satisfying "many so-called 'practical men,'" Mac-
Kaye made a case for the increased utility of play and leisure in contemporary
American society. "Recreation does not produce material *things*," he admitted, "it only produces welfare; and you can not measure accurately the utility
of a thing in terms of welfare." People's support for the forestry movement
was not based on dry technical, economic, legal, or scientific grounds, he added. "The main thing that really arouses their enthusiasm, that really 'gets
them' is not the navigability of water for commerce, but the accessibility of
land for recreation."[37]

Benton clearly depicted one public issue that would inevitably arise from
this growing demand for recreational forestland. All the national forests and
national parks throughout the country then constituted approximately 143
million acres, he noted. But two-thirds of the nation's population lived within
a day's ride of the eastern mountains, only about four million acres of which
were then in public ownership. The East's "mountain land," he asserted, "being needed for timber production and stream protection, and including the
main scenic features of the country, is the main recreation ground of the Nation. And the people will require, for a healthful and properly balanced life, all
of this mountain land that it is possible to place at their disposal. . . . The forests should be linked up."[38]

MacKaye followed Chamberlain's vision of linked public lands within one
American region to its logical continent-spanning conclusion:

> It would seem as if this system of connecting forests and camping grounds might
> come to be applied on a national scale. It would be following somewhat the evo-

lution of the American railway system, which was largely a matter of "linking up." Some day we are going to improve our navigable inland waters and link them up so as to make a system. A system of navigable rivers would dovetail intimately with a system of mountain chains skirting the edges of the various watersheds. Here would be the basis of a land-and-water transportation system which could connect and unify a possible national recreation ground which would reach from ocean to ocean.[39]

For the first time he broached the idea for a trail from New England and New York "to connect through the New Jersey and Pennsylvania highlands with the Blue Ridge of Virginia and thence quite readily through the Southern Appalachian Range"—the future route of the Appalachian Trail. And he imagined a network of "scenic trails" from the eastern mountains connecting "with the heads of navigation of the various interior rivers flowing into the Ohio and Mississippi valley," which would be serviced by a network of nonprofit shelters, buses, and motor boats. By the protection of rivers as "scenic highways," MacKaye insisted, "a continuous connecting link would be provided between the playgrounds of the east and west." The "national recreation ground" he envisioned would comprise "not a series of unrelated tracts, but a single thing—a thing to grow and be developed apart from our more hideous *commercial* development."[40]

Appearing in an obscure forestry journal of modest regional circulation, MacKaye's essay on "the great *problem of recreation*" had little immediate public impact. Yet, in combination with the creation of the National Park Service, propositions such as those he and Chamberlain offered began to spur a Forest Service response. In 1917, Graves commissioned a leading landscape architect, Frank A. Waugh, to review recreational practices and policies in national forests across the country. Waugh's report, *Recreation Uses on the National Forests,* urged the Forest Service to acknowledge forthrightly that every acre of national forest could not in fact be managed by the tenets of multiple use. Waugh emphasized, "On the principal areas of the National Forests recreation is an incidental use; on some it is a paramount use; on a few it becomes the exclusive use."[41]

The Forest Service was not yet prepared to implement his idea that some areas within national forests be reserved and developed exclusively for recreational use, but Waugh's report revealed a paradox that would come to frame the debate over the Forest Service's recreation and wilderness preservation policies in the years ahead. Landscape architects such as Waugh and Arthur Carhart urged the agency to accommodate planned recreational development within national forests. They sought to segregate recreation from timber production and to protect scenic values. But some of their plans for intensive recreational development, as historian Paul S. Sutter has observed, "solved the problem of privatization of recreational opportunities by pushing public programs that were equally destructive of wilderness."[42] By contrast, foresters associated with the Forest Service, including Aldo Leopold, Robert Marshall, and MacKaye, promoted a variation of Waugh's idea—the zoning of some areas of the national forests and other federal lands exclusively for wilderness preservation—as a means "to save portions of the natural world *from* the juggernaut of modern recreation and automobile access."[43]

Benton had been pondering the destiny of America's wild terrain since his own youthful experiences in the New England mountains at the turn of the century. By 1916, public ideas and political forces were converging to demonstrate growing public support for recreation and preservation as uses of America's wildlands. To many, the American wilderness was no longer a place to fear and subdue; instead, it represented an environment worthy of protection for its own sake and useful for enjoyment and recreation. The priorities of wartime would temporarily blunt the momentum of the incipient public movement based on these ideas. During the 1920s, though, the outdoor-recreation crusade would erupt with activity and results, and MacKaye would be one of its intellectual and spiritual leaders.

MacKaye's 1916 article represents a solitary and somewhat enigmatic landmark among his published writings, as well as his surviving personal papers, in the years before the publication of his pathbreaking 1921 article proposing the Appalachian Trail. The key elements of recreational possibilities of public forests—the linking-up of undeveloped public and private lands, the creation of a nationwide outdoor recreational network connecting natural features such as mountain ranges and rivers, the priority of play and leisure as purposes of life and government policy—reveal clearly the mental template and vision that he fully developed in his later writing about long-distance trails, regional planning, and wilderness preservation. Except for his audience of New York foresters, however, Benton in 1916 had no official forum for his nascent recreational and preservationist views. Nor is there any indication that such issues or policies had been part of his government duties. Nonetheless, he had begun sketching in his imagination a novel vision of the American landscape.

Indeed, Benton's sense of professional identity as a forester already encompassed issues that could not find official expression under the auspices of any single government agency. His simultaneous, but divergent, endeavors in these years manifested his equivocal official and professional role. On the one hand, after more than a decade of professional life as a trained forester, he had established a foot in two federal agencies—as an employee of the Forest Service and a collaborator with the Department of Labor—and his National Colonization Bill depicted an ambitious alternative government program for transforming the relationship between labor and resource production on America's public lands. On the other hand, his views on the recreational possibilities of public lands emphasized leisure, play, and recreation as forms of resource use and consumption.

In the turbulent crosscurrents where his own ideas, larger social forces, and official policies intersected, MacKaye sought links among sometimes conflicting ideas about the uses of the nation's human labor, natural resources, and public lands. By conventional historical accounts, these were years when the American conservation movement could still be neatly divided into two traditional factions described as "utilitarian" and "preservationist." Benton MacKaye, in his effort to "humanize" the nation's conservation policies, embodied a complex intellectual perspective that strove to encompass and unite both camps.

⁜ 6 ⁜

Employment and Natural Resources

1917–1919

At the end of July 1916, Benton and Betty MacKaye journeyed to Shirley Center to join his siblings—Hal, Percy, Jamie, and Hazel—and their mother Mary for a reunion at the modest yet beloved family dominion, comprising the Cottage, the adjacent Grove House (which Mary had purchased just after her husband's death in 1894), and a small meadow across Parker Road. It was an auspicious interlude in the family's earnest and unceasing pursuit of their "larger work." Benton and Betty were busy with their political tasks in the nation's capital. Percy and Hazel were both at the peak of their careers as authors and directors of pageants, a theatrical movement then at its popular apogee. Only a few weeks earlier, Percy's *Caliban,* staged with a cast of 2,500, had attracted 130,000 spectators to ten outdoor New York performances. James, though still employed as a chemist, had recently published *The Happiness of Nations: A Beginning in Political Engineering* (1915), the latest effort to popularize his ideas. Hal's artistic efforts as a novelist and librettist were mostly behind him, but the eldest of the MacKaye siblings had established himself as a capable New York patent lawyer.[1] Benton and Betty could not know, however, that this midsummer idyll in Shirley would mark virtually the last peaceful respite in a short marriage.

Alice Paul called upon Betty to serve the suffrage cause once again during the 1916 elections. That June, Paul had instigated the formation of the National Woman's Party (NWP), which would later merge with the Congressional Union. The NWP's single objective was to campaign *against* candidates who did not support a federal suffrage amendment, in those states where women were already enfranchised. This controversial strategy, which was opposed by less militant suffrage groups, necessarily entailed opposition to President Woodrow Wilson and Democratic congressional candidates: the Democratic platform supported suffrage—but only through the adoption of separate state amendments. Throughout a grueling fall, Betty campaigned first in her native Illinois, then in Colorado.[2]

⁜

Her husband was headed west, too. Louis Post had at last won approval for a joint land colonization study by the Agriculture and Labor Departments. MacKaye, still a Forest Service employee, traveled to the Pacific Northwest to investigate conditions in the region's forests, mill towns, and logging camps.[3] Arriving in the region in late September, he was on the scene as the embattled economy of the lumber industry exploded in one of its bloodiest and most notorious episodes.

The forest economy of the Pacific Northwest differed dramatically from the conditions MacKaye had observed in New England or the cutover lands of the Great Lakes forests. Although timber companies and railroads owned substantial landholdings in the region, the federal government still held vast tracts of relatively undeveloped and unsettled land, in national forests and the rest of the public domain. MacKaye planned to survey the possibilities for implementing his colonization ideas on this "patchwork of public and private holdings."[4] What he found was a forest economy locked into a painful cycle of overcapacity, competition, low prices, low wages, poor working conditions, and bitter conflict between timber workers and their employers. "The 'timber-mining' industry is essentially migratory," he explained after his trip, "hence employment therein is essentially unstable. The lumber jack must live in a camp, and the man with a family is excluded as a worker. 'Timber mining' being itself a tramp industry, it is a breeder of tramps; it is an industry of homeless men."[5]

During his travels that fall he investigated forests and rangeland in Montana, Idaho, Oregon, and Washington. By early November he was in the Snoqualmie National Forest in the heart of the northern Cascades, where he stayed with Forest Service ranger J. R. Bruckart and his family at Darrington, Washington, "a little bum back town" located at the divide marking the headwaters of the Sauk and Stilaguamish (North Fork) rivers. "I am spending my time now interviewing settlers in the valley and getting their 'sob stories,'" he reported sardonically to his mother. "It is fascinating to see this wonderful west developing under the ungainly capitalist system."[6]

When the two federal foresters received word of violence at the Puget Sound lumbermill town of Everett, less than forty miles to the southwest, they immediately made their way to the scene. On Sunday November 5, the day before MacKaye and Bruckart arrived in Everett, the city's waterfront had been the setting of a deadly confrontation that came to be known as the "Everett Massacre." Five men had been shot dead, two lay dying, and at least fifty were wounded in a wild shootout that had erupted when a citizens' army deputized by Sheriff Donald McRae attempted to prevent two boats carrying 250 members of the Industrial Workers of the World from landing at the city dock. The "Wobblies" had come from Seattle to participate in a free-speech demonstration, in response to another violent incident exactly a week earlier. Then, McCrae's viligantes tossed forty-one IWW demonstrators out of Everett after running them through a brutal gantlet in a remote city park.[7]

The consecutive "bloody Sundays" were the culmination of a tense and violent summer and fall in the lumbering center at the mouth of the Snohomish River. The intensity of the bitter, ongoing struggle between millowners and the AFL-affiliated shingle weavers union was linked directly to the fluctuating price of the red cedar shingles that were the principal product of the Everett mills. The town was "among the most highly unionized in the country," according to Everett's perceptive historian, Norman Clark. The extent of labor and radical activism was such that it "was not uncommon for a shingle weaver to carry his union card, his Socialist card, and his IWW card in the same pocket."[8]

The violent events in Everett were the climax of labor struggles that had occurred up and down the West Coast that year. In 1916, thanks to the increased wartime demand for agricultural and industrial products, the region's timber industry was recovering from several years of depression. When the shingle weavers union had walked out, in May, demanding the restoration of a wage scale that the owners had abandoned during the economic downturn, Everett's determined millowners had imported strikebreakers and demanded an open shop. By midsummer, the strike was effectively broken. The eleven-year-old IWW, trying to exploit the failure of the AFL shingle weavers in Everett, sought to revive its own flagging efforts in the Northwest to build one all-encompassing industrial union.

Everett was a bitterly polarized community in 1916. The city's Commercial Club, controlled by the millowners, served as the headquarters and arsenal of a vigilante army composed of several hundred Everett citizens. Sheriff McRae and his deputies were not reluctant to use increasingly brutal tactics to keep the IWW from spreading its message among the city's millworkers.

Benton MacKaye had a unique perspective and opportunity in the aftermath of the deadly events at Everett. At once federal official, Harvard gentleman, softspoken radical, and forester conversant in the argot of the lumber trade, he talked with a representative cross-section of the city, including witnesses to the bloody waterfront shootout. At the Everett Commercial Club, Fred Baker, a mill operator and club officer, righteously defended the violent means used to crush IWW agitators. The local deputies, said the smiling

Baker, "don't use them [the IWW men] as if they were china." Such tactics would continue, he told MacKaye, because "peaceful methods . . . don't work."[9]

On his rounds of the divided community, MacKaye also talked with a self-described "laboring man," carrying a gun and a lead pipe, who could barely articulate his disgust with the IWW's rhetoric and allegedly foreign origins. A less blustery, more bourgeois perspective was provided by two Everett tradesmen: A dentist and a tailor explained in detail the events leading up to and surrounding the harborside shootout. Unable to abide the violence and intimidation countenanced by the millowners, they had both resigned from the Commercial Club. Many Everett citizens, the two men insisted, shared their own nonviolent views. At Everett's Socialist headquarters, a member described the coroner's report on the shootout as "a joke," because testimony had been gathered only from the deputies on the dock; none of the ferry passengers was questioned.[10]

MacKaye soon pieced together his own record of the violent events. His sympathies were clear enough. Writing to his sister, he appended a lengthy narrative of the Everett situation for his Hell-Raising friends in Washington. "It is a sort of city drainage project," he explained, borrowing a metaphor from the conservation cause:

> As fast as the I.W.W. eddies and currents come in, they are led out in rivulets and sudden pressure applied at the city limits. Splendid engineering perhaps when applied to water and inorganic substance. But IWW's are organic with a nervous system, and this may be the fatal error in the calculation of the C. Club engineers. . . .
>
> . . . They say there is going to be more trouble in Everett. The press are telling their usual half truths. To my mind—the need is clear, for Everett and every other vent hole, now and the many times this sort of thing is going to blow out. If the press tries to hold in this force something is going to burst.[11]

Except for the low-key presence of the two Forest Service men, the federal government took little notice of the tumultuous events at the Puget Sound community. Neither Sheriff McCrae nor any of his deputies was ever charged for their role in the shootings, though five of the dead and more than half of the wounded were IWW members who had been on board the ferry. On the other hand, seventy-four passengers of the vessel were arrested, when the *Verona* returned to Seattle, and charged with the first-degree murder of the two deputies killed on the dock. The IWW men were jailed throughout November.

Five months later, after twenty-two jury ballots, they were all acquitted. It was never conclusively determined whether the first shot on that violent Sunday had come from the dock or the ferry.[12]

Before heading back east at the end of November, MacKaye stopped in Seattle, Portland, and San Francisco, interviewing IWW members in each city. The itinerant government forester was not reluctant to promote his ideas publicly, notwithstanding the region's volatile mood. "Uncle Sam Planning a Substitute for Hobo Camps," read a headline over his interview in the *Bellingham Herald*. He also described his program for permanent logging camps in the region's national forests to newspapermen in Seattle and Portland.[13] "I was collecting data for designing one of my 'Utopias,'" MacKaye recalled of this excursion almost forty years later.[14] What he had learned firsthand in Everett in 1916, though, demonstrated that, in the Pacific Northwest, utopia was yet a distant vision. Instead, in MacKaye's righteous eyes, the region was a hotbed of violence, death, social unrest, and environmental destruction.

While Benton pursued his work in the Northwest, his wife returned to Washington after the 1916 election campaign to their new apartment at The Ethelhurst on 15th and L streets. Despite the efforts of the National Woman's Party to dramatize the suffrage issue, the European war had overshadowed American politics during the election. Wilson, in fact, had won ten of the twelve suffrage states in his narrow reelection. And the NWP campaign had taken a heavy physical and emotional toll on some of the militant group's best organizers. Alice Paul herself suffered a collapse. Betty was afflicted with a "miserable pain," perhaps involving her heart, which impaired her physical and mental health. She struggled against loneliness while her husband investigated forest conditions thousands of miles away. "Let us pray the God's—he will not be going on such big costly jobs often—for I cannot *eat* here alone, and I am so lonely without that *darling* boy—I could shed buckets of tears," she lamented to Hazel, anticipating a "dreadful" Thanksgiving without her husband. "I can manage breakfast and a bite of luncheon alone—but dinner is too much for me. I should starve to death *alone*."[15]

When Benton returned from the West in December, the couple soon joined many of their progressive and left-wing friends in opposing the growing pressure for American involvement in the European war against Germany. Betty was engaged to publicize a Washington meeting of the Women's Peace Party in early December and the Conference of Oppressed or Dependent Nationalities immediately succeeding it. She also lent her name and her efforts to several ad hoc antiwar organizations, such as the Rational Defense League and the Emergency Peace Federation.

MacKaye, as a federal employee, assumed a lower profile than did his wife, but his antiwar sentiments were just as intense. "Am rushed every minute, between my regular job and trying to do 'my bit' toward steering off these warhounds from demolishing what is left of American liberties," he wrote his mother in February.[16] To some close acquaintances he sent copies of an Emergency Peace Federation appeal supporting a congressional investigation "to *establish* the well known fact that the press is owned body and soul by the big interests" who stood to benefit from the war.[17] As it became increasingly clear that President Wilson would seek a congressional declaration of war, Benton, writing to his brother Percy, privately staked out his antiwar position. "To my mind, the desire for war is instigated by those who would Prussianize America. I shall fight the war talk to the limit—fight *against* Wall St. not *for* it." He also railed against military conscription and passage of "espionage and censorship bills to kill Article I of our bill of rights providing for freedom of speech."[18] "Well, well dear old Ben," Percy replied calmly to his brother's screed, "it's hard for me to imagine myself on the other side of the fence from you, politically, . . . but that's where I seem to be."[19]

The serious but civil disagreement between Benton and Percy over the war issue was a familial version of the political fissure the war opened among socialists, leftists, and reformers. A seasoned progressive like Chicago's irascible Harold Ickes, for example, wrote to the Emergency Peace Federation to scorn "the emotional drunkenness of the professional pacifist who is thinking less of his country and of the future of our civilization than of whether the calcium light is showing to the audience in clear and bold outline, his pseudo-heroic and tragic features."[20] Benton's friend Walter Lippmann, then an editor of the liberal *New Republic,* enthusiastically urged American intervention. On April 6, Congress overwhelmingly approved Wilson's request for a declaration of war.

By June Benton had completed a 272-page draft of his study, now titled "Colonization of Timberlands." The prospects for his colonization project were uncertain, though, as government agencies turned their attentions to wartime priorities. "The plans are all chaotic," Betty wrote Percy's wife Marion, who had inquired about government employment prospects for her son Robin, "and no one has the least idea of what may be in store for the summer."[21] Despite his personal opposition to the war, Benton carried on conscientiously and loyally with his official tasks.

Early in the winter of 1917–1918, for example, a national fuel emergency loomed. The combination of increased industrial demand for coal and an overburdened railroad system threatened to cramp production of war matériel—and, not least, to leave many Americans literally in the cold. For some progressives, the wartime experience suggested the possibilities of rational, centralized federal planning.[22] For MacKaye, the wartime fuel crisis seemed to offer an opportunity to test his ideas for government development of natural resources.

Continuing to work with Zon during 1918, he developed several Forest Service plans for exploiting wood as a substitute for coal. One scheme called for cutting fuelwood on a government-owned tract within the District of Columbia. For another project, planned for private forestlands in southwestern Michigan, he envisioned a "permanent organization for the cooperative marketing of woodlot products."[23]

For two full years MacKaye had worked closely with Post of the Labor Department on his colonization study, though the Forest Service remained his formal employer. But both of his official mentors realized that the ideas and perspective of their restless, imaginative subordinate were increasingly at odds with the trend of Forest Service policies. When MacKaye requested a formal transfer from his old agency, Zon came to his support, noting that "it would be very difficult indeed to assign him a project in the Forest Service in which he would be able to accomplish his best work." In January of 1918, he was transferred to the Labor Department as an expert and assigned to the Bureau of Labor Statistics.[24]

An article published that February in the *Journal of Forestry*, "Some Social Aspects of Forest Management," represented MacKaye's parting shot at the forestry profession and the Forest Service. Foresters, he charged, had failed to fulfill the promise of stable forest employment and a "permanent community life" for those who struggled for their livelihoods in the American forest. He portrayed his fellow foresters as agents of an industry "conducted as one of harvesting or 'mining' timber and not of reproducing and cropping timber." The sustainability of stable, prosperous forest communities, by his account, was related directly to the sustainability of the forest's natural resources. Forest management, MacKaye insisted, needed to address head-on "the problem of the lumberjack."[25]

He linked the practical considerations of the forest worker's daily life directly to the physical arrangement of forest labor. "If he can reach the same spot at the end of each day, he can establish a home there and a family life," MacKaye wrote, "but if he cannot reach the one spot, he must live in a camp and will then soon establish a hobo life." Invoking the forester's notion of the "working circle"—"the area within whose radius an annual cut is continuously maintained"—he suggested how a permanent "sawmill community" would be the hub of a "logging community," which still "might have to be relocated from time to time" to follow a sustained rotation of timber growth and harvesting. "Improved transportation" in the forest would increase the "daily working radius," thereby reducing the frequency of the logging community's relocations. "This is simply the commuter's problem as applied to forest management," he observed. While incorporating such human communities into a "forest working plan," foresters would also have to address the need for "self-government" and other civic aspects of community organization. "Educational facilities also go without saying," MacKaye urged. "The school-house, as well as the voting booth, should follow the community."[26]

MacKaye acknowledged that such a scheme for developing "integral working units" within forests, encompassing both the social and the "timber cropping" aspects of forest management, would require "public ownership or control" of forests. But he concluded his article by warning his forestry colleagues that if they continued to ignore "the social aspects of forest management . . . we must not be surprised in future times of crisis if the labor situation, in the industry which is ultimately in our charge, becomes acute and grows worse instead of better; for fundamentally it will be 'up to us' for our failure to prepare."[27] MacKaye's notion that foresters, rather than businessmen, were "ultimately" in command of the timber trade revealed both his basic skepticism about capitalism and his blithe faith in the benefits of government regulation of industry. Only five pages long, "Some Social Aspects of Forest Management" was nonetheless the manifesto for a radical alternative to the current system of management of America's public forests. MacKaye had sketched out a program for the integrated environmental, economic, and social planning of permanent human communities on public lands.

In August, Louis Post directed MacKaye to "survey and report on the possibilities of using forest lands in the Northeastern States for establishing wood working factories suitable for the employment of crippled soldiers."[28] (Two months later, as he worked on the fuelwood projects, Benton was transferred directly to Post's office, at a substantial increase in pay.) Working with Forest Service, Labor Department, and Fuel Administration colleagues, he developed a detailed proposal for creation of the U.S. Wood Fuel Corporation. President Wilson approved a $100,000 allocation for the project from his wartime emergency funds. Immigration Commissioner Frederic Howe, who was designated to organize and administer the corporation, dispatched MacKaye and the Forest Service's Austin F. Hawes to the White Mountain National Forest to prepare a detailed report for a pilot project. The two men, who would remain friends throughout their lives, located an abandoned but well-equipped logging camp in the Wild River valley along the Maine–New Hampshire border. Their plan called for the employment of forty men, beginning with twenty in the winter of 1918–1919, to harvest "turnery" wood for a local bobbin factory, fuel wood, and pulp wood on the 30,000 acres of national forest lands within the Wild River watershed. Once established on a permanent basis, they predicted, the project could sustain "an ultimate forest community of about 200 people."[29]

By early November, the project organizers were ready to lease the logging camp and to hire a superintendent for the operation. But soon after the armistice was announced on November 11, the project was suspended and the Fuel Administration ceased operations. Thus faded the idea of the U.S. Wood Fuel Corporation, which would have provided an experiment in MacKaye's notion of planned, government-financed communities supported by sustained development of natural resources on federal lands.

Earlier in 1918, Benton's work had been interrupted by the physical and emotional collapse of his wife. During this ordeal, by his later account, Betty had exhibited signs of "suicidal mania" and been hospitalized. She regained a measure of equilibrium, but the episode provided her husband with a powerful warning of the depth and volatility of her emotional problems.[30] It was perhaps of some compensation for him that about this time his ideas began to win some public recognition. The worrisome prospects of postwar unemployment, plunging industrial production, and inflation prompted much debate about "reconstruction"—a term elastic enough to accommodate almost any legislative proposal or ideological viewpoint. Now, as MacKaye described in "The Soldier, the Worker, and the Land's Resources," an article in the January 1918 *Monthly Labor Review,* the Labor Department was recasting its colonization and land settlement program as a way to put returning soldiers to work building their own farms and communities in planned colonies on government-owned land. By late April, MacKaye had completed yet another draft of his colonization study, titled "After-the-War Colonization." The Labor Department distributed summaries of the report, which received sympathetic attention from such publications as the single-tax *Public* and the influential

New Republic. The latter journal touted MacKaye's "scientific diagnosis" of the conditions that impelled the Wobblies to revolt "against the intolerable by-products of timber mining and crass speculation in public lands."[31]

But MacKaye's report was not greeted with "sympathy and cooperation" by the secretary of the interior, Franklin K. Lane, as Louis Post had hoped it would be.[32] Rather, the Labor Department initiative prompted Lane to accelerate development of his own soldier resettlement program. Soon the Labor and Interior departments were engaged in an escalating battle over colonization and soldier resettlement policies. In part, they were simply fighting over political turf. But a philosophical difference became the fundamental obstacle to any compromise between the proposals offered by each agency. Lane was a staunch supporter of fee-simple title to land, believing that long-term, low-interest purchases would foster qualities of patriotism and productive citizenship among settlers. The Labor Department (in alliance with at least that part of the Agriculture Department represented by the Forest Service) adhered to the single-tax principles articulated by Post and MacKaye. The government must retain permanent title to settlement lands, they vehemently insisted, in order to promote stable communities and to prevent speculation.[33]

Throughout the spring and summer of 1918 the two agencies competed to win President Wilson's support for their respective soldier resettlement programs. Ironically, some of the key Interior Department staffers who developed that agency's resettlement policy were philosophically more in tune with the Labor Department's plan, especially on the question of land title, than they were to their own chief's views. Colonization expert Elwood Mead, for instance, had recently joined the Interior Department. And Lane's principal assistant on resettlement matters was none other than MacKaye's former roommate and Gifford Pinchot's confidant, Harry Slattery.[34]

Indeed, it was Slattery who in late May drafted Lane's letter to President Wilson seeking to preempt the Labor Department program. That may be one reason that the Interior Department's rhetoric sometimes echoed the propaganda MacKaye was penning for his agency. Colonization, Lane told the president, "should be done upon a definite planning basis." In fact, Lane called for federal acquisition and development of arid lands, cutover lands, and swamplands throughout the country. He requested funds for surveys and studies of those lands, noting that the "era of free or cheap land in the United States has passed. We must meet the new conditions of developing lands in advance—security must to a degree displace speculation."[35]

Simultaneously, Wilson received an appeal from a frequent correspondent and adviser, U.S. Tariff Commissioner and former California congressman

William E. Kent. A progressive reformer, conservationist, and devotee of Henry George, Kent worked closely with the Labor Department to promote its colonization plan. He urged the president to issue an executive order assigning to the secretary of labor the personnel, from all departments, already at work on postwar resettlement. Such a group, he recommended, should include himself, Mead, Slattery, Forest Service chief Graves, War Labor Policies Board chairman Felix Frankfurter, and the "experts" on Post's staff, such as Mac-Kaye. Subsequently, Kent suggested, legislation could be sought to make such a commission a separate government agency.[36] Invited by Wilson to elaborate on this proposal, Kent vigorously attacked the traditional approach to American land settlement. He and colleagues in the Labor Department, Kent reported to the president, were unanimous in their opinion that any "farm rehabilitation" policy must make "security of tenure dependent upon proper use and that the only way of being able to exercise such control of use would be by retention of the fee simple title in the Federal Government or in the State."[37]

Kent noted the contributions of Elwood Mead, who had "practically solved" the problems of settlement in Australia, and of the Forest Service's Raphael Zon. He brought to the president's attention MacKaye's "extremely valuable report" on colonization, which "shows a splendid vision of things to be considered"; and he mentioned MacKaye's draft legislation for realizing that vision.[38] The preoccupied president was reluctant at first to commit himself on either side of the quarrel among his subordinates. Indeed, he confessed befuddlement about the various wartime tasks and postwar plans already assigned to various departments. Wilson's indecision and his weakening influence over Congress left the field open for a proliferation of resettlement proposals. By one estimate, at least 100 such bills were filed during this period.[39]

The November 1918 congressional elections gave Republicans control over both the House and the Senate. The armistice followed a week later. These two major political developments lent urgency to a resolution of the political impasse over soldier resettlement policy. MacKaye drafted a new version of his original colonization bill within "only a few weeks" after the armistice.[40] Filed December 17, 1918 by Representative Clyde Kelly of Pennsylvania, H.R. 13415 included the major elements of the 1916 National Colonization Bill but now provided as well for a large-scale public-employment and public-works program to be directed by the United States Construction Service. The Construction Service, which would be administered by the National Emergency Board for Soldier Employment, again comprising the secretaries of labor, agriculture, and the interior, would build (1) planned agricultural settlements on the public domain or on private lands purchased by the government; (2) "permanent forest and coal-mining communities" on national forests and in Alaska; and (3) roads, waterways, and other projects related to creating the communities. The bill authorized development of communities in cooperation with state governments on a cost-sharing basis, provided that either federal or state governments retained permanent title to the land. And instead of the $50 million proposed two years earlier, the Kelly bill requested ten times that amount.[41]

However, the president publicly swung his support behind his interior secretary's program. Secretary Lane had called for a $100-million initiative, under his direction, to develop a settlement project in each state. He proposed to sell 25,000 developed farms to settlers, financed by four percent loans payable over forty years. The Lane approach was flawed, MacKaye asserted, partly because it did not provide for a "community form of development or cooperative colonization" with regard to marketing, schools, churches, or town planning. Its most serious drawback, he maintained, was that the fee-simple sale of farms would inevitably lead to "the evils of speculation."[42]

Hearings on the Kelly bill were held by the House Labor Committee in January 1919, during a brief session of the lame duck Congress. Committee members were eager to support substantial public works programs but exasperated at the administration's inability to speak with a single voice. The congressmen grilled the measure's advocates, including Post, Kent, and Ferdinand A. Silcox, a future Forest Service chief temporarily assigned to the Labor Department. Post expressed his department's willingness to compromise with Lane but cited the sticking point: land title. Kent conceded the necessity of supporting the Lane measure, even with the fee-simple provision, if that was the only way to initiate a colonization scheme quickly. Echoing the social concerns of his fellow forester MacKaye, Silcox testified that "the lumberjack has a lack of community life as barren as most of the isolated farmers." But he also acknowledged the "very definite practical difficulties in establishing permanent forest communities" of the sort MacKaye envisioned.[43]

❖

In the midst of this political maneuvering, MacKaye, at Post's urging, undertook one more field investigation to gather information for his still unfinished colonization report. By early December, he had embarked on a month-long trip through eastern Canada to examine recent initiatives concerning the development of public lands, resources, and government-planned communities. In his journey across Ontario, Quebec, and New Brunswick, MacKaye met with Canadian federal and provincial officials involved with forestry, agriculture, railroads, and soldier resettlement.[44]

In Ottawa, MacKaye talked at length with Thomas Adams, the town-planning adviser to the Commission of Conservation.[45] Adams's considerable reputation as a planner was based principally on his experience in Wales and England, where he had served as executive secretary of the acclaimed "garden city" Letchworth. Moreover, Adams had recently produced an extensive report for the commission, *Rural Planning and Development* (1917), a thoroughgoing treatise on community planning principles and techniques. Adams had followed the progress of the 1916 Crosser National Colonization Bill, and though he shared with MacKaye many concerns about land speculation, haphazard development, and the need for land classification, his comprehensive proposals were more cautious than was MacKaye's evolving colonization plan.[46]

MacKaye did not refer to himself as a town planner or city planner, nor had the concept of regional planning yet achieved much professional recognition; but his meeting with Adams was, as he later put it, "'the bridge' to my regional planning efforts in the 1920's."[47] By that time, however, he and Adams would attach significantly different meanings to the purpose and the practice of regional planning.

During his 1918 Canadian excursion, MacKaye spent much of Christmas week at a new community under development by the Ontario government, at Kapuskasing, a solitary location on Canada's transcontinental railroad in the province's northern "clay belt." Conceived as a soldier settlement, the Kapuskasing colony comprised sixty-some families at the time of MacKaye's visit. Each soldier had been granted one hundred acres, ten of which were cleared at government expense, along with the promise of a free house. While MacKaye praised the spirit and the purpose of the enterprise, he also noted several serious deficiencies and drawbacks.

Not only, he lamented, had the settlers been granted their farms in unrestricted fee simple ("The same old mistake of the homestead laws!"); they were allowed to choose their farm lots without any form of soil survey, land classification, or town planning. As a result, the settlers' lots were strung out for seventeen miles along the railroad line. The sprawling, unfocused settlement took a particular toll in the solitude and loneliness of Kapuskasing's women. "Without we can get together to a card party, or a lecture, or something else," one young farmwife complained to MacKaye, "we might as well be dead."[48]

For MacKaye, Kapuskasing provided a cautionary lesson in colonization. Without "three vital things"—land classification, town planning, and "land tenure based squarely upon use"—he predicted that "the Kapuskasing colony, or any other colony, is doomed to ultimate failure in providing for the happiness of its members."[49]

When he returned to Washington in January 1919, Benton was called upon to revise the Kelly bill to meet the political challenge of the Lane program. The economic climate had changed dramatically as well. During December 1918, he reported, "a net shortage of labor in the country turned into a net surplus."[50] The new bill cut the funding request to $100 million, matching Lane's request. The U.S. Construction Service's governing board would now include the secretary of war and the postmaster general as well as the original three Cabinet members. The Construction Service would also undertake a wider variety of public works projects, particularly the first links in a 15,000-mile federal post road system that had been authorized the previous year. The bill also added provisions for another program dear to progressive hearts, the development of public water power.[51] The legislation now included so many controversial, even radical, elements that it had no prospect for passage—especially in view of Wilson's public support for the Interior Department's proposal. The 65th Congress adjourned in March without voting on any significant resettlement or public works bills.

MacKaye drafted yet another version of the Kelly bill for the 66th Congress. A revision of the competing Lane program also appeared, produced with the assistance of Elwood Mead and sponsored by Wyoming representative Frank W. Mondell.[52] In "Making New Opportunities for Employment," an article that appeared in the *Monthly Labor Review* just before 66th Congress convened in May 1919, Benton compared the various bills dealing with public works and land development. "Congress thus far has taken no action affecting the employment of returned soldiers and workers," he observed. "What these unemployed men are going to do about it remains to be seen. But whatever they do their rights are clear." MacKaye's exasperation mirrored the dejection of liberals and leftists in the immediate aftermath of the war, but he could barely contain his personal frustration over the lack of any concrete results from his own efforts of more than three years. "The choice by Congress of an inadequate or backward-looking policy in this matter may well prove worse than no action at all," he charged. "The imperative need is a program that goes to the roots of this problem—not a makeshift that irritates its surface."[53]

Enthusiasm for large-scale public-works and resettlement programs evaporated during the spring and summer of 1919. Republican control of Congress, a resurgent economy, and an increase in strikes and other labor agitation undermined support for such legislation. Furthermore, colonization advocates—whether the single-taxers of the Labor Department or the supporters of private ownership in the Interior Department—were themselves looking backward. They failed to grasp the social and economic reality of the times. Most returning soldiers and displaced workers, notwithstanding the prevalence of "back-to-the-land" rhetoric and programs, simply did not want to be farmers. Nor did they share the community ideal envisioned in various forms by MacKaye, Mead, Post, Lane, and the other government proponents of new approaches to settling the nation's undeveloped lands. "It was somewhat late for this kind of legislation," MacKaye admitted many years later, "because little remained of the public domain which was available for settlement. Congress had waited for nigh forty years before even considering the Powell doctrine. Had Congress acted on this doctrine at the time of its proposal in the *Arid Lands Report* (1878) the saga of public land settlement since that date would have had happier endings."[54]

Despite Congress's failure to enact postwar colonization legislation, Benton worked intensely during that spring and summer to document his own work on the subject. He faced a July 1 deadline, because the Labor Department had not been able to obtain an appropriation to continue his position. Nonetheless, Louis Post secured funds to publish MacKaye's 144-page report, *Employment and Natural Resources,* that September. The result, in the words of historian Roy Lubove, constituted "a scheme which surely ranks among the most mature and memorable fruits of the American conservation movement."[55]

The report's presentation in the sober tone and terse style of a government publication camouflaged its sheer utopianism. The product was a provocative work of wide range and scope, a substantial achievement for the forty-year-old MacKaye. But *Employment and Natural Resources* was no mere armchair intellectual exercise. MacKaye had walked the terrain. He had talked not just with his professional peers, but with the loggers, farmers, millworkers, and developers who were attempting to wrest their livelihoods from the land and its resources. Environments that he had observed firsthand—New Hampshire's White Mountains, the cutover lands of the Lakes States, the valleys and mountain slopes of the Cascades, Canadian soldier colonies—were his principal case studies. His descriptions of colonization programs and planned communities throughout the world—the English "garden cities" of Letchworth and Welwyn, the agricultural settlements of Australia, the Durham farm colony recently initiated by California's Land Settlement Board—were based on information gathered directly from the leading authorities on those projects, such as Thomas Adams and Elwood Mead.

The purpose of the report remained to promote the colonization and public-works policies advocated by the Labor Department. Farming settlements and permanent forest communities continued to be the central elements of that vision. But he also described his ideas for a federal construction service, government-sponsored mining communities in Alaska and elsewhere on the public domain, development of hydroelectric power and other river projects, and the postal highway and motor truck system.

The "ultimate source of all employment," MacKaye maintained, was land, "in its broad sense," including what he and other conservationists liked to call the "big four" natural resources: soils, forests, minerals, and water. The worker's "true access to land"—that is, the ability to earn a living from it—"is achieved only through industrial processes. Industry . . . is the 'pipe line' from land to men. . . . Any obstruction in this pipe line is an obstruction to men's access to the sources of life."[56] For MacKaye, such hydraulic imagery—pipelines, flows, dams, levees—provided a key to visualizing the operation of the modern industrial economy.

MacKaye proposed to stop further disposal of the public domain. Six principles, in his scheme, should govern federal agricultural settlement policy: community cooperation, land reclamation, "ready-made farms," a government rural credit program, limitation of farm areas according to local land and market conditions, and land tenure that would be "dependent on use."[57]

"Under the cooperative principle the unit of development is not the farm but the community," he wrote. For the citizens of such communities "not only their material welfare but their social and spiritual welfare, demands an end to needless rural isolation. Hence the demand for the development of the rural community as a definite thing—as a concrete organism and not an assemblage of conflicting interests." Schools, churches, adult education, and "training in self-government" were the "requisites of a true community life." And not the least of these community considerations, for this member of the leading family of American pageantry, was recreation, including "community theaters and dramatic presentations."[58]

To create an efficient, fair, and economical alternative to traditional homesteading practices, MacKaye insisted, the government was obligated to reclaim unproductive lands and to provide "ready-made" farms for settlers. "The settler or agricultural worker should no more be expected to go upon the farm before it is made and equipped than the manufacturing worker should be expected to go into the factory before it is made and equipped," he asserted. Such measures would not only conserve land and natural resources; most importantly, they would promote "the conservation of human energy."[59]

He recommended as well that individual farm areas not be limited to the fixed sizes of 160 or 320 acres set by various federal laws without regard to the land's physical character. "What seems to be needed is not a farm unit of rigid and arbitrary area," he suggested, "but one of flexible area." In addition, farmers should be permitted to cooperate in organizing farm units, as long as safeguards prevented exploitation of farm laborers, tenants, and "the food-consuming public."[60]

Finally, he asserted in unequivocal terms the fundamental aim of the Labor Department's land tenure policy, which was to depose what he came to call "the fee-simple despot." "Any colony or community settlement based on individual fee titles is doomed at the start," he wrote. "It is only a question of time—and no long time—when it will disintegrate into individual earldoms and tenancy enabled to get a foothold. The way to preserve its integrity and maintain a uniform system of individual use is for the colony itself or the State to hold the fee and thus control the individual use."[61]

Drawing on his own fourteen-year experience as a working forester, MacKaye framed three principles to govern the development of new communities on America's forest lands: "(a) Timber culture as against 'timber mining,' (b) permanent forest employment, and (c) stability of the forest community." The "main purpose" of the American lumber industry, MacKaye charged, had been "to shovel out the timber." By human nature and economic necessity, the timber companies were compelled to cut and move on. "Timber culture," by contrast, represented the "essence of forestry," which entailed "sustained-yield" management under "some form of public ownership or control."[62]

Benton described with firsthand authority the need for permanent forest employment in the three Pacific states, where an estimated 100,000 workers lived in labor camps. The transience and uncertainty of the lumberjack's trade made community and family life virtually impossible. "There is no more familiar sight in this region than the 'blanket stiff' wandering the coast from camp to camp seeking a better job." In MacKaye's vision, national forests would no longer be managed as uninhabited timber plantations, whose resources were leased or sold to loggers, miners, corporations, and ranchers. Instead, he saw the federally owned and managed forests becoming the sites of "real communities and not mere shack towns."[63]

In one series of seven detailed maps, Benton illustrated concrete possibilities for permanent forest communities in a million-acre tract along the western slopes of the Cascades, where he had visited in 1916. The fundamental unit of his development program would not be defined by the arbitrary units and limits of a rectangular survey but by organic boundaries: the mountain divides separating the region's river basins. The national forests, in MacKaye's vision, would serve as the habitat for human communities. He imagined that his communitarian, sustained-yield vision for the Northwest's forests "could be repeated in any number of western mountain valleys."[64] He pictured river basins, each with a fixed carrying capacity, as cells of a larger geographic and social organism. However, his ideas for a steady-state watershed economy essentially ignored population growth, business cycles, technological change, and other dynamic forces that might upset the environmental, economic, and social balance he sought to achieve. His program for the development of national forests did not even acknowledge such uses as recreation and wilderness preservation. Although these issues may have been on his mind, they were hardly priorities of the Labor Department, the sponsor of his work.

Employment and Natural Resources dropped into bureaucratic and political oblivion as soon as it was published, but Benton MacKaye could take some satisfaction from the warm reception his work received from his own family and some of his professional colleagues. The report was "acclaimed by a Fordful of MacKayes" when it arrived at the Shirley post office, Percy reported.[65] Jamie, the source of many of Benton's own ideas, congratulated his younger brother for his "'larger work.'. . . It is a real monograph on the science of land use—a vital contribution to political engineering in the concrete."[66] From the University of Michigan, Filibert Roth, a respected veteran forester, extolled MacKaye as "a *pioneer in truth*"—and expressed delight that "the government press of the most money-mad people in the world" would publish so blunt an attack on the speculation and fraud fostered by federal land practices.[67]

MacKaye's efforts had failed to reform public policy, but the report represented the culminating statement of the first substantial chapter of his career as both a forester and a social activist. As an exercise in regional planning, *Employment and Natural Resources* illustrated the techniques, principles, and ideas that would be the foundation of all his later works: imaginative maps and charts depicting alternatives to the social, political, and legal conditions

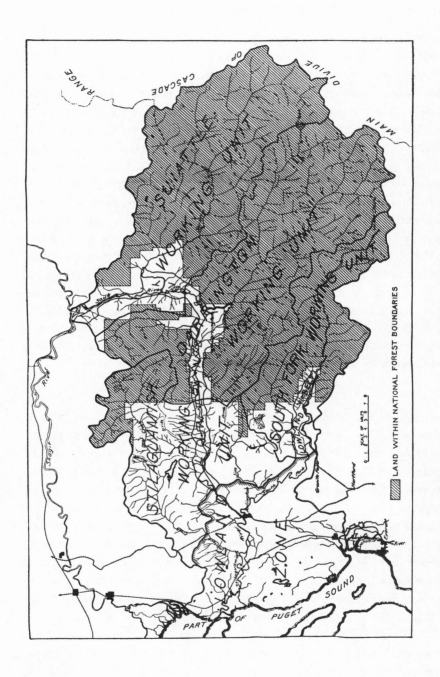

SEUATTLE WORKING UNIT

WASHINGTON WORKING UNIT

SOUTH FORK WORKING UNIT

SKAGADAM'S WORKING UNIT

TUK WORKING UNIT

SOUTH CREEK WORKING UNIT

ZONE

RANGE

OP CASCADE

MAIN DIVIDE

Skagit River

Sauk River

Stillaguamish River

Snohomish River

Snoqualmie River

Columbia River

PART OF PUGET SOUND

SCALE OF MILES
0 1 2 3 4 5 6 7 8

▨ LAND WITHIN NATIONAL FOREST BOUNDARIES

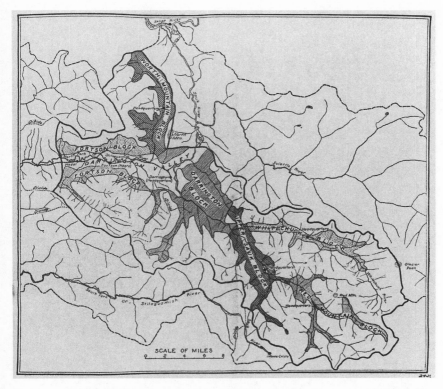

This map from MacKaye's Snoqualmie National Forest plan shows one of the working units (the Darrington unit) divided into six "cutting blocks," which would be logged in rotation over fifty-year cycles. (*Employment and Natural Resources* [1919]. Author's collection.)

FACING PAGE

In *Employment and Natural Resources,* his 1919 report for the Department of Labor, Mac-Kaye used a region on the western slope of the Cascade Range in Washington, including and surrounding the Snoqualmie National Forest (which he had visited in 1916), to illustrate in a series of maps his vision of a "permanent forest community" based on the forester's notion of the "working circle." He defined a region by the drainage basins of certain tributaries of the Skagit and Stilaguamish rivers, which he further divided into forest "working units," not all of which were within the boundaries of the national forest. (Author's collection.)

that already existed on the ground; the river drainage basin as the basic unit of settlement and development; the classification of land-use zones by altitude, slope, and vegetation, separating an undeveloped mountain zone from a settled lowland zone along the rivers; small planned communities, focused around social as well as economic activities—all of this coordinated by a ra-

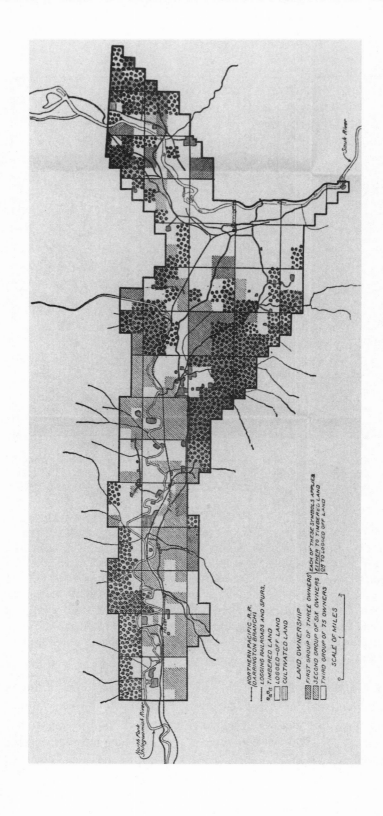

NORTHERN PACIFIC R.R.
(DARRINGTON BRANCH)

LOGGING RAILROADS AND SPURS.

°₀°₀ TIMBERED LAND

LOGGED-OFF LAND

CULTIVATED LAND

LAND OWNERSHIP

FIRST GROUP OF THREE OWNERS

SECOND GROUP OF SIX OWNERS

THIRD GROUP OF 75 OWNERS

EACH OF THESE SYMBOLS APPLIES EITHER TO TIMBERED LAND OR TO LOGGED OFF LAND

SCALE OF MILES

0   1   2

North Fork Stillaguamish River

Sauk River

tional plan developed by omniscient technical experts. But *Employment and Natural Resources* also revealed an expansive social vision that set MacKaye apart from many other conservationists and preservationists of his era. He had contemplated how the nation's public lands could provide opportunities for creating permanent human communities where working men could live with their wives and children, where those families would have a measure of control over their economic destiny, and where natural resources could be used and sustained without being extinguished.

In certain significant respects, MacKaye's work for the federal government during the 1910s could be regarded as a failure. Rather than rising through Forest Service ranks to more influential positions, he had left the agency to join the Department of Labor. There he had worked himself out of a job, when the political tide turned after the 1918 elections. The colonization work and ideas to which he had devoted the better part of four years had produced almost no concrete results in the form of legislation or policy. The political forces of progressivism, socialism, and reform had lost their momentum; and the Hell Raisers had exhausted much of their influence, if not their indignation.

Nonetheless, many of MacKaye's professional colleagues and reform-minded friends had recognized his idealism, energy, and intellect, which could not always be fitted comfortably into the demands and constraints of workaday bureaucratic routine. MacKaye had demonstrated his talent for visualizing the landscape on a large-scale. There were plenty of pragmatists, compromisers, and "practo-maniacs," but precious few visionaries and idealists. His working odyssey from forester to community planner had carried him from the frigid summit of New Hampshire's Mount Washington to the bullet-riddled docks of Everett, Washington, and from the eroded hills of eastern Kentucky to the wasted, stump-strewn cutover lands of Minnesota and Wisconsin. After the publication of *Employment and Natural Resources* in 1919, however, MacKaye could not be sure what he would discover at the next unexpected bend in his life's path.

FACING PAGE

Another map from the Snoqualmie National Forest series depicts land-ownership and land-use patterns in an agricultural settlement in the Darrington Valley. (*Employment and Natural Resources* [1919]. Author's collection.)

❖ 7 ❖

Turning Point

1919–1921

In 1919, Benton and Betty MacKaye faced an important personal turning point. The forty-year-old MacKaye left his position with the Department of Labor on July 1, 1919; and the Nineteenth Amendment, guaranteeing women the right to vote, was moving inexorably toward ratification. For virtually the first time in their four-year marriage, they could now make decisions about their future without the compulsion—or the excuse—of professional duties and political events.

At the beginning of September, the activist couple moved to 1622 19th Street N.W., which would be their last home in Washington. They shared the rowhouse with the family of economist and author Stuart Chase, then on the staff of the Federal Trade Commission, as well as with a few politically and personally congenial bachelors. And they called the place "Hell House"—partly, it appears, in recognition of their self-styled role as "Hell Raisers," partly as a play on the name of another well-known institution of reform, Jane Addams's Hull House.[1] Intentionally a "cooperative house," in which residents shared living expenses and domestic duties, it was also a salon, a seminar, a soviet, a commune, a meeting place where political schemes were hatched and debated. "The usual run of 'reds,' freaks, reformers, Non-Partisan Leaguers, Chinamen et al. continue to come to us," Benton wrote to his mother, describing the stimulating, close-knit household near Dupont Circle.[2]

MacKaye continued to organize daylong walks in the countryside surrounding Washington. With his wire-rim glasses, ever-present pipe, hawklike countenance, lean and wiry build, and shock of dark hair, he presented an intense and professorial bearing. Chase later recalled:

> The upper reaches of the Potomac were our favorite exploring ground. As we ate our picnic luncheon, Benton would dilate on the inequities of the lumber barons, on how a valley should be planned for maximum "habitability"—one of his favorite words—and what a pest the automobile was rapidly becoming. It was pungent and salty discourse, not always strictly fair. He hated, down to the depth of his soul, all despoilers of the land.[3]

Recollecting his married years, MacKaye later wrote that the "gatherings at these walks, together with the impromptu evening talk-fests in our house, be-came our family institution."[4]

For Betty and Benton, the decision to live with other friends and couples was not only a frugal response to postwar housing shortages and inflation. It was also an intentionally political act, a statement in daily life of their ded-ication to communal and cooperative principles. "We are simple souls, full of busy ideas and not in a notion to fuss about with the fashionables!" Betty had once written to her mother-in-law. "All of our friends are the journalistic crowd who never know where the bread and butter are coming from and do not care. They are a jolly, happy group who know life and live it!"[5]

When funds and political support for MacKaye's work in the Labor Depart-ment ran out in mid-1919, Louis Post helped him secure a temporary appoint-ment with the Postal Service. During the fall of 1919 and into the spring of 1920, MacKaye dutifully conducted a study of the postal delivery system as the basis for a cooperative agricultural marketing scheme, linking planned farm communities in the country with organized purchasing districts in the city.[6] But his real ambition in the latter half of 1919 was to be a writer—or more pre-cisely, a journalist. Benton had already made a few tentative forays into print beyond the pages of forestry journals and official government publications. During early 1919, he had turned his attention to a favorite and always contro-versial cause, the fate of Alaska. He proposed to Post that the agency send him to Alaska to investigate colonization possibilities, but his superior had neither the authority to launch nor the funds to support such an initiative. Debate about Alaska centered, as it always has, on the disposition of the territory's abundant resources and spacious lands. As Congress considered legislation to ease the terms of leasing and purchasing Alaska's publicly owned treasures, MacKaye inveighed in such political journals as the single tax–promoting *Public* and the short-lived *Reconstruction* against "commercial despotism" and speculation, and he called for colonization of the territory's lands by means of public ownership and administration. "We can seize the world-given chance to put through the first clean cut project in social engineering," he exhorted *Reconstruction*'s readers, "and build, in our northern land, a nation within a nation."[7]

Through his journalist friends, MacKaye that summer turned his hand to a series of syndicated articles for the Newspaper Enterprise Association. Fred Kerby, a hero of the Ballinger-Pinchot Affair, coached him with columns about the high cost of living and the union-conceived Plumb Plan to nationalize the railroad system. "I think you have the best mind in the U.S.A. on fundamental economics," Kerby encouraged his friend—perhaps the first and last time that anyone credited MacKaye with an understanding of the "dismal science."[8]

Benton and Betty reveled in the sociable and socialist atmosphere of Hell House, but their living arrangement also had a more troubling aspect. The Red Scare reached directly into the household in late 1919 when Chase and another of its residents, Aaron Kravitz, also a staffer at the Federal Trade Commission, were named prominently in a congressional investigation of alleged socialist influence in the agency. Their hard-hitting report on the monopolistic price-setting practices of Chicago meatpackers had raised the hackles of midwestern congressmen. Chase and Kravitz defended themselves and their affiliations, but they were dismissed from the FTC a year later.[9]

The visitors to the 19th Street house constituted an intriguing cross-section of the era's left-wing personalities. Sinclair Lewis, then in the midst of writing *Main Street,* entertained Hell House parties ("after a couple of drinks," Chase recalled) with his witty improvised songs.[10] MacKaye's pocket diary for 1920 comprises a terse catalogue of left-wing activists searching for a role and a purpose in a nation transformed by the war experience. Hell House callers included the likes of Louise Bryant, wife (soon to be widow) of journalist John Reed; Roger Baldwin, just out of prison for refusing to comply with the draft law, who had returned to his post as director of the American Union Against Militarism, predecessor of the American Civil Liberties Union; Dean Acheson, the future secretary of state, then the young secretary to Supreme Court Justice Louis Brandeis; reformer Frederic Howe, U.S. Commissioner of Immigration during the war years; pro-Soviet journalist Lincoln Colcord; and the controversial author and economist Scott Nearing.[11]

As the Red Scare climaxed in late 1919 and early 1920, these self-styled Washington radicals took an increasingly apocalyptic view of the nation's immediate fate. Attorney General A. Mitchell Palmer's minions arrested four-thousand suspected subversives across the country in the first days of 1920. The indignant Hell House crowd organized the Citizens' Amnesty Committee, with Stuart Chase and Betty MacKaye among its officers, to protest the arrest and threatened deportation of aliens rounded up in the Palmer Raids.[12] By midwinter, as evening conversations began to focus on the possibility of financial panic, Benton, Betty, and the circle of their activist friends began to pursue the idea of volunteering their services to the Bolshevik government of Russia, not yet recognized by the United States. One late January day, Benton and Betty visited with Albert Rhys Williams, a Congregational minister and writer who had been in Russia in 1917 and 1918, where he had worked in the Soviet foreign affairs commissariat; now he was promoting the Soviet cause in the United States. MacKaye began studying Russian with his housemate Kravitz, a native of the country; he read Williams's admiring biography of Lenin; and he pored over maps of the vast Soviet republic, which depicted inviting open spaces to colonize and watersheds to manage. Over cups of cocoa, the Hell House residents "sat around embers in the fire place in the dark, and talked of going to Siberia."[13]

Within a matter of days after his visit with Williams, MacKaye was meeting with Ludwig C. A. K. Martens, self-proclaimed representative of the American bureau of the Russian Socialist Federal Soviet Republic.[14] Since his arrival in the United States in late 1918, Martens had been the focus of intense controversy and investigation concerning his allegedly subversive activities on behalf of the Bolshevik government. In fact, at the time MacKaye first met Martens, the "American Lenin," as one anticommunist legislator labeled him, was already embroiled in legal proceedings that would eventually lead to his deportation. MacKaye also met with Martens's secretary, Santeri Nuorteva, who had been influential in forming the Communist Party of America in 1919; later, Nuorteva would be accused of being the conduit for Russian funds purportedly intended for financing revolutionary propaganda efforts in the United States.[15]

By mid-March, after Benton and his close friend Herbert Brougham met with Martens again and attended the first anniversary celebration of the Soviet Bureau's establishment, several of the Hell House habitués came to a decision about their plans. On March 24, he and his compatriots presented their résumés to Martens, offering to go the Soviet Union as technicians. The names of MacKaye, Brougham, and Charles Harris Whitaker, editor of the *Journal of the American Institute of Architects,* were on the draft letter to Martens explaining the group's intentions. Their "dossiers" also included the résumés of Betty, Stuart Chase, and a number of other friends, many of them federal employees, some of them Russian immigrants, all emphasizing their socialist affiliations or sympathies. Their aim was "to take an active and sympathetic part in the new life of Russia," they wrote Martens. "In this resolve we act voluntarily and without seeking the official support of the American government, well knowing that its disposition is not at present favorable to undertakings of this sort. We have no capitalistic backing, and we must continue to support our families through the employment of our skills and abilities, just as we have had to do here in the United States."[16]

Benton gave as references his former government mentors, Louis Post and Russian-born Raphael Zon of the Forest Service, as well as American activists known to be on good terms with the Soviet government, including Albert Rhys Williams, Isaac McBride, and Lincoln Colcord. "Am particularly interested in the opening of new country and resources through railroad development," he wrote in describing the sort of work he hoped to pursue in Russia, "and in the utilization of land—farming, timber, and mineral land—so as to avoid, from the start, the needless exploitation of labor which has been the history of land concessions in America. Should like to take a constructive part in the plans for colonizing territory to be opened by such projects as the 'Great Northern Railway.'" He hoped, in other words, to carry out in Russia the ideas for the rational, planned development of worker-managed communities on publicly owned lands he had described in *Employment and Natural Resources*. Betty, who included Louise Bryant among her references, offered her help "in the field of public health—social hygiene, medical services, dietetics."[17]

Martens, preoccupied with his own legal troubles, acknowledged but did not act on the group's proposal.[18] In any event, the enthusiasm Benton and his friends shared for the Russian experiment provided a measure of their optimistic, even starry-eyed, expectations for the Soviet revolution. Their willingness to leave the United States also indicated their pessimism about the prospects for socialist reform in their own country.

Feeling rebuffed by the Russians, the Hell House group immediately turned its attention to developing a domestic project for reform they called "The Northwest Program." A tier of northern states, including Wisconsin, Minnesota, the Dakotas, Montana, and Washington, had long been at the forefront of American progressivism, socialism, and populism. MacKaye had traveled this country extensively during his colonization research. He had made a wide range of contacts and acquaintances, including Industrial Workers of the World in Washington, hard-pressed Wisconsin farmers, Nonpartisan League members, Socialist Victor Berger, and intellectuals such as University of Wisconsin economist John Commons. Working closely with Chase, Whitaker, and Brougham, who grandly called their kitchen-table think tank the "National Volunteer Committee on Industrial Administration," MacKaye helped to sketch out a prospectus for an elaborate survey of the region's natural resources and its economic framework. "How far is the Northwest area self-sustaining?" they asked. "How far, economically, can it be made self-sustaining?" The Northwest Program, they maintained, would lay the groundwork for "a new social order the keynote of whose productive system shall be service—not profits." Their manifesto proposed "to destroy the present industrial feudalism" that controlled the region's resources and economy. They foresaw a nonviolent, technocratic revolution that would meet the challenge of capitalism "not in terms of moral indignation, but in terms of social engineering." Although MacKaye and Brougham attempted throughout the year to launch the Northwest Program with some of their Wisconsin acquaintances, their idea for a comprehensive regional survey led nowhere.[19]

By the end of June, MacKaye had completed his Postal Service project. He and Betty returned to Shirley Center to take stock of their future—and to transact some MacKaye family business. Mary MacKaye was seventy-five years old, the Grove House mortgage was coming due, and she and her daughter Hazel had no secure source of income. Benton, on whom Mary had depended for much of her support over the previous fifteen years, was without a job. And her other sons, Harold, James, and Percy, none of whom was particularly well off, all had family obligations of their own. Nonetheless, when the members of the family gathered in Shirley that summer of 1920, they created the MacKaye Family Association, whose stated purpose was "to acquire, hold, and preserve property and records of all kinds connected or associated with Mary Medbery MacKaye . . . and her descendants." The Cottage was deeded to the Association, the Grove House and its mortgage to Hazel.[20] It was an almost comical estate: a few acres, and two unimproved, rundown houses, stuffed with papers, books, photograph albums, and furniture, the property mortgaged beyond the family's realistic prospects of repayment. Yet they had held to this one common cherished spot for more than thirty years.

During late June and July, Benton read his boyhood diaries, retraced some of his expeditionary walks, helped his Shirley neighbors with their haying, and sketched out ideas for articles on "land management." In the middle of July, he heard from Herbert Brougham, who had gone west several weeks earlier to take the post of managing editor at the *Milwaukee Leader*. Brougham offered Benton a job as an editorial writer for the paper, a daily edited by the prominent Socialist Party figure Victor Berger. He quickly accepted the position in a new trade at one of the most steadfast outposts of American socialist journalism.[21]

❧

Benton had regularly voted for Socialist Party candidates, but he remained at a distance from the events that had riven the Socialist Party during the war years. Even though his new journalistic position was with a semiofficial Socialist Party organ, he considered himself a "radical," not a partisan. "Strictly speaking a radical is one who goes to the root of things, who seeks to change society not by petty reforms on the surface, but by a drastic review of its whole body," he had written not long before his arrival in Milwaukee:

> Radicals, generally speaking, are agreed that the existing capitalistic, competitive organization of society has outlived its usefulness—if ever it had any usefulness. They are agreed that only a society founded on cooperation and mutual aid is a tolerable one for the average man and woman. They are agreed that production for profit must give way to production for use. They are agreed that humanity precedes property in the scale of social values. They are agreed that an international association of peoples must displace the arrogant anarchy of sovereign and imperial States. They are agreed that society cannot be kept wholesome without the widest possible freedom of speech, of press, and of assemblage. These are specific ends to which most people who call themselves radicals are in accord.
>
> It is regarding the means to be used rather than the ends desired that radicals usually find their differences. They disagree on tactics rather than principles.[22]

MacKaye's new boss Victor Berger was embroiled in a controversial campaign for Congress—hence the paper's need for another editorialist. Through a bizarre series of political events, including a conviction under the Espionage Act for his antiwar statements, Berger was pursuing his fourth congressional campaign in only two years. (He had served one earlier term in the House of Representatives after, in 1910, becoming the first Socialist ever elected to that body.) At the time Benton took up the pen for the *Leader,* moreover, the paper was still without the postal privileges of which it had been stripped by the postmaster general in 1917. Berger, although considered by Congress and the courts to be a threat to the republic, was "the most conservative of all Socialists," who took a patient and nonviolent approach to the promotion of socialism; his attitudes suited MacKaye, ideologically and temperamentally.[23] As MacKaye wrote in an election-day editorial for the *Leader:* "Evolution is slow revolution. . . . The ballot is not the only weapon. There's the strike. There's co-operative enterprise. All these things spell 'Revolution,' though not the bullet kind."[24]

MacKaye, although a capable writer who had turned out many articles, reports, and memoranda over the years in his own dogged fashion, had no background as an editorialist. Brougham, drawing upon his two decades of newspaper experience, tutored his friend in the "method of 'dragnetting' the news for editorials."[25] Benton spent long hours studying and clipping other news-papers in search of data and ideas for the several front-page editorials he was expected to turn out for six editions a week. His pieces, while hewing to the *Leader*'s ideological line, ranged widely in topic, from the instability of postwar global politics to the dangerous habits of the ever-increasing number of auto-mobile drivers on local roadways.

And he added a new feature to the *Leader* editorial columns: his dynamic, distinctive maps, replete with large arrows and bold type, depicting a geopolit-ical scene in a dramatic state of flux. "Less knowledge of rhetoric and more of geography might have 'kept us out of' it," he wrote of the "real reasons" for the war, which he saw as the world struggle for control of resources and com-merce. "So let's, for God's sake, get busy now and study geography while the studying is good. 'Map Knowledge' to fight 'pap knowledge' is what we need for 'preparedness,'" The "best way to keep out of hell (like any other place)," he added, "is to study its geography."[26]

While most of his editorials were concerned with matters of the moment, others revealed his comprehension of how the rapid pace of technological change was transforming modern society. Noting the inauguration of com-mercial passenger air service out of New York City and Florida, he observed that "commerce has penetrated all three geographic 'spheres'—the land, the waters, and the atmosphere. . . . The deep harbor on the sea, former maker of cities, will be joined by the flat 'harbor' on the land, rival maker of cities." He also wrote optimistically of the possibilities of harnessing geothermal power in Hawaii and "wave power" from water as alternatives to "the world's rush for oil." In an editorial about the 1920 census figures, he observed that, for the first time, the nation's urban population was greater than its rural population and that the nation's population center had moved eastward. "In a word, the crowded places are more crowded while the lonely ones continue lonely," he wrote. "Capitalism craves population."[27]

Protesting a plan to build a dam in Yellowstone National Park that would flood 8,000 pristine acres, he defended the social importance of recreation on America's wild lands. National parks and national forests represented one of the government's greatest commitments to public ownership, according to this veteran of high-level battles over federal conservation policies, a socialist foot in the door of the nation's commercial interests:

> We should be far less interested in "playgrounds" were it not for just one thing—the future. Those who at present have the time and means to play can find plenty of space to do it in. "We should worry" about them. But the time is approaching

when workers will be players, too. That is the very time we "radicals" are planning for. Therefore we should not forget to do our part right now to hold open a place—and a spacious one—for the coming "players of the world." . . .

. . . In the public forests, as well as in the parks, recreation should be recognized as a primary "utility"—one at par with the supposedly more "useful" utilities of logging, grazing, and water conservation.[28]

In his plea to protect America's keystone national park from development, MacKaye had offered an explicitly class-based justification for preserving expansive tracts of public recreational land. While not suprising in the pages of a socialist newspaper, MacKaye's inclusive social perspective encompassed a rationale for wilderness conservation that was anything but elitist.

❖

While MacKaye burrowed into his editorial work, Betty threw herself into the cause of disarmament and peace. (On August 18, the day before she joined her husband in Milwaukee, Tennessee finally ratified the Nineteenth Amendment, thus securing women's right to vote.) Within a few weeks of her arrival in Milwaukee, Betty had joined the Socialist Party. "Thousands of suffrage workers will vote the Socialist ticket this year," she predicted, "and we shall find them leading their sisters all over the nation into the battle for the co-operative commonwealth."[29] Benton and Brougham accompanied her to a party meeting a few weeks later. "Herbert and I got our red cards," he noted in his diary, "and became 'comrades.'"[30]

One enigmatic aspect of the relationship between Benton and Betty MacKaye was the persistent presence of their friend, colleague, and sometime housemate Herbert Brougham. An experienced newspaperman, Brougham, after graduating from Yale, had worked for the *New York Times* and the *Philadelphia Public Ledger*. Dismissed from the *Ledger* for his antiwar stance, he eventually found a job with the Labor Department, which may be where he first met MacKaye. By 1920, he was associated with Glenn Plumb, counsel for the Railway Brotherhoods, as an editor of the union's publications. Though married and the father of two children, Brougham apparently did not always live with his family. He sometimes resided at the 19th Street house.[31]

Throughout 1920, the lives of Benton, Betty, and Brougham were intertwined. During this period, the three arrived at all key decisions about jobs and living arrangements on a mutual basis, first in Washington, then in Milwaukee, and finally in New York. It is clear that Betty and Brougham had a relationship that was close and complex. She was apparently as close to him in certain respects as she was to her husband; she certainly seems to have spent as much, or more, time with Brougham than with Benton, at least from the evidence of her husband's own diary. Singlemindedly grinding out his editorials, Benton was likely to keep to the apartment in the evening while his wife accompanied Brougham to a political rally, a committee meeting, or a play or concert. There is no real evidence that MacKaye himself was in any way jealous or suspicious of the time his wife and friend spent together—indeed, the personal arrangement among the three was apparently open and mutual. By 1920, the strain of being a pioneer feminist was taking its toll on Betty, who had had physical and mental difficulties in the past. It is difficult not to conclude that Benton, so wrapped up in his own work, had his head in the clouds, or in the sand, when it came to understanding his wife's increasingly agitated emotional condition. Brougham may have provided Betty with the companionship and intimacy that Benton did not, or could not.

In October 1920, while the three were taking one of their regular long Sunday walks in the countryside surrounding Milwaukee, they discussed a letter recently received from Hazel, describing the familiar plight she and her mother faced: no money, no job prospects for her, no comfortable place to live for the winter. Benton and Betty invited Hazel and Mary to Milwaukee to live with them, and they found a more spacious apartment to make room for the enlarged household.[32] All five were in residence by November 22, but this new family arrangement was disrupted within only a few weeks. The complexities of the relationship among Benton, Betty, and Brougham climaxed in a public controversy that would have fateful personal consequences for the three.

It had not taken long for Betty to gain visibility, and a measure of notoriety, for her peace and disarmament work in Milwaukee. She immediately organized a local branch of the Women's Peace Society. Headed by her former suffrage colleague Fanny Villard, the group was devoted to immediate and universal disarmament, free trade worldwide, and the abolition of mob violence. "We wonder only whether armaments will fall before capitalism," Benton cheerfully wrote in an editorial about the society's efforts. "Neither can crumble too soon to suit us."[33]

But Betty's provocative comments in early December before a disarmament mass meeting at Milwaukee's Pabst Theatre totally disrupted the MacKaye family circle. The source of the trouble was her impassioned call for a "bride strike." Betty MacKaye's logic was impeccable—if, as some of her critics soon pointed out, totally oblivious to the realities of human nature. War required men as combatants; men required women as mates, providers, bearers of children. Hence, if women withheld themselves as wives, perhaps men would, finally, take notice and halt war. "The women of the world will disarm the males in every country, since the men do not seem able to do it for themselves," she announced. "They will refuse to bear children to be slaughtered by the tanks and odorless gases made by men."[34] Betty had demonstrated one of the unspoken, perhaps subconscious, concerns of many who had feared the consequences of woman suffrage. Sex was a weapon after all.

Her defiant suggestion, besides alienating some of her former allies, may also have reflected considerable private anguish concerning her own status as a woman and a wife. Childless, now in her forties, always on the brink of mental and physical collapse, living in cramped quarters with her husband's sister and mother, and closely involved with another man—such personal circumstances may have had as much to do with her demand for a sex strike as did her politics.

In fact, there is reason to wonder whether sexual relations had ever been a part of her marriage to Benton—or, at the very least, a significant part. She was an outspoken opponent of "mechanical birth control," taking the view that contraceptive methods other than abstinence were, as one of her closest friends put it, "the last gasp of sensualism," a subtle way of enslaving women to men's appetites and control.[35] "So long as the claim is made of the necessity of the male," Betty had told a reporter only a few weeks before her marriage to Benton, "women will give themselves to prostitution both in and out of marriage."[36] In any case, Betty's austere views on the subject of sexuality were well known to the public—and, presumably, to Benton—at the time the two were married. The couple must have reached some sort of premarital understanding consistent with her forbidding standard of sexual self-control. Benton would claim, after her death, that they "never planned a family." His wife, he recalled, "was deeply impressed with the way that civilization had distorted the design of nature—especially as to the relation of the sexes. . . . Whatever else, not until the relation of the human sexes became normal could there be lasting happiness."[37] Betty, early in their marriage, had written somewhat cryptically to her sister-in-law Hazel that she and Benton "lived an ideally happy life. So ideal that most humans could not grasp it."[38] As matters developed, the worldwide politics of war and peace became, for Betty MacKaye, somehow confounded with the unresolved enigma of her personal sexuality.

The politics of sex and feminism had reared its head early in MacKaye's tenure at the *Leader*, when Berger had requested that Benton hold up publication of an editorial discussing birth control.[39] And it was a woman, Elizabeth Thomas, president of the worker-owned company that published the paper, who now refused to run Benton's editorial and Brougham's article about Betty's bride-strike proposal. "Herbert told her that he hereby resigned," Benton recorded in his diary on December 7. "First day that no ed of mine appeared in paper."[40]

Betty and Brougham exploited the furor surrounding the incident to promote her campaign. Within a few days, the two traveled to Chicago to get the "'bride strike' story going." There she predicted "announcements of celibacy" across the country "by men and women who are determined to do something tangible as a protest against the absurdity of war."[41] Yet the recent events left Benton in an untenable situation. His mother and sister, only recently arrived, depended on him for their support. His wife and his friend Brougham had, he felt, been unfairly treated at the hands of his own employer. His own career as a journalist was just getting started.

Nevertheless, when the unorthodox and unsettled household talked over their plans and troubles the next Sunday, December 12, all five "decided we'd return to Wash[ington] and the east."[42] In a resignation letter to Miss Thomas, Benton explained that he had "no share whatever in the difference which urges Mr. Brougham's withdrawal" but that his resignation, effective at the end of the year, was necessary because "my own plans are so intimately bound up with his, which we had hoped to put in operation through the Leader. . . . Briefly, we hope to try some other medium whereby to effectuate the economic purposes in which we are interested."[43]

Hazel, writing to Percy, tried to explain this latest family crisis. "Well, Betty felt the 'call.' She felt she might be a real force in this great movement for Peace—which, as you know, is the thing nearest her heart. Benton had come out here to work out the Leader's policies *with* Herbert. Incidentally, he was getting his training as an editorial writer, but he felt that he had already got his 'stride' in that regard, and the only interest for him that lay in continuing the job was that he and Herbert could, together, do some 'political engineering.'"[44]

Hazel made plans to secure a short-term position in January at the National Woman's Party to produce an upcoming pageant; this job, and lodgings for her and her mother at the Woman's Party headquarters in Washington, would see the two of them over the immediate crisis. Betty, having departed for New York with Brougham on December 14, reported back that she had found a position with the Federated Press—then wrote a few days later that the job had fallen through but that she had located a post for Benton and an apartment for the two of them in New York. Benton conscientiously attended to his final duties at the *Leader* and celebrated "an old-fashioned" Christmas with his mother and sister in Milwaukee.[45] "There is a new confident note about him nowadays," Hazel wrote of Benton as they prepared for their trip back east on the eve of 1921. "He frankly admits he's found that he can write—and has a real belief in himself. That was worth coming to Milwaukee for—if for nothing else!"[46]

❖

At their austere one-room apartment at 145 West 12th Street Betty and Benton lived "tenement fashion," as Benton related to Percy.[47] He immediately went to work for the new but short-lived Technical Alliance. The group was inspired by the iconoclastic economist and social critic Thorstein Veblen, who had been teaching most recently at the New School for Social Research. Veblen's recent book *The Engineers and the Price System* had called for the organization of a "soviet of Technicians," a "general staff of engineers driven by no commercial bias." Such technocrats, Veblen envisioned, would, in effect, take over the industrial system and install a new regime dedicated to production for use rather than for profit.[48]

Among those who responded to Veblen's manifesto for a peaceful revolution by engineers was the charismatic, controversial Howard Scott, "a real Napoleonic type," in the recollection of Stuart Chase.[49] Scott, as self-styled "Chief Engineer," was the key figure in organizing the Technical Alliance. He offered MacKaye a job "on a tentative basis for a couple of months," joining Chase as a member of the group's small staff. The Technical Alliance's fifteen-member organizing committee included other acquaintances of Benton, among them architectural critic Charles Harris Whitaker, economist Leland Olds, physicist Richard Tolman, and educator Alice Barrows Fernandez. (Some of the members, including architects Frederick L. Ackerman and Robert Kohn, would join MacKaye, Chase, and Whitaker in forming the Regional Planning Association of America in 1923.)[50]

The nonprofit Technical Alliance had been established in late 1919 with the stated goal "of applying the achievements of science to social and industrial affairs."[51] The group seemed to provide a perfect vehicle for the practice of MacKaye's special skills and for the pursuit of his ambitious ideas about an alternative American landscape of dispersed, worker-controlled communities. During the first months of the year, he worked on a report for the IWW on waste in the lumber industry, in which he argued for the extension and implementation of scientific forestry throughout the country. Set loose to conceive broad reformist schemes, he also drafted two extensive reports, "The Need for Social Engineering" and "A Plan for Developing a Program for Industry in the U.S."[52]

On March 12, MacKaye delivered his final report to Scott, who offered no additional work. As it happened, the Technical Alliance was already struggling to survive, as a result of Scott's domineering personality and the difficulty of finding paying clients for the organization's services. By May, the organization had collapsed. For MacKaye, the brief stint with the Technical Alliance was intellectually fruitful, however. He had been introduced to an expanding circle of like-minded reformers centered in New York. And he continued to pursue his own evolving ideas about social engineering, mulling over a specific, original project that he thought might serve as a flank attack on the established industrial system. "Doping out plans for Appal. trails," he jotted in his diary on March 15, an entry that represents the first written documentation of his best-known idea. MacKaye saw the unraveling of the Technical Alliance as an opportunity to begin what he called, in his family's term, "my larger work." Soon he responded to the entreaties of Rotus Eastman, an old "comrade" from Harvard Socialist Club days, who lived a rustic farmer's life in Glen Sutton, Quebec, just over the border from Vermont. Intending to stay with Eastman for a month or so, Benton left New York by train on April 3, arriving in time to help gather the last maple sap from his friend's sugarbush. "This is the red letter day," he noted in his diary—in bold red pencil. He was celebrating the reunion with his old friend, a return to the invigorating landscape of the northern woodlands, and what he hoped, once again, would be the start of a new vocation, this time as a self-styled social engineer.[53]

While MacKaye's imagination teemed with visions of hiking trails and new recreational communities up and down the Appalachian range, his wife was feeling the full burden of the world's troubles. The focus of her activities at the beginning of 1921, after the bride strike furor had abated, was a campaign to secure passage of a congressional resolution calling for an international disarmament conference. As legislative chairwoman of the Women's Peace Society, she carried the ultimately futile fight to Capitol Hill. In early April, a Women's Peace Society delegation met with Warren Harding and received a tepid response from the new president. The disarmament conference seemed, for the moment, to be an idea without a future.[54]

Discouraged and exhausted, Betty began planning her own temporary retreat from the war against war. Before Benton departed for his rustic Canadian retreat, she tried to rent a house along the Hudson River for her, her husband, and Herbert Brougham. She thought of traveling to Europe for the summer. She attempted, without success, to secure a position on the Chautauqua circuit as a lecturer on "social hygiene." And she sketched an outline for a book to be titled "The Sexual Revolution." Intended partly as an attack on "the fallacy of the mechanical birth control movement," her proposed book would conclude with chapters on "Enforced Motherhood," "Free Divorce," and "The Future of Marriage." Her crusade to "devote her life to the solution of sexual problems," as she wrote in an autobiographical note appended to her book proposal, had apparently failed to provide solutions to her own troubles, however.[55]

When Benton left New York for Quebec, he was concerned enough about Betty's condition to make certain that she had company during his absence. He could remember her breakdown in 1918, when for almost two months she had suffered "sleepless nights and mental disturbance." He arranged for Kathryn Lincoln, a young New York acquaintance, to move in with Betty, who tried to cope with her troubles by keeping busy.[56] In a subdued, affectionate letter to her husband dated April 5, she described her publicity work for the Women's Peace Society, her excursions with Kathryn, and the efforts of Herbert Brougham to find a job. "We will be talking of you and your conversation with Rotus," she concluded. "We can see you tipping in a chair telling Rotus the story of your life."[57]

The next day she wrote another letter, reflecting a more restless state of mind. Preparing for a Philadelphia peace meeting that night, she was preoccupied with domestic concerns: She had just visited Staten Island in search of a house with space enough to add Benton's mother and sister back into the household. She complained of not having a comfortable dress for the hot weather. She had worked on the apartment until it was "as clean as a fiddle. I went over it yesterday with a hoe, and put it to right." She seemed happy with her companion, writing, "I can see that Kathryn and I will get on famously. She is much of a child, of course, but a sweet one. I feel a thousand years old beside her."[58]

That evening, as her husband was settling in with his friend in Quebec, she was sharing the platform at a Philadelphia peace meeting with Socialist Party leader Norman Thomas. The title of her talk was "Can Women Stop War?"[59] Her own response to that awesome question must have given her no reason for optimism, for upon her return to New York she soon fell into a deep depression. On April 15 Benton received a letter at Glen Sutton from Betty, followed by an urgent telegram from Kathryn Lincoln and Brougham describing her deteriorating condition. He returned to New York the next day and discovered that his wife was experiencing an "attack of nervous depression."[60]

❖

The tragic and dramatic events that ensued were front-page news in New York, Chicago, and Washington, three cities where Betty had lived and worked. "MRS. MACKAYE GONE, THREATENED SUICIDE" read the headline in the *New York Times* on April 19. The *New York Journal,* in bold type emblazoned across the top of the front page, announced: "NOTED SUFFRAGIST VANISHES IN QUEST OF DEATH." When Betty had collapsed three years earlier, her recovery had been aided by a stay at the Croton-on-Hudson camp of her friend, Mable Irwin, a Universalist minister and a lecturer on eugenics. Hopeful that such a retreat might once again produce similar recuperative results, Benton made plans to take his wife back to Irwin's house. On the 18th, Benton, Betty, and Irwin went to Grand Central Station to catch a train for Oscawanna, near Croton. While Benton waited in line to buy tickets, the two women made their way through the busy station to a rest room. The detailed newspaper accounts of what happened next were lurid and sometimes contradictory. But the outcome was grim and unequivocally final.

"I'm going to end it all," Betty had told her companion, according to one newspaper report. "I am going to kill myself," she had exclaimed, according to another.[61] Benton would later deny that she had said any such thing. But in the event, she bolted from the older Mrs. Irwin, mingled with the teeming crowd, and hurried out of the station and into the street. When Benton, in the confusion, learned of his wife's escape, he checked the station's exits and the nearby streets, but to no avail. It was not until later in the day that he reported his wife's disappearance, at a police station near his apartment, where he claimed, according to the *Times* account, "that his wife often said she intended to end her life by jumping into the Hudson or East River."[62]

Later that afternoon, a woman's body was recovered in the East River. Benton did not come to the undertaker's to identify the corpse, however. Not yet ready to face the fact of his wife's death, he spent the night contacting city hospitals and his friends in the hope that Betty had turned herself in for treatment. The next morning, Mrs. Irwin and Charles Whitaker identified Betty's body at a Brooklyn funeral home.[63] Whitaker telephoned Benton's apartment, where Hazel was keeping vigil with her beloved brother, to give his friend the sad yet certain news. And it was Whitaker who received Betty's ashes after her cremation.

Benton gathered himself together to attend the brief outdoor ceremony on Staten Island to scatter her remains. "The place my brother has chosen for the service in memory of his wife is out of doors—for it seems to him that the open air is the only fitting place to scatter the dust of those who have gone from this world," Hazel wrote in her memorial comments. There, she continued, next to a lily pond, the "young frogs are piping their spring song. That song which seems to him to be nearer to what the race calls 'spiritual' than any other sound in the world. It is the sound of eternal spring—of the pain and ecstasy of that ever-recurring time." For the rest of his years, no matter where he lived, Benton always noted the first springtime sound of peeping frogs, the "Carboniferous chorus" that symbolized for him the ancient and hopeful continuity of life on earth.[64]

In the aftermath of Betty's suicide, Benton's family and friends rallied to his support. Emotionally spent, suffering his recurrent intestinal troubles, he moved in for a time with his brother Hal's family in Yonkers. Letters of consolation poured in. In an attempt to allay Benton's possible sense of guilt for his wife's death, one of Betty's long-time friends described a previous—and previously unknown to Benton—suicide attempt by his wife. Other letters recounted the unhappy circumstances of her childhood, the tragedy of her first husband's death after their brief marriage, her restless quest for a career and a cause. Betty's "mind was failing her," wrote her sister. "In fact Gurney, her first husband, told me so before he died. . . . Jessie [as her sister called her] made many friends yet she was always a puzzle to all of them."[65]

One colleague, in a public tribute, described her late friend as a martyr to the suffrage cause, "who has paid with her life a part of the price of women's freedom. . . . It is well that men should know something of the fearful price that women have paid for their freedom. It is well, too, that women who have never worked for their own freedom but have gratefully accepted that liberty, without cost to themselves, should realize how some women have poured out their very lives in the cause of freedom."[66]

Other acquaintances, who had shared Betty's interest in spiritualism, claimed that they were still communicating with her. Her friend Theresa Russell, suffragist and wife of the muckraking journalist Charles Edward Russell, reported receiving "messages" from Betty from the "other side." Herbert Brougham also wrote that he was still hearing from her.[67] Of all MacKaye's unresolved emotions, his feeling about Brougham, unexpressed in any surviving records, may have been among the most confusing and troubling. In any event, except for a "good talk" a few weeks after Betty died, the two men apparently had little contact from that point on.[68] Brougham, who later married Kathryn Lincoln, died twenty-five years to the day after Betty's suicide.

MacKaye found his solace, as time passed, in the knowledge that Betty's fragile psychological condition and periods of depression were chronic, recurring, and might well have become more severe had she lived. "There was something wrong with her mind and that is all we know," he wrote to one of Betty's relations:

> This winter she had been working with all her soul in the cause of disarmament. She sensed the danger of another war. Apparently she was more stirred up than any of us realized. . . .
>
> . . . She used to say also that she thought she was coming to the change of life. She thought, at the time, that this might be the cause of her illness three years ago. She was repelled at the thought of dealing with the doctors. So we just tried to get out of our minds the experience we went through in 1918.[69]

For MacKaye, his wife's suicide opened an emotional wound that never really healed. He passed the rest of his years as a bachelor; and he rarely mentioned, even to his closest acquaintances, that he had ever been married. In his own effort to recover from the soul-scorching circumstances of Betty's death, however, MacKaye discovered and nurtured untapped personal resources of his own. His wife's death was the turning point of a lifetime, the event that set him on the path to his greatest social contributions. "The philosophy of utility fits every occasion," wrote his brother James in a characteristically reserved and calculating note of consolation, "and I am applying the principle in urging you to begin at once to plan for the future—that is the useful thing to do."[70] Benton assiduously applied his brother's philosophy. He earned a measure of compensation for his private loss and pain when, within only a few months of Betty's passing, he publicly proposed his pioneering idea for a hiking trail along the ridgeline of the Appalachians, the full length of that ancient mountain chain.

His ambition would soon be precise and intense, whereas just a few months earlier it had been only vague and rhetorical. On the last day of 1920, as he had boarded an eastbound train for New York to face an uncertain future, his final editorial had appeared in the *Milwaukee Leader*. Under the headline "Happy New Year," his commentary spelled out several reasons for hope and optimism on the barren landscape of American capitalism usually depicted in the newspaper's pages.

"You're in the wilderness and desolate—and scared perhaps," MacKaye had written, "but the quest of 'getting out,' ignites a spark that shows there is somewhere a heaven. If man cannot—just yet—HAVE happiness he can at least PURSUE it." Invoking Jefferson's vision of the supreme American purpose, the editorialist's valedictory concluded: "May the new year bring for you the zest of the Pathfinder in the wilderness—the thrill of him whose only thought is the pursuit of the happiness of his fellowmen."[71]

MacKaye had no way of knowing that during the year after he wrote this editorial—and despite the most harrowing of personal setbacks—he would fulfill the very rhetorical challenge set forth in his own hopeful words. He would lead his fellow countrymen on the trail out of the confused, capitalistic wilderness of urban America into the primeval natural community preserved in the forests and mountains of what he came to call the "Appalachian Domain." Henceforth, he would become best known as a prophet of play, "Moses in reverse," as his old friend Sturgis Pray called him—a Pathfinder in the wilderness.[72]

❖ 8 ❖

First Steps along the
Appalachian Trail

1921–1923

T he direction of MacKaye's life and career at this precarious moment
in the spring of 1921 was set by the generous and supportive Charles
Harris Whitaker. MacKaye and Whitaker shared common ideological
and political convictions, but they also had in common a New England upbringing, which instilled in both a yearning to recreate communities modeled
partly after the region's rural villages.[1] While MacKaye slowly regained his
emotional and physical equilibrium at his brother Hal's home in Yonkers,
Whitaker invited his distraught friend to spend some time at his modest farmstead in the northwestern New Jersey township of Mt. Olive. "Come on out
and live there for a while and be my aide de camp," he wrote. "It is grand beyond words!"[2]

By early June, Benton had settled in at Whitaker's rural retreat, where he returned to his writing. "Began work—doping out gen'l industrial plan," Benton
noted in his diary, "blocking out recreation project." A few days later he began writing a "Memo. on Regional Planning." As Whitaker saw MacKaye's
schemes taking shape on paper, he also had in mind a plan to broadcast those
ideas to others. By the end of the month, as MacKaye was "working out Appal.
Trail" in his memorandum, Whitaker wrote to another friend, Clarence S.
Stein, about MacKaye's work. Stein, an urbane, progressive New York City architect, headed the Committee on Community Planning of the American
Institute of Architects. In the AIA's journal, Stein had been reporting for several years on the community planning movement. In Stein's eyes, community
planning encompassed not only the development and rehabilitation of existing cities, but also the creation of entirely new towns and cities on previously
undeveloped land. As Whitaker recognized, MacKaye's embryonic ideas suggested a novel type of community—entirely different from anything yet conceived by urban-oriented architectural thinkers like himself and Stein.[3]

MacKaye was now free from political constraints and the muddled ideology
that so often characterized group efforts. So he outlined a far-reaching plan
for the transformation of modern American industrial society that empha-

sized play as a first priority. His handwritten, sixty-page memorandum was divided into two sections: "Regional Planning as a Reconstruction Policy," which outlined his personal philosophy of a new discipline; and "Projects in Regional Planning," in which he described several practical planning tasks. His diagnosis of the nation's social and economic ills constituted an unveiled attack on the profit system. If "surplus wealth" was the underlying cause of war and unemployment, as socialist orthodoxy proclaimed, then regional planning and a "unified industrial system" organized on nonprofit, socialized principles provided the antidote. English-style "garden cities" and what he called a "recreation plant" were the two key components of his blueprint for the evolutionary takeover of the industrial system. Writing explicitly about regional planning for the first time, he asserted that the subject "to be of full use, must deal not alone with man's work but with man's life." "Given time," he optimistically wrote, "the cooperative principle will replace the competitive one."[4]

MacKaye described three specific projects, "with a view to securing tangible employment for the regional planner," perhaps "in the form of a series of articles in popular magazine style." The first proposal called for a survey of industrial localities throughout the Appalachian region, but he concluded that the time was "probably not yet ripe for obtaining a client for such a project." The second was a detailed, six-point industrial survey of Vermont. He sketched out a vision of the state that divided it into self-contained, "self-owning," "factory-ized" agricultural communities, organized according to watershed boundaries.[5]

It was MacKaye's third proposed planning project, titled "Survey and Plan for an Outdoor Recreation System throughout the Appalachian Mountain Region," which piqued Whitaker's enthusiasm. "In view of . . . the fact that outdoor recreation makes instant appeal to all classes of humans," he wrote, "it is suggested that the most popular approach to a comprehension of regional planning might be made by presenting some big bold conception in public recreational life."[6]

"Engineering projects which can be visualized geographically, and in their entirety, are well adapted to seize the popular imagination," he continued. "Witness the Panama Canal and the 'Cape to Cairo' railway." Citing parallels with Vermont's 210-mile Long Trail, which the Green Mountain Club had begun clearing and blazing a decade earlier but which would not be completed the full length of the state until 1930, he proposed "the building of a 'long trail' over the full length of the Appalachian skyline—from the highest peak in the north to the highest peak in the south—from Mt. Washington to Mt. Mitchell."[7]

That was the backbone of his grand scheme. Then he elaborated on its rationale and purpose. "It would make a man sized project in regional planning and engineering, laying the foundation for a socialized outdoor life for the workers of the nation," MacKaye imagined. "Putting regional engineering on the map in this line—recreation—should lead to its comprehension and application in other lines—those of industry. It might well prove to be our route at least toward a socialized *industrial* life for the workers of the nation."[8]

The "clients" for a "regional survey and plan" for what he called Appalachian Skyline Trail might, he speculated, include mountain clubs, hotel operators, sporting-goods businesses, newspapers and magazines, railroads, chambers of commerce, women's clubs, social workers, and "the more far seeing labor organizations." A plan for publicity and popular education would include a series of articles for newspapers and magazines describing the geography of the trail, "section by section." The first such article "would outline the enterprise as a whole, being largely a popular essay on recreation in its relation to industry," he suggested. "It should contain a map showing the proposed route and its branch lines."[9]

His recreation proposal also included an element that foreshadowed his later ideas on "townless highways" and his battle in the 1930s over the "skyline" highways planned and built on some Appalachian mountain ridges. The construction of a road the length of the Appalachians, he suggested, might "tend to follow the valleys rather than the skyline. An essential part of the scheme should be the establishment of inns where board and lodging can be had without profit; also the running without profit of auto-stages for touring."[10]

With such detailed plans on paper, MacKaye, on July 10, accompanied Whitaker to meet Clarence Stein for the first time at the Hudson Guild Farm, near Lake Hopatcong at Netcong, New Jersey, only a few miles from Mt. Olive.[11] The farm was maintained as a recreational retreat by New York's Hudson Guild Settlement House, which was affiliated with the Society for Ethical Culture, a movement founded by educator Felix Adler that sought to combine personal improvement with social reform. Stein, long active in the Ethical Culture movement, had designed some of the buildings at the Hudson Guild Farm.

This July 1921 meeting, at which Whitaker offered to publish an article concerning the Appalachian recreation plan in the AIA journal and Stein agreed to promote it through his AIA Committee on Community Planning, launched the Appalachian Trail; and it launched MacKaye's self-designed career as a regional planner. The meeting also initiated a close friendship between Mac-Kaye and Stein that lasted until the year they both died, 1975. Stein would become MacKaye's most significant professional and financial patron.[12]

In a new draft of his regional planning memorandum, MacKaye elaborated on the rationale for pursuing a recreational project as a more palatable first step in a broader plan for "social readjustment." Were they to mount a "frontal attack" on the industrial system, involving the construction of worker-owned industrial communities, it would meet stiff opposition from "ultra conservatives" and big business, he predicted. "If they did not call it 'visionary' they would say it was 'bolshevistic' and 'dangerous.'" By contrast, a project combining a mountain trail and a series of recreational camps and communities "would make a flank attack on the problems of social readjustment. This fact if understood, would lose for the project the support of the ultra conservatives among the recreation group. But it would retain the support of the liberal minded and of the radicals therein. And these together would form a majority of the recreation group." The Appalachian project, in other words, would provide an indirect route to his conception of the ideal American society—"to live, work, and play on a non-profit basis."[13]

From this memorandum, MacKaye distilled his article "An Appalachian Trail: A Project in Regional Planning," which appeared in the *Journal of the American Institute of Architects* in October of 1921 (and is included as an appendix to this book).[14] He and Whitaker toned down the political thrust of the article by weeding out any direct references to its essentially socialist underpinnings, but the ideological implications of the proposal were still distinct. If the Appalachian Trail that eventually took shape did not incorporate all the elements of MacKaye's original proposal, the trail that evolved nevertheless reflected that plan's essential, even subversive, spirit. "The camp community is a sanctuary and a refuge from the scramble of every-day worldly commercial life," MacKaye wrote. "It is in essence a retreat from profit," an amateur, noncommercial, communal project in which "cooperation replaces antagonism, trust replaces suspicion, emulation replaces competition."[15]

MacKaye's article did not stint on grandiose rhetoric and bold imagery. "Let us assume the existence of a giant standing on the skyline along these mountain ridges," he wrote, "his head just scraping the floating clouds. What would he see from this skyline as he strode along its length from north to south?"[16] From this giant's-eye view, MacKaye led his reader from New Hampshire's Mt. Washington to North Carolina's Mt. Mitchell.

MacKaye addressed not only the nature of recreation, but the whole "problem of living," which he saw as "at bottom an economic one." The complications of industrial society, he pointed out, included not only war, unemployment, environmental contamination, high prices, threats to personal liberties, and labor-capital conflict. The forces of industrial efficiency would reduce "the toil and chore of life" and increase time devoted to "leisure and higher pursuits," he predicted. "The coming of leisure in itself will create its own problem."[17]

As an answer to the leisure "problem," MacKaye plotted a campaign to put a volunteer army of 40,000 to work developing along the Appalachian skyline and in its river valleys "a camping base strategic in the country's work and play." Assuming his omniscient viewpoint, "standing high on the skyline," he surveyed the countryside from Mt. Washington south, noting the "smoky beehive cities" as well as "picturesque Allegheny folds," the iron plants of Pennsylvania and the "wooded wilderness of the Southern Appalachians." From this skyline perspective of a vast region, he discerned the landscape's latent opportunities: recreational facilities, sanitariums for those who have suffered "the grinding-down process of our modern life," and employment in agriculture and forestry. Most important, though, the view from the mountaintop provided "a chance to catch a breath, to study the dynamic forces of nature and the possibilities of shifting to them the burdens now carried on the backs of men. The reposeful study of these forces should provide a broad gauged enlightened approach to the problems of industry. Industry would come to be seen in its true perspective—as a means in life and not as an end in itself."[18]

The Appalachian Trail was to be but one component in MacKaye's grand plan for "a series of recreational communities" that would constitute "a base for a more extensive and systematic development of outdoor community life. It is a project in housing and community architecture." The whole project comprised the hiking trail, shelters along its course, community camps providing less strenuous and solitary activities for families, and food and farm camps in the valleys offering employment and a "tangible opportunity for working out by actual experiment a fundamental matter in the problem of living." His plan was intended in part as a constructive response to what he saw as the age's militaristic spirit. "It appeals to the primal instincts of a fighting heroism, of volunteer service and of working in a common cause."[19]

The text of MacKaye's article, however inspirational, may not have had as great an impact as did the accompanying map. Drafted in his characteristically impressionistic and large-scale cartographic style, the map depicted the route of the proposed trail very much as it would eventually be built. The map's features included the main trail itself, a thick black stripe reaching from Mt. Washington to Mt. Mitchell; branch trails, incorporating existing paths as well as possible extensions of the main trail north to Katahdin and south into Georgia, the trail's eventual terminal points; railroads intersecting with the trail; the locations of cities with populations over 100,000 within a day's drive of the trail; and the mountainous areas the trail would traverse.

It was a measure of the intellectual distance MacKaye had traversed from his professional origins as a forester that his trail proposal did not appear in a journal of forestry, conservation, or recreation. Like the more reform-minded foresters and conservationists, many progressive architects and planners were ambivalent about the institutions and interests their professions served. In the *Journal of the American Institute of Architects* these architectural reformers found an outlet and a leading intellectual forum of their profession. Indeed, the October 1921 issue of the journal, in which MacKaye's article appeared, was devoted mostly to community planning. As an explicit "retreat from profit," MacKaye's "project in regional planning" struck a chord with Stein, Whitaker, and other friends and colleagues in the most visionary and radical wing of the era's evolving American planning movement. These were people interested not just in architecture, community-building, and other physical aspects of planning, but in overarching social and economic objectives as well.

Among the readers who responded enthusiastically to the sweep of MacKaye's vision was his erstwhile boss, Gifford Pinchot. "I have just been over your ad-mirable statement about an Appalachian Trail for recreation, for health and recuperation, and for employment on the land," wrote the former Forest Serv-ice chief. "Your giant certainly sees the truth."[20] Pinchot's endorsement was but one example of the intellectual recognition and personal support Mac-Kaye began to receive after introducing the Appalachian Trail idea.

Indeed, the setting and events of the summer had proved to MacKaye the validity of his brother James's philosophy of utility. "They say one's body starts anew every seven years," he wrote his mother in July, recalling the moment in the summer of 1914 when he had first realized the depth of his love for Betty. "Well my soul started anew then. . . . I am 42 years old—just seven times six. This last cycle of my life has been obviously the biggest one. . . . It is no mere 'death' I think of hourly—it's the coming of a philosophy—a something that I never had before."[21] Out of the ruins of his personal life in the aftermath of

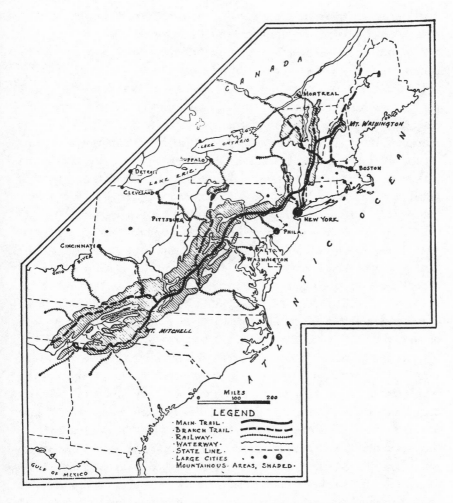

MacKaye's original map of the Appalachian Trail appeared in "The Appalachian Trail: A Project in Regional Planning," in *The Journal of the American Institute of Architects* 9 (October 1921). The article and the map were reprinted later that year in a brochure titled "A Project for An Appalachian Trail," produced by the AIA's Commmittee on Community Planning, with an introduction by the committee's chairman, Clarence S. Stein. MacKaye's map depicted a main trail from Mount Washington in New Hampshire to Mount Mitchell in North Carolina, as well as various existing or proposed branch trails. The map's caption read, "Area shown contains more than half the population of the United States and over one third the population of Canada." (Author's collection.)

Betty's death and his own career travails Benton had begun to construct a coherent, hopeful vision of a future for himself and for his fellow citizens.

"I am not at all sure whether the proposition is worth a hoot or not," he confessed to Percy, enclosing a copy of his Appalachian Trail article. "Neither am I worried." During late summer and early autumn, MacKaye was "doping plans" for writing about natural resources on a global scale, meeting on several occasions in New York with journalist Lincoln Steffens, and planning his return to Shirley Center.[22] The Shirley properties were customarily vacant during the winter months, and as a bachelor again, with a head full of ideas and without other immediate responsibilities, he viewed Shirley as the economical and logical place to resettle and work. By October, as he sorted through family papers from thirty years of MacKaye occupation, Benton began laying a special claim to the Shirley property. It became the principal bench mark for his own lifelong intellectual survey of the globe. From this time on, though the Cottage was jointly owned by the family association and the Grove House became Hazel's legacy, Benton would oversee the "Empire," as he came to call the modest family holdings. His dominion over the property would cause some tension within the family, especially between Benton and Percy. Both brothers romanticized Shirley Center, but Benton truly made it his home.

In Shirley, MacKaye set to work at two tasks that he saw as closely related: a campaign to promote the Appalachian Trail project, and the development and refinement of his general approach to regional planning. His arrival in Shirley coincided with the publication of his article in Whitaker's magazine. Clarence Stein arranged for his AIA Committee on Community Planning to sponsor the production and distribution of reprints of the article, to which Stein added his own introduction. Stein immediately grasped the scope and underlying purpose of his new acquaintance's vision. "He would as far as practicable conserve the whole stretch of the Appalachian Mountains for recreation," Stein wrote. "Recreation in the biggest sense—the re-creation of the spirit that is being crushed by the machinery of the modern industrial city—the spirit of fellowship and cooperation. . . . It is a plan for the conservation not of things—machines and land—but of men and their love of freedom and fellowship."[23]

MacKaye compiled an eclectic mailing list of reformers for Stein's publicity campaign. "In general I suggest avoiding—at least at the start—persons who are merely officials," he advised Stein. "The officials I have listed are personalities also." Besides some of his "recreationist" acquaintances, the list included town planner Thomas Adams, forester Gifford Pinchot, social worker Jane Addams, pacifist Fanny Garrison Villard, and a variety of labor leaders ("including liberals and radicals"). "The subject should appeal to a wide range of them," MacKaye predicted.[24]

As he plotted both the trail's potential route and the means by which the trail might be built, MacKaye met in Shirley Center and in Cambridge with his earliest tramping companions: his brother James; his friend Horace Hildreth, with whom he had explored Vermont in 1900; and landscape architect and Appalachian Mountain Club activist Sturgis Pray. In late November, Benton also met in Boston with his friend Allen Chamberlain, the conservationist and writer. On Chamberlain's urging, MacKaye in early December attended the two-day meeting in Boston of the New England Trail Conference, where he was able to talk with the most influential figures in the regional trail community. His proposal struck a chord with these outdoor activists. The Appalachian Trail would soon be transformed from pipe dream to literal pathway.[25]

Like Chamberlain, other eastern outdoor enthusiasts, especially in New England and in the New York–New Jersey region, had, over the previous decade, offered various proposals to connect existing trails and to expand the regional trail network into new mountain terrain. But none of these plans matched the boldness and breadth of MacKaye's. The burgeoning use of the automobile provided hikers quick access to more trailheads and tramping possibilities. In New Hampshire's White Mountains, for instance, trail circuits that had previously been clustered around railroad terminals and resort communities were rapidly being linked. James P. Taylor, the enthusiastic Vermonter who in 1910 had first proposed the Long Trail, running the length of the Green Mountains, four years later promoted a trail reaching from New York City to Quebec. Others had sketched maps for trails connecting the White Mountains to Maine's Katahdin. And in 1916 Chamberlain and Philip W. Ayres, who headed the Society for the Preservation of New Hampshire Forests, had taken the lead in creating the New England Trail Conference (NETC), a federation of outdoor clubs whose purpose was to "promote cooperation in the creation and maintenance of connecting trails in New England."[26]

One Forest Service official, William L. Hall, at the NETC's first meeting in 1916, had spoken "of the desirability of completing a system of trails to traverse all the New England mountain ranges, and be linked ultimately with the southern Appalachians." The NETC members soon began sketching out the details of long-distance trails all over the New England states, as well as throughout New York's Adirondack and Catskill ranges. Inspired by the example of their New England neighbors, a group of New Yorkers had met earlier in December 1921 to form the Adirondack Mountain Club.[27]

As Chamberlain explained to MacKaye after reading his Appalachian Trail article, the NETC and recently organized trail workers in New York and New Jersey were thinking about consolidating to form the Northeastern Trail Conference; they already had the active support of the Forest Service and state forestry officials for their trail-linking efforts. "The plan really grew out of a desire to thread various public lands and reservations on a trail system so as to arouse wider interest in public forests," Chamberlain wrote.

It was pure propaganda. It took.
Now the aim is to work for the public acquisition of the right of way for the connecting trails—the trunk lines. Otherwise the day will come when many important links will be lost where they now cross over private land on sufferance merely. . . .
The motorists own the roads now. Why isn't the walker entitled to a road of his own at public expense?[28]

So when MacKaye appeared at the NETC in December 1921 to describe his project, he was preaching to enthusiasts who were already thinking and working on precisely the same lines, if not on so vast a scale. At the same meeting, in fact, a New York trail builder, J. Ashton Allis, offered a plan to extend the Long Trail south through Massachusetts, Connecticut, New York, and New Jersey as far as the Delaware Water Gap.[29]

MacKaye's article, and his own efforts over the next several years to launch and organize the trail project, were catalysts for a new burst of activity based on this existing, but unfocused, energy and enthusiasm. He managed, through his own network of personal acquaintances, to locate and bring together the strategic individuals and organizations that would launch the project. "It will be comparatively simple to push on the trail proper portion of our program," he wrote to Stein immediately after the conference. "The main problem will be how to handle the community feature."[30] As the years went on, MacKaye's assessment of the project's prospects proved to be altogether accurate. Some of the recreationists, as he had predicted, had a more modest agenda and a less ideological rationale for their efforts than did he and his planning associates Stein and Whitaker; for many in the hiking community, a trail was simply a trail.

Many of the parties associated with the origins of the Appalachian Trail, in the NETC and otherwise, were not mere hobbyists. Ayres and Chamberlain, for instance, were shrewd, experienced political activists who had been instrumental in securing passage of the 1911 Weeks Act. Other NETC members, such as Sturgis Pray, Arthur Comey, Harlan P. Kelsey, and John Nolen Sr., were leaders in the fields of landscape architecture and city planning. And many other figures in the regional outdoor recreational movement were trained foresters, schooled in the Pinchot conservation doctrine of "the greatest good for the greatest number." These men (it would be a few years before many women assumed significant roles in the Appalachian Trail effort) shared a broad vision of the social utility of a protected and open public landscape. "It presents a planning project of real significance," Nolen wrote MacKaye after reading his Appalachian Trail article.[31] New Hampshire forester Philip Ayres predicted, "The interesting thing about your plan is that it is inherently likely to be carried out. The trend of the times is likely to enforce it."[32]

"Here is a project to be dramatised," MacKaye had concluded his 1921 article.[33] As a social inventor, he hoped to build not merely a physical trail along the eastern mountain ridgeline. In his appeal to the public imagination, he aimed as well to create a community of productive citizens, whom the trail would link in both geography and spirit.

As 1921 came to a close, MacKaye remained in and around Boston, visiting with his friends and family. On New Year's Day, Clarence Stein arrived in Boston, where MacKaye introduced him to acquaintances who shared their conception of the trail, including Pray, Chamberlain, and Nolen. Stein reported to MacKaye that the National Federation of Settlements had endorsed the trail project; the organization planned to map the location of its affiliated summer camps in relation to the proposed trail route.[34] When MacKaye returned to Shirley after his round of discussions in Boston, he had formulated more precisely the role he envisioned for himself. He proposed to write a book and a series of articles on the Appalachian Trail project. The task would require that he travel the length and breadth of the region he now perceived as the "Appalachian Domain." He pictured himself as the scout who would visualize and plot a regional plan for the entire mountain chain.

As the weather turned frigid and he settled in to a routine in Shirley, MacKaye felt confident and composed about his new life. He wrote to his mother in late January describing in comic terms his daily ritual: At 5:30 a.m., the "alarm clock chops a dream in two." The domestic chores required to maintain an unheated house without running water and electricity—stoking stoves, carrying firewood, filling kerosene lanterns—occupied his early morning hours. "And then—then—with the industrial and economic problem completely solved—for one small daylight sojourn, I take up the work of the unpestered and the uninvaded," he wrote of his daily writing routine "on the

Acropolis of Shirley Center." His afternoons were often devoted to Thoreau-vian walking excursions across the frozen landscape of Shirley, when he traded tales with local woodchoppers, farmers, and general store "wise men" about the rigors of surviving the below-zero weather.[35] The bracing New England weather had instilled in MacKaye a new physical and emotional zest, as well as a renewed sense of purpose.

January 22 was a solemn date for the MacKayes. Benton wrote Percy, recalling how, exactly thirty-three years earlier, he had received from his Aunt Sadie, right there at Shirley Cottage, the grim news of their cherished brother Will's death. "All things have been different since that moment," he wrote. "They have been 'easier.' The falls have not been so high—I've not been high enough to have them." He nonetheless conveyed his own sense of optimism and hope:

> This earth is a small flea bite but who knows but what it is strategic in the universe? This earth scrubbed up has the potentialities for the beginnings of a corking heaven. Anyhow it's the only place we have to reclaim for such a purpose. And so, as you say, each to his magnum opus!
>
> And how the air has cleared regarding this since my talks with you in Cambridge! I've come up here to do "the biggest thing that's in me." My first task is a regional plan for the Appalachian country. I've just received 136 topographic quadrangles from Washington. I spread them out in groups before me. It is as if I were, according to scale, about 100 miles above the country in an aeroplane. I am thus flying over the region from north to south. I started at Mt. Washington and am bound for Lookout Mountain, Tenn. I've today reached northern Virginia on this my preliminary trip. A couple of days will fetch me to Lookout Mtn. I'll take a side trip through Harlan County, Ky. I'm making notes en route and drawing a sketch map. With a knowledge of the country (and of human nature) I am endeavoring, in these winter days, to blaze at least one trail toward that "civilization" that we talked about.[36]

As he traveled along the Appalachians from an imaginary aerial perspective, MacKaye sketched watershed boundaries, traced routes for the ridgeline hiking trail, and searched out sites for the camps and communities he had described in his article. By the beginning of February, he had completed this intense and solitary mind's-eye reconnaissance. The contours of the Appalachian landscape imprinted in his imagination, he began his book, with the working title "The Trail Out: An Outdoor Survey of Our Industrial Wilderness."[37]

❖

The appeal of the Appalachian Trail project paralleled developments in the co-alescing American wilderness preservation movement. By the early 1920s, the fate and the uses of America's remaining undeveloped lands were subjects of intense debate among a small but expanding circle of foresters, conservation-ists, and land-use activists. A month after the publication of MacKaye's Appa-lachian Trail proposal, the *Journal of Forestry* carried an article titled "The Wil-derness and Its Place in Forest Recreational Policy," written by the Forest Service's Aldo Leopold.

"By 'wilderness,'" Leopold wrote, "I mean a continuous stretch of country preserved in its natural state, open to lawful hunting and fishing, big enough to absorb a two weeks' pack trip, and kept devoid of roads, artificial trails, cot-tages, or other works of man." He challenged his fellow foresters to ask them-selves "whether the principle of highest use does not itself demand that repre-sentative portions of some forests be preserved as wilderness."[38]

Leopold's eloquent voice gradually redefined the terms in which the whole question of wilderness protection in the United States was perceived and dis-cussed. His early plea to forestry colleagues was a protest against the Forest Service's approach to wilderness preservation and recreational management in the national forests. Historian Paul Sutter explains that what Leopold meant by wilderness preservation was "preservation from certain forms of ad-ministrative and recreational development," especially road building and the leasing of Forest Service lands for summer homes and other commerical uses. Leopold believed "that the Forest Service had, in many cases, been too gener-ous to those seeking recreational access." And Leopold did begin to influence the agency's priorities. In 1924, at his prodding, the Forest Service would es-tablish the 574,000-acre Gila Wilderness Area in New Mexico's Gila National Forest. The nation's first designated wilderness area (although not afforded complete legal protection until passage of the Wilderness Act forty years later), the Gila Wilderness exemplified a principle Leopold had articulated in his momentous article. "It will be much easier to keep wilderness areas than to create them," he had written in his 1921 article. "In fact, the latter alterna-tive may be dismissed as impossible."[39]

Another Forest Service staffmember directly attacked the threat posed by national-forest roads. Arthur Carhart, the first landscape architect employed by the agency, in 1919 proposed that a tract surrounding Trapper's Lake in Col-orado's White River National Forest be used for wilderness recreation rather than for vacation home sites; his proposal was adopted the next year. After conducting a recreational survey of Minnesota's Superior National Forest,

Carhart also proposed that some national forest areas be managed as wilderness. However, as historian David Backes has observed, Carhart's "idea of wilderness focused on scenery" and emphasized "making such areas available to the masses." Indeed, Carhart's plan for recreational uses in what would later become the Boundary Waters Canoe Area called for "motorboat highways" and a string of eight rustic lakeside hotels.[40]

MacKaye, the eastern counterpart and contemporary of Leopold and Carhart, also emphasized the recreational benefits of wilderness, but only as part of his larger social agenda. In fact, he identified a social dilemma exacerbated by the nation's geography and the distribution of its population. "Camping grounds, of course, require wild lands," he wrote in his article proposing the Appalachian Trail. Most of the nation's wild lands, however, were in the national parks and national forests of the west. These "playgrounds of the people," he continued, "are for the Western people—and for those in the East who can afford time and funds for an extended trip in a Pullman car. But camping grounds to be of the most use to the people should be as near as possible to the center of population. And this is in the East."[41]

MacKaye came to envision a reconstituted wilderness along the Appalachian Mountain range, where the original wilderness no longer existed—at least not on the scale or in the character that Leopold had experienced on federal lands in the Southwest. The Appalachian Trail would represent a conceptual wilderness, traversing numerous political jurisdictions, environmental habitats, and human cultures across thousands of mountainous miles. "The region spans the climates of New England and the cotton belt," MacKaye observed; "it contains the crops and the people of the North and of the South."[42]

When Leopold, Carhart, and MacKaye used the term *wilderness* in the early 1920s, as scholars such as Backes and Sutter have carefully elucidated, these legendary figures of the wilderness movement who shared some of the same concerns did not necessarily mean the same thing. In practical terms, however, on the American landscape the two concepts introduced respectively by Leopold and MacKaye—the extensive wilderness *area* and the regional *linear* wilderness represented by the Appalachian Trail—would gradually, but never entirely, be connected.

The crusades for wilderness preservation, wildlife protection, and outdoor recreation were joined by an increasing number and variety of Americans during the 1920s. The character of the American conservation movement was transformed, veering away from the utilitarian priorities emphasized by Gifford

Pinchot and other influential government officials. The movement grew in size; its agenda expanded; its organizational structure diversified and fragmented. The Izaak Walton League, for instance, founded in January 1922 as an organization of sportsmen devoted to protecting wildlife, grew dramatically during the decade. An authentic nationwide mass conservation organization, the league dwarfed the memberships of other specialized and regional groups like the Sierra Club, Appalachian Mountain Club, Save-the-Redwoods League, American Game Protective Association, and Audubon Society.[43]

Scientists, for their part, were promoting the value of undisturbed lands not for recreation or "social readjustment," but as ecological benchmarks and settings for research. In 1921, the Ecological Society of America (of which Raphael Zon had been a founding member in 1915) and the American Association for the Advancement of Science both proposed the creation of wilderness reserves, essentially as laboratories of scientific study. In the same years, federal and state officials sought to expand the lands and facilities available for outdoor recreation and wildlife protection. At the instigation of National Park Service director Stephen Mather, a January 1921 conference in Des Moines, Iowa, at which twenty-five states were represented, resulted in the creation of the National Conference on State Parks.[44]

MacKaye was aware of some of these contemporary strands of amateur, scientific, and political activity in the areas of conservation and outdoor recreation, and he sensed the trend of the moment and the promise it offered for his own career. "The whole project grows vaster (as I knew it would) the more I get into it," he wrote in a letter to Stein from Shirley in early February 1922. "I want if possible to launch it in a way so that I can give my future to the kind of work which I think I see here."[45] After mapping out possible routes for the trail, he set to work outlining and writing a manual for those who would actually "scout" the trail on the ground. Already, though, he was struggling to reconcile the mundane physical task of building the hiking trail itself with his more ambitious—and abstract—program for controlling what he saw as the uncontrolled and insidious growth of modern industrial civilization. He arrived at an elaborate metaphor that turned the American conception of "wilderness" on its head. Working on the manuscript for "The Trail Out," he attempted to link the surveying of a hiking trail with what he characterized—possibly for the first time—as a "new exploration": an intellectual and analytical quest for a path out of the "industrial wilderness."[46]

He was racing ahead of his old friends and new acquaintances in the outdoor clubs. For him, the scouting and building of an Appalachian Trail was merely the first step in a grassroots investigation to reveal the workings of the "industrial clockworks" of natural resources and financial capital. He envisioned "Industrial exploration as an outdoor recreation"—an idea that surely would have perplexed many hiking enthusiasts. Then he sketched out a scheme for series of regional surveys, from the White Mountains of New Hampshire to the Great Smokies of Tennessee, that would investigate not only the industrial potential of each locale but also the prospects for services to provide forest fire prevention, flood control, and outdoor recreation.[47] Attacking his one large undertaking from two extremes—the nitty-gritty of trail building on one hand and a general philosophy of planning and reform on the other—MacKaye was not finding it easy to get his evolving, inchoate ideas onto paper. The task of developing this new variety of applied geography, he began to think, might require two books. In any case, the first order of business was to get out on the ground, to begin the survey himself by traveling the length of the Appalachians, perhaps by train or automobile.[48]

Clarence Stein urged MacKaye to come to New York to promote both the trail project and the book proposals.[49] As it happened, that city was just one way station on an excursion through the Northeast, from March through June of 1922, that proved to be perhaps the most effective and important missionary work MacKaye ever accomplished for the Appalachian Trail. Reprints of his AIA journal article in hand, he traveled from city to city, an apostle of outdoor life. By the sheer force of his idea and his personality, MacKaye began stitching together the network of enthusiasts and public officials that would eventually compose a permanent community of trail builders.

On March 16, he met in Boston with Allen Chamberlain, Philip Ayres, and Harvey N. Shepard, president of the Massachusetts Forestry Association. He also addressed a gathering of Boston landscape architects, to whom he confessed that his self-appointed mission as a conservationist and planner was "to provide an incentive for people to leave the large cities, [and] thus obtain a socially healthier distribution of the population."[50]

In Hartford several days later, he called on Austin F. Hawes and Albert M. Turner. Hawes, while in the Forest Service, had collaborated with MacKaye on the fruitless plan to create the U.S. Wood Fuel Corporation; now he was Connecticut's state forester. Turner, field secretary of the state's Park and Forest Commission, was also serving a term as president of the New England Trail Conference. Both men (especially Turner) would remain close friends of MacKaye's and key promoters of Appalachian Trail, which would traverse Connecticut's northwest corner.

MacKaye spent almost a month in New York, visiting some of his acquaintances from the Washington hell-raising days and his stint with the Technical Alliance, such as Stuart Chase, Charles Ervin, Harry Laidler, Frederick Ackerman, and Howard Scott. During these weeks he also met with Daniel Beard, national commissioner of the Boy Scouts; planner Thomas Adams, who had left Canada to assist in the development of a regional plan for New York sponsored by the Russell Sage Foundation; and Isaiah Bowman, director of the American Geographical Society.[51]

Probably the most significant of his New York meetings was a March 21st lunch at the City Club with Stein and Raymond H. Torrey. For several years Torrey had edited a feature page for the *New York Evening Post,* in which he detailed the activities of the many outdoor clubs in the New York metropolitan region. No mere reporter of those activities, Torrey was the "supreme ombudsman in the boiling consortium of New York hiking clubs," laying out trails, writing and editing guidebooks, organizing clubs, and lobbying for greater political support of public parks and forests.[52]

A few weeks later, on April 6, Torrey set up a meeting with several other movers and shakers in New York–area hiking circles, including Major William A. Welch, general manager of the popular Palisades Interstate Park along the Hudson River, and J. Ashton Allis, the banker and outdoorsman who had already proposed a trail from the Delaware Water Gap on into New England. It was at this meeting, as MacKaye later recollected, that Torrey recommended the formation of the New York–New Jersey Trail Conference, modeled on the New England club federation.[53]

Torrey's contributions to the trail project were just beginning, though. His lengthy column in the *Post* the next day, titled "A Great Trail from Maine to Georgia," provided an enthusiastic description of MacKaye's proposal. Including a version of MacKaye's trail map, Torrey's article represented the first extensive public description of the Appalachian Trail project to an audience other than New England trail club members and the readers of Whitaker's architectural journal. "Some mighty big things are coming out of this trail movement in the next few years if its development grows at the pace it now shows," Torrey predicted. Other "dreamers of a super–Long Trail for the benefit of outdoor folk" had been "surpassed in imagination" by MacKaye's Appalachian Trail, he continued. The Maine to Georgia project seemed "a large order," Torrey admitted, "but as it is presented by the author, the difficulties turn into opportunities, which enthusiastic trail workers ought to solve with proper centres of organization for the work."[54] Torrey himself spearheaded the job of trail building across New York and New Jersey, from the Connecticut border to the Delaware Water Gap. He and MacKaye would remain friends and allies through the internal feuds that were to split the Appalachian Trail community in the mid-1930s.

Moving on to Washington, MacKaye called on some of his influential hell-raising friends, such as Fred Kerby, Harry Slattery, Hugh Reid, and Louis Post; Secretary of Labor James J. Davis; former Forest Service colleagues Raphael Zon and Franklin W. Reed; and two outdoorsmen to whom he had been referred by Chamberlain: L. F. Schmeckebier, an economist and historian with the Brookings Institution, and Francois E. Matthes, a geologist with the U.S. Geological Survey. Some of these individuals provided the initial membership of the short-lived Appalachian Trail Committee of Washington, which was organized at Washington's Penguin Club on April 13 and was a precursor of the important Potomac Appalachian Trail Club.[55]

During these busy months, MacKaye also corresponded with trail enthusiasts and public officials who were knowledgeable about conditions in the southern Appalachians, where the sort of organized amateur trail organization common in the Northeast had not yet taken root. Franklin Reed, serving as district forester for District 7, which encompassed the southern national forests, cautioned MacKaye that it would "be particularly difficult in the Southern States to popularize the idea as the recreationists are not given to hiking as a means of recreation." Nonetheless, Reed endorsed the project and promised the Forest Service's cooperation in the South. A few months later, describing the status of completed trails along the trail's projected route through the Natural Bridge, Unaka, Pisgah, and Nantahala National Forests, Reed could report "that the idea of the Appalachian Trail seems popular with the Forest Supervisors, and the popularity of the idea with the general public is predicted."[56]

MacKaye located other enthusiasts in the southern states. Halstead S. Hedges, an ophthalmologist and outdoorsman from Charlottesville, Virginia, agreed to scout possible routes along the Blue Ridge, including areas that would later be incorporated into the Shenandoah National Park. MacKaye's long-time acquaintance Harlan Kelsey, a landscape architect and nurseryman with extensive business and personal connections from New England to the Great Smokies, also provided advice on a southern trail route. Paul M. Fink, a Tennessee hiker of near-legendary exploits, described in careful detail a route through the little-explored mountain terrain along the North Carolina and Tennessee borders and proposed a southern trail terminus at Lookout Mountain, near Chattanooga. Fink immediately recognized the reinforcing link between the trail and the growing southern sentiment for national parks.

"Speaking from the viewpoint of a Southerner," he related, "this trail is need-ed, for we have no routes for a long trip anywhere in our mountains, at least none marked either by signs or on the maps, and each tramper must lay his own itinerary."[57]

MacKaye's pathbreaking mission was not done yet. He also called on Robert Sterling Yard, president of the National Parks Association, and Arno C. Cam-merer, assistant director of the National Park Service.[58] Yard and Cammerer had been long-time allies and operatives during the Park Service's Mather-Albright regime, but they would soon part ways philosophically over the agency's policies. A dozen years later, MacKaye and Yard would be among the eight founders of the Wilderness Society.

MacKaye headed back north. In New York, he was on hand on April 25 at the first meeting of the New York–New Jersey Trail Conference, where Major William A. Welch was named the group's first chairman. A few days later, MacKaye and Stein paid another visit to Isaiah Bowman of the American Ge-ographical Society. MacKaye appealed for funds and institutional support for his proposed Appalachian survey and guidebook. Bowman tentatively prom-ised to raise $1,000 to fund the project. MacKaye immediately went to Charles Whitaker's Mt. Olive farm to begin work, but within days Bowman wrote to say that only $500 would be available for the project. The pace and intensity of Benton's efforts began to take a toll. After experiencing another attack of se-vere intestinal maladies, he spent a month recuperating at Whitaker's farm and at Hal's home in Yonkers. As during previous such episodes, enforced im-mobility provided an opportunity for reading, writing, and reflection.[59]

Although the Appalachian Trail idea had generated enthusiasm, MacKaye's more practical concern was to secure some financial support for his immedi-ate work. To that end, he spun out a series of documents. A detailed one-page outline, "Suggestions for Scouting the Appalachian Trail," was designed to as-sist self-appointed trailblazers in gathering information. "Make maps. Take photographs. Collect literature," he advised. Though terse, the brief document provides perhaps the most concrete depiction of MacKaye's early conception of the trail; it also reveals his appreciation of the practical challenges the pro-ject's creators would confront. The "essential information" to be gathered about trail sections, he suggested, should include data about location, access, gradient and other physical characteristics, and the legal status of land trav-ersed. (*Take pains not to antagonize,*" he urged, of dealings with landowners.) Campsites, scenic vistas, historical sites, and "items of botanical, zoological or geological interest" were among the features he suggested that trail scouts in-ventory. Finally, and not least important, he sought estimates, "in money and in man-days," of the cost of trail building and maintenance. In this first year of trail work, such reports were to be forwarded to the New England Trail Conference, the New York–New Jersey Trail Conference, Schmeckebier in Washington, and Hedges in Charlottesville.[60]

Simultaneously, MacKaye worked on drafts of his proposed trail handbook and his more ambitious "opus," which he was now referring to as "The New Exploration." He also drafted a constitution for the "Appalachian Trail, Inc.," hoping that a more organized entity would have better chances of securing financial and institutional backing.[61]

❖

In late June 1922, Benton returned to Shirley Center, where he found his mother, his sister, and Percy's daughter Christy installed at the Cottage. Every member of the family, it seemed, was hemmed in by sickness, poverty, or both. Nonetheless, he received encouragement from his brothers. "You must guide the torrent you have loosed," urged Percy. "So far from being the cork on the waters you describe, you are the superengineer who must chart and conduit the inundation!"[62]

Benton began work on an article he had proposed for publication in the American Geographical Society's *Geographical Review.* Determined to produce a concrete example of his proposed method of trail scouting, he outlined a survey of a prospective trail section in southern Vermont. And with his old trail companion Horace Hildreth, he returned to the mountain terrain they had traversed twenty-two years earlier during their "Classic of the Green Mountains." It had been many years and numerous bouts of poor health since Benton had headed into the woods with a pack on his back. On July 15, retracing parts of their 1900 route from Shirley via train, trolley, and taxi, he and Hildreth arrived at a camp northeast of Bennington, adjacent to the Long Trail. The cleared, blazed trail and overnight shelters they found had not been in place during their original hike. There were other dramatic signs of change on the landscape. Two dams—one completed, one under construction— created two substantial reservoirs on the upper Deerfield River. Unlike some nature lovers, MacKaye was not necessarily disturbed by the presence of the hydroelectric dams. Indeed, since his days as a Washington bureaucrat, he saw such dams as logical, potentially beneficial components of a multiple-purpose conservation policy. What worried a progressive like MacKaye, however, was the ownership and control of these natural resources by a private corporation, the New England Power Company, instead of a public power authority.

He and Hildreth worked their way north on the trail, across the summit of

Stratton Mountain, where in 1900 they had ridden trees swinging in the wind to survey the skyline. Hildreth departed after four days on the trail, but MacKaye stayed in the region, to recuperate from his mountain exertions; he visited with long-time family friends, talked with old-timers about changes in the logging business, and took notes in local libraries about the region's history.[63]

During the rest of the summer, he incorporated the Vermont data into his newest manuscript, "Making Geography: A Conservation Survey on the Appalachian Trail." Like another seminal conservation philosopher, Vermonter George Perkins Marsh, he viewed humanity as a purposeful natural agent, capable of altering the environment and nurturing the conditions for the good life. "The goal of our geography-making," he declared, "is conservation: of two things—human energy and natural resources." He emphasized a utilitarian view of conservationist goals with which a Marsh or a Pinchot would have readily agreed. "The primary object of the Appalachian project is human," he continued. "The human biped comes first. Quadrupeds, trees, waters, soils, coal—these are *for* man, not man for them. Man's energy should be spent, not in serving them, but in making them serve him."[64]

As the fall advanced, MacKaye fell into a more somber mood, reading Thoreau and remembering his wife. He also worked on his "New Exploration," the broader philosophical counterpart to his trail-scouting handbook. "I am writing these days as I never wrote before," he reported to Percy, "the biggest stuff that's in me." By mid-November, he noted in his diary, he had finally "crashed through," connecting the two projects.[65]

He explained to Stein the gist of this breakthrough in his thinking. "The big quest of the future . . . is the probing and untangling of the world's industrial cobweb," he wrote. The conservation movement and technocrats such as Howard Scott lacked the support of "the people at large":

The amateur must be aroused as well as the technician. The amateur is the hobbyist, the "nut," the man with a cause, the man who has "got religion." Hence I turn my guns on *him,* for without him there is no impelling public will-to-do.

. . . Our job is just one wee slice of this—though it is a strategic slice. It is to provide the *outside space,* the "exterior" facilities, for this extra time from toil. . . . Our job is to chart and plan the country's (exterior) *recreational* development.

Or rather this is *half* our job. The other half is to chart and plan the country's (exterior) *industrial* development—to discover the possibilities of straightening the industrial tangle so as to reduce the present waste of time spent in useless industrial motions—to reduce the time of toiling and increase the time for recreation. In short, we must seek not only more *space* for recreation but more *time* for it. For to build a greater recreation ground and not provide a greater chance to use it would be a brainless undertaking, or else a grossly selfish one.[66]

MacKaye described to Stein several projects and reconnaissance trips he hoped would provide opportunities to develop the two major aspects of his work: an evolving philosophy of "neo-recreation"—"something which is both work and play (not both *toil* and play)," and the techniques for conducting "a regional survey of the industrial wilderness," throughout the "Appalachian domain." Not for the last time, MacKaye appealed to his friend for support. "As soon as I complete the book I shall have to do something, quickly, to make a living," MacKaye concluded.[67]

Stein joined Benton and James MacKaye at Shirley over the Thanksgiving holiday. During their long walks and intense discussions, they assessed the progress of the trail project and plotted the next stage of the campaign.[68] Stein and MacKaye outlined a four-point agenda to build on the progress thus far. First, they envisioned a federation of thirteen regional groups, including those already under the umbrella of the New England and New York–New Jersey trail conferences, to formally endorse and carry on MacKaye's trail plan. Second, Stein would ask Alexander Bing, a New York businessman with whom he had collaborated on innovative housing projects, to form and finance "a group to study the planning of the industrial wilderness." Third, they would approach Charles Harris Whitaker and publisher Horace Liveright about publishing MacKaye's book. Finally, they would seek assistance for their efforts from state governments, including that of Pennsylvania, where Gifford Pinchot had just been elected governor, and of New York, where Stein was affiliated with a state housing and planning committee. It was an ambitious program, but the idealistic and intellectual pair also had a pragmatic side, a reasonable sense of the possible. Although it would take several years, they would succeed in achieving, in some form, all four of their objectives.[69]

❖

At the end of 1922, just over a year after MacKaye's Appalachian Trail project had first been publicly broached, the scheme had taken hold. Already, as he reported in that December's issue of *Appalachia,* individuals, outdoor clubs, and public officials were at work from the White Mountains of New Hampshire to the Great Smokies along the Tennessee–North Carolina border, "exploring and scouting the chief links" in each of eight trail "divisions." A third of the 1,700-mile trail he originally proposed, according to his estimates, was already in existence, principally in such states as New Hampshire, Vermont, New

York, and New Jersey, as well as in the national forests of the South. "In almost every locality along the Appalachian ranges a greater or less amount of trail-making is going on anyhow from year to year," he observed. "The bright idea, then, is to combine these local projects—to do one big job instead of forty small ones."[70]

MacKaye could report on the progress of the "one big job" when the New England Trail Conference met in January 1923. The meeting, as he put it, "was basic in clinching the start" made in the region. The Appalachian Trail, by Raymond Torrey's account, "was the principal subject considered" by the 1923 NETC. In his enthusiastically received address, MacKaye promoted dimensions of the project he had downplayed in his original proposal. He envisioned the Appalachian Trail as the backbone of a publicly owned "super national forest" stretching from Maine to Georgia. The trail itself, MacKaye suggested, could be built by local organizations in a series of links, "each link to be sufficient of itself and to serve for local use." He also floated his idea for a "central organization" to oversee the trail's creation and maintenance, taking up the tasks that Stein's AIA Community Planning Committee had been performing.

He likened such a federation of local groups to "the original Thirteen States of the Union." But such an organization, MacKaye cautioned, was "something which should grow and ripen rather than be suddenly created."[71] Indeed, he and Stein harbored some concern about losing control and leadership of the trail project, although the control they held was already tenuous at best. More important, as MacKaye saw it, was for would-be trailblazers first to comprehend the philosophy—his philosophy, that is—behind the project:

This is *not* to cut a path and then say—"Ain't it beautiful!" Our job is to open up a realm. This realm is something more than a geographical location—it is an environment. It is the environment, not of road and hotel, but of trail and camp. It is human access to the sources of life. The first hand survey of these sources, through the free efforts of the amateur, would secure a yield of immediate fun and promise of ultimate welfare. It would stimulate *vision* in the public mind—a vision of constructive National development as against a medley of destructive notions. The times are ripe for such a program. Here is a job for young and old. The task at hand is an Appalachian Trail; the goal is an Appalachian Domain.[72]

MacKaye's rousing and vivid "messages" became traditional and eagerly anticipated features at meetings of eastern trail clubs in the years ahead. They were, essentially, sermons, designed to exhort and inspire. In later years, some leaders of the Appalachian Trail effort charged that MacKaye had not paid sufficient attention to the detailed, practical tasks of locating, building, and

Appalachian Trail Completed (in part)
"   "   Scouting Organized
"   "   Not yet Organized
Automobile Road
Main Peaks
Cities and Towns
Numbers refer to Divisions of Trail.

In a December 1922 article for *Appalachia,* the journal of the Appalachian Mountain Club, MacKaye showed the Appalachian Trail's progress during the year after he proposed the project. The route of the trail's southern section had already been shifted westward, to the Great Smokies. Another map (*facing page*) portrayed possible side trails in New England.

maintaining the physical trail. Such criticisms tended to arise from those who had not been involved in the trail project during its earliest few years—the very years, in fact, when MacKaye made his greatest contributions to the small details as well as the overarching concept of the trail project, and at the greatest personal sacrifice. He succeeded in establishing the concept of the Appalachian Trail. As importantly, but harder to measure than the miles of trail blazed, he located and linked together other dedicated and influential trail en-

thusiasts throughout the region spanned by the project. His own prospects were still unsettled and his efforts had been carried on with little financial reward. Through his writings, correspondence, speeches, and travels, however, MacKaye inspired the creation of the "camp community" his 1921 article had called into action. His vision of the Appalachian Trail spoke to human needs and aspirations that most mainstream institutions—political, commercial, educational, social—had failed either to see or to address.

A self-styled radical, MacKaye had envisioned the Appalachian Trail project as "a flank attack on the problems of social readjustment." In the project's earliest years, most trail enthusiasts probably did not suspect that they were following a roundabout path toward social and political reform. The Appalachian Trail quickly won support as a recreational project and would eventually become emblematic of the American wilderness ideal. MacKaye himself, however, did not soon abandon his original vision of the trail as a regional "project in housing and community architecture," encompassing many of the ideas for colonization and employment he had developed during the previous decade.

The Regional Planning Association of America
and the Appalachian Trail Conference

1923–1925

MacKaye soon found himself at the center of a movement to develop a new approach to American community building. At Clarence Stein's urging, he came to New York City immediately after the January 1923 New England Trail Conference. Stein had pressed forward in New York with their idea of forming a group to survey the "industrial wilderness," a version of a concept Stein called the "City Planning Atelier" which the architect had envisioned while returning from Europe the previous year.[1]

MacKaye holed up in the Columbia University library to work on his would-be book about "industrial exploration." Settled in New York again, his network of stimulating friends and associates quickly expanded. "These weeks in N.Y. have been tremendous ones for me—imbibing ideas from my wondrous group of friends here," Benton breathlessly reported to Percy at the beginning of March, as an important new professional, intellectual, and personal chapter opened in his life.[2]

On February 17, Stein had hosted a lunch for a small group of reform-minded colleagues at the City Club at 55 West 44th Street. The upshot of this gathering, where the discussion included MacKaye's Appalachian project, was the creation two months later of the Regional Planning Association of America (RPAA). It may well have been at this affair that MacKaye first met Lewis Mumford. The young New York author was fast gaining a reputation for his trenchant yet stylishly expressed writings on architecture, housing, and other contemporary social and aesthetic concerns. At twenty-eight, Mumford had been a contributor for several years to Whitaker's AIA *Journal,* among other magazines. His first book, *The Story of Utopias,* had been published just the year before. MacKaye's concrete program for restoring a balanced regional and community life on the American landscape instantly struck a chord with Mumford.[3] "I well remember the shock of astonishment and pleasure that came over me when I first read this proposal," Mumford recalled, many years after seeing for the first time MacKaye's 1921 article describing the Appalachian Trail project. "But even the most sanguine backer of MacKaye's idea could hardly have guessed that this was such an *idée force* . . . that MacKaye would live to see the Trail itself and some of the park area, as in the Great Smokies, finished before another twenty years had passed."[4]

The cosmopolitan New York writer and the exuberant Yankee forester soon became productive professional collaborators. Mumford discovered in MacKaye an inimitably American practitioner of the communal and regionalist ideals he was espousing in his own writing. MacKaye, though sixteen years older than his new friend, found in Mumford a sounding board, advocate, editor, and intellectual disciplinarian for his own unruly ideas and literary efforts. ("Why should I ever write myself," he once confessed to Mumford, "when you can portray my ideas so much better?") Over the next decade, in their individual and joint writings, MacKaye and Mumford produced the most comprehensive expressions of the RPAA's regionalist ideas and ideals. Just as important, the two men in these years cemented a strong, if sometimes guarded, friendship that flourished for more than half a century.[5]

On the day after Stein's City Club dinner, the Sunday *New York Times* carried an extensive article by MacKaye, adapted from his recent NETC speech and titled "Great Appalachian Trail from New Hampshire to the Carolinas." He now reiterated his call for the creation of a "Super National Forest" the length of the Appalachians. His original 1921 article had been more cautious about the government's role in achieving his grand scheme for preserving the mountains' wild character. "This does not mean that all the forest and mountain land should literally be in Federal ownership," he still maintained. "Much might be state or town owned. But enough should be in public hands to make of the Appalachian region a national, or people's 'sphere of influence.'"[6]

With his brother Percy's introduction, Benton made the rounds of a few New York publishers, hoping to stir interest in his proposed book. In March, he retreated to Whitaker's New Jersey farmstead to revise his manuscript investigating what he called the "industrial wilderness." He completed a ten-chapter draft later that month and submitted it to an editor at the Macmillan Company, who rejected the work. This literary project would occupy him for another five years.[7]

On April 18, the Regional Planning Association of America held its organizational meeting at the architectural office of Robert D. Kohn at 56 West 45th Street. Over the next decade, this small, informal, intellectually fertile group fashioned a penetrating critique of prevailing American trends in architecture, housing, transportation, and land-use planning. During the RPAA's short heyday, the group's provocative vision for the regional development of the nation's physical and social landscape achieved only limited public, professional, and commercial acceptance. The RPAA, recounted historian Carl Sussman, "sought to replace the existing centralized and profit-oriented metropolitan society with a more decentralized and socialized one made up of environmentally balanced regions." That objective represented a direct challenge both to established economic interests and to the conventional practices and ideas of the city planning profession. But it has held a persistent appeal for later generations of scholars, architects, planners, and reformers.[8]

Students of the RPAA's history all support Mumford's assertion that "essentially this little group was a society of friends."[9] While the RPAA's membership consisted of about thirty members, the group's major initiatives, projects, and ideas were usually produced by an inner circle of five or six. Charles Harris Whitaker, in his role as *JAIA* editor, was an intellectual gatekeeper of new ideas about architecture, planning, and housing. He had introduced most of the figures who composed the initial membership of the RPAA. Architect Henry Wright, businessman and philanthropist Alexander M. Bing, and writer-economist Stuart Chase also made substantial contributions to the RPAA's work. But the special character of the relationship—and enduring friendship—among MacKaye, Stein, and Mumford was perhaps the major source of the RPAA's intellectual and organizational energy between 1923 and 1933, the years of its greatest activity. The RPAA's stimulating personal chemistry was the result of a "providential conjuncture . . . that produced a kind of fusion reaction that released energy and light," Mumford later testified. "Where two or three are gathered together something happens that no individual ego, however inspired, knowledgeable, constructive, or imperious, can bring into existence."[10]

At the RPAA's first meeting, Alexander Bing was elected president. The other RPAA officers included first vice-president John Irwin Bright, a Philadelphia architect who was Stein's predecessor as chairman of the AIA's Committee on Community Planning; MacKaye as second vice-president; and Stein as secretary and treasurer.[11]

Besides Stein, other socially minded architects and planners who joined the group included Wright, Kohn, Bright, Frederick L. Ackerman, Frederick Bigger, Sullivan Jones, and Henry Klaber, many of whom had worked on the staffs of federal housing programs during World War I. Whitaker, Mumford, Chase, and *Survey Graphic* editor Robert Bruère were among the talented literary figures who helped publicize the RPAA's ideas and projects. Public-housing advocates Edith Elmer Wood and Catherine Bauer sometimes participated in the group's activities as well. All were critics of land and real estate markets based on speculation; and they likewise doubted the efficacy of the contemporary housing-reform movement, which sought to improve housing conditions with stricter building, zoning, and sanitary codes and which RPAA members saw as "restrictive."[12]

Clarence Stein "sat indisputably at [the RPAA's] center."[13] He provided the group's meeting places, maintained its organizational framework, and raised funds to support its activities. A "rare combination of artist and organizer," according to Mumford, Stein had cultivated many contacts with New York politicians, social workers, and other civic activists involved in housing reform and other social causes.[14] After studying at Columbia University and Paris's École des Beaux-Arts, he joined the firm of architect Bertram Grosvenor Goodhue in 1911, quickly rising to the position of chief draftsman. In 1919, he started his own New York practice, in association with Charles Butler and Robert Kohn, another Ethical Culture enthusiast. Two years later, he began a productive collaboration with Henry Wright, designing such innovative community projects as Sunnyside Gardens in Queens, New York, and Radburn, New Jersey. But Stein's efforts on behalf of the RPAA demonstrated the rare, unteachable talent of nurturing the conditions that enable others to do their best work.

For MacKaye, creation of the RPAA provided a title, a letterhead, and a certain professional identity. Beyond that, and more important from a practical point of view, the organization provided a modest degree of financial support for his independent work. Indeed, the RPAA's principal expenses, for much of the group's existence, were periodic stipends for MacKaye.[15] The rest of the RPAA inner circle had other professional positions and opportunities. Mac-Kaye, by contrast, for a time looked to the fledgling RPAA as the principal sponsor and patron of his career.

As a working forester who had traveled extensively throughout the country on his government duties, MacKaye brought a unique perspective to this otherwise urban-oriented group. Among all the RPAA members, he was the most knowledgeable about the problems and prospects of the mountains, forests, farmlands, and small towns toward which the powerful influence of the American metropolis was rapidly spreading. His perspective was genuinely indigenous, a product of his American training, experience, and observations. So it was no surprise that MacKaye's innovative Appalachian Trail project, which spanned an extensive geographical region and addressed a variety of social and land-use concerns, became the group's "first rallying point."[16] Mac-

Kaye's idea encompassed key elements of the RPAA's aim to "deliberately plan for better domestic and industrial development of a whole region," in the words of an early statement of its methods and principles.[17]

In an earlier attempt to describe the new group's mission, Stein had proposed that it be called the "Garden City and Regional Planning Association."[18] The RPAA's founders were quick to drop the term *garden city* from the name, but they owed a substantial debt to the garden city movement and tradition. Members of the RPAA circle were well acquainted with the English garden city movement, which had originated with the publication of Ebenezer Howard's 1898 book, *To-morrow: A Peaceful Path to Real Reform* (republished in 1902 as *Garden Cities of To-morrow*). Howard, a British court stenographer driven to make a place for himself as a social reformer, tirelessly promoted his vision of new "garden cities," designed to control the relentless expansion of industrial cities like London, with all their social, economic, and environmental troubles. His idealized diagrams portrayed new "social cities" for "a vast army of workers" on undeveloped community-owned land, comprising 6,000 acres and some 32,000 people. The area occupied by homes would be confined to 1,000 acres; the community would be bounded by greenbelts of agricultural land and interwoven with parks and open spaces.[19]

Howard's ideas inspired creation of two English garden cities, Letchworth and Welwyn Garden City. If they fell short of Howard's strictest garden city principles, Letchworth and Welwyn nonetheless provided a benchmark for measuring progress toward the garden city ideal. Indeed, MacKaye in *Employment and Natural Resources* had pointed to garden cities, and to Letchworth in particular, as a model for controlling the "inefficient" and "unsightly" expansion of cities.[20] Other RPAA members, like Stein, Whitaker, and Ackerman, had made firsthand acquaintance with important figures in the English garden city movement, such as Howard and architect Raymond Unwin, who with his partner Barry Parker had designed the site and many buildings at Letchworth.

The RPAA's program attempted to adapt garden city principles to the fast-changing social, economic, and technological American landscape of the 1920s. The innovative design and funding of housing and new communities for middle- and working-class Americans, the decentralization of population and industry, the protection of open space and recreational greenbelts, and the exploitation of new technologies such as the automobile, hydroelectric power, long-distance electrical transmission, and electronic media—all were encompassed by the RPAA's evolving conception of regional planning.

❖

During the spring of 1923, MacKaye made his headquarters at the Hudson Guild Farm. In mid-May, a small RPAA contingent, including Mumford and Chase, joined him there for a weekend of discussion and socializing.[21] As it happened, the Scottish polymath and planner Sir Patrick Geddes had just arrived in the United States for a series of lectures at the New School for Social Research in New York. Mumford, a disciple of Geddes and his "unofficial, part-time secretary," invited him to the meeting at the farm, where he was greeted as "the authentic Father of regional planning."[22] The sixty-nine-year-old Geddes had begun his extraordinary and diverse career as a botanist under the tutelage of Thomas Henry Huxley, then had transformed himself into a world-traveling sociologist, town planner, and ecologist.

For MacKaye, this encounter with the prophet of regional planning had the power of revelation. Geddes stayed on at the farm after most of the other conference participants had departed. The two planners—the bearded Scotsman and his lean, pipe-smoking new follower—took several long walks together. Geddes read and pronounced his benediction on MacKaye's book manuscript. Mumford recognized the "special link" between MacKaye and Geddes, and years later recounted how the two, running late, had had to sprint the last stretch of a six-mile walk to Geddes's train for New York, so engrossed had they been in their conversation about MacKaye's true vocation.[23]

MacKaye described to Geddes his geographical studies under William Morris Davis, his experiences as a forester, and his own ideas about regional planning. During one of their walks, he recalled, "Geddes rounded on me in the path," and declared,

> "None of those! Not conservation, not planning, not even geography. *Your subject is geotechnics.*"
>
> And then with a lunge he resumed speed, but only for a few strides. Again he stopped short.
>
> "Geography," said he, "is descriptive science. . . ; it tells what *is*. Geotechnics is applied science. . . ; it shows what *ought to be*." And on he bounded.
>
> "But what's the matter with 'conservation,'" I pleaded, "or 'regional planning?'"
>
> "Nicknames," he retorted. "Of course you 'conserve,' and of course you 'plan' just as you do in building; but verbs like these stand for operation and make no term for your comprehensive science any more than 'nail-driving' is a substitute for 'architecture.'"[24]

Soon after their perambulatory conversations, MacKaye attended Geddes's New School lectures, "Talks from the Outlook Tower," which distilled much of his life's work and thinking. Geddes described his debt to French regionalists like Frédéric Le Play, from whom he determined that his own studies should be based "on the regional *place,* the *work* that goes on it, and the *folk* that inhabit it."[25] He outlined "The Valley Plan of Civilization," his approach to charting geographic processes, historical progress, and social evolution by studying the "Valley Section, . . . that general slope from mountains to sea which we find everywhere in the world."[26]

In his 1922 *Story of Utopias,* Mumford had enthusiastically depicted his mentor's concept of the regional survey. Geddes's approach, Mumford explained, was "to take a geographic region and explore it in every respect."[27] Before their meeting, MacKaye had been unaware of Geddes or his ideas. But the regional survey was an almost exact counterpart of the "conservation survey" he had already envisioned for the Appalachian region.

Geddes's stimulating, sprawling 1915 book, *Cities in Evolution,* had anticipated many of the elements that MacKaye and Mumford would meld into their own regionalist perspective during the 1920s and 1930s. There Geddes urged the preservation of forests and moorlands "between the rapidly growing cities and conurbations of modern industrial regions." Such "constructive conservation," Geddes proclaimed, "is more than engineering: it is a masterart; vaster than that of street planning, it is landscape making; and thus it meets and combines with city design."[28] MacKaye's Appalachian Trail project represented just such an exercise in "landscape making" and "constructive conservation."

At the Hudson Guild Farm gathering, MacKaye, Mumford, Stein, and Chase were designated the RPAA's program committee. Their detailed memorandum bore the heavy stamp of MacKaye's efforts and influence. Indeed, this first formal RPAA program consisted principally of the adoption of his entire Appalachian project, which now had Geddes's imprimatur. With an eye toward the possible formation later that year of an all-Appalachian trail conference to assume the trail's administration, the memo's authors proposed that the RPAA adopt "the regional planning features of the project" from the AIA Committee on Community Planning. Specifically, they suggested the "reconnoitering and surveying of a series of unit valley-sections (or small regions) within the Appalachian Domain," preferably in the region of the Appalachians from New Jersey through New England, in association with state and federal government agencies. They also proposed the "scouting and organizing," with hiking clubs and other amateur groups, of several key links of the Appalachian Trail, to make these "conveniently and inexpensively accessible for walkers and campers living in the neighboring cities." Finally, they suggested the publication of a book or monograph reporting the results of these efforts.[29]

Although the memo did not specifically identify MacKaye as the person to carry out the program, it was clear enough that he alone was qualified for the tasks proposed. On Stein's assurance that funds would be available to carry out some of the proposed RPAA initiatives, MacKaye set to work from his base at the Hudson Guild Farm for much of that summer. Though his ambitious plans for a series of valley-section surveys were not fulfilled, he surveyed, mapped, and tramped northwestern New Jersey, sometimes alone, sometimes with groups of young people from the Hudson Guild Farm.[30]

Mumford, writing in *The Freeman* that summer, described the subversive essence of his new friend's vision and method. "Unlike so many reformers, who urge people to desert their pleasures and recreations and 'consider things seriously,'" Mumford wrote, MacKaye was proposing

> simply to call people, especially young people, out into this Appalachian region; he asks nothing more, at first, than that the folk who are cribbed, cabined, and confined by the great cities along the coast should camp out in the open spaces of Appalachia, scramble over its hills, make themselves at home in its woodlands, fight the forest fires when need be and guard against them at all times. In short, he wants them to possess the whole landscape, not by act of legislature, but by the process of use and wont, whereby the people of England once upon a time acquired their right to the common lands, and to this day keep their title to the common footpaths that run across the fields. In short, he does not propose that this domain should be given to the "people," as our national forests are given; he proposes that it should be conquered.[31]

MacKaye balanced Mumford's romantic rhetoric with the practical task of creating a peaceable, landscape-conquering corps of trail makers. Working with Stein, Major Welch, and Raymond Torrey, he prepared for an autumn conference on the trail project. Co-sponsored by the Palisades Interstate Park and the New York–New Jersey Trail Conference, the meeting convened October 26 through 28 at the imposing Bear Mountain Inn. The conference brought some of MacKaye's planning friends, like Stein and Mumford, together with Welch, Allen Chamberlain, Albert Turner, Harlan Kelsey, the state foresters of New York and New Jersey, and others from the region's conservation and hiking communities.[32]

The approximately thirty "not-too-serious people" who attended shared information and thoughts about the progress of the project; some hiked stretches of the trail in and around the park. The group also adopted Welch's proposed design for a uniform trail marker: a copper monogram incorporating the crossbars of the letters A and T, a variation of which was later approved as the official trail emblem. The Bear Mountain meeting brought yet more people and interests under the trail project's umbrella. The gathering, MacKaye later said, "did for the Hudson-Delaware section what the January NETC meeting of 1923 did for the New England section—clinching work already done, crystallizing plans for next steps."[33]

The Bear Mountain conference provided encouraging evidence of the trail project's progress, but MacKaye's own immediate prospects remained tenuous and unsettled. Robert Bruère, of *Survey Graphic,* commissioned him to write an article for an issue on "Giant Power." The term had been coined by engineer Morris L. Cooke, who served under Gifford Pinchot, then governor of Pennsylvania. The patriarch of utilitarian conservation was now espousing the social and economic virtues of new advances in electricity-generating technology. Giant Power proponents envisioned huge powerplants located near reservoirs or coal mines (Pinchot's state, of course, was rich in coal), connected to an extensive grid of long-distance transmission lines. More than the power plants themselves, it was the power grid's range that appealed to progressive planners and conservationists. The efficient distribution of electricity by publicly owned utilities, they insisted, could promote the decentralization of industry and communities. Modern industrial technology, they hoped and believed, need not necessarily lead to unbridled metropolitan growth.[34]

MacKaye's article sketched out a possible layout for communities and industry in the Appalachian Domain. An "efficient environment," he contended, was composed of communities based on seven fundamental principles: "self-comprehensible" work, the use of modern technology, a community population of "democratic proportions," "distribution attaining compactness as against both isolation and congestion," access to "natural attractive environs," modern sanitation, including "smokeless factories," and recognition that the "community should be a unit."[35]

MacKaye's prescription for sound regional development was short on specifics. Just how big was a community of "democratic proportions"? By what standard could a community's "compactness" be measured? In any case, the potential for community development in the Appalachian region in the era of Giant Power only reinforced, in MacKaye's eyes, the prospects for his ridgeline trail proposal. "We have the tradition, the folk, the land, and the resources left over from colonial days," he concluded. "We have the mechanics of the modern day. We have the power of a giant wherewith to turn the wheels. And we have a choice."[36]

Staying at Mt. Olive in December 1923, then moving on to Hal's in Yonkers at year's end, MacKaye incorporated ideas about the Appalachian Trail and Giant Power into an extensive report for the RPAA titled "A Suggested Policy for Approaching the Problem of Regional Planning in the United States." "It's the biggest blame thing I've tackled yet," he reported to Stein, "but I'm determined to 'surround' it." While finishing the latest version of his ever-evolving treatise, Benton received word from his sister in Washington that their mother's health was failing fast. He arrived in time to see Mary MacKaye before she died, on May 14, in her seventy-ninth year.[37] While her death temporarily reduced the number of people needing Benton's financial support, Mary MacKaye's passing only exacerbated Hazel's dependency on her beloved brother.

MacKaye's trip to Washington coincided with the first National Conference on Outdoor Recreation. About 130 recreation and conservation organizations were represented at the gathering, which was chaired by Colonel Theodore Roosevelt, son of the late president and patriarch of the conservation movement. The Appalachian Trail represented just one of the many initiatives and organizations to which Americans flocked during the 1920s as they sought to enjoy and protect the nation's wild and scenic landscapes. The era's economic prosperity and the widespread ownership of dependable automobiles provided more Americans with both the leisure time and the mobility to explore previously remote places. One official response to such forces was the appointment by the Secretary of the Interior in early 1924 of a five-member Southern Appalachian National Parks Commission. The panel, which included Appalachian Trail leaders Welch and Kelsey, issued a report that led, a decade later, to establishment of the Great Smoky Mountains and the Shendandoah National Parks, both traversed by the trail.[38]

The increased pressure for recreational use of the nation's public lands created tensions both within and between the National Park Service and the U.S. Forest Service. At the 1924 conference, a committee was established, including the respective agency heads Stephen T. Mather and William Greeley, to resolve "territorial disputes" between the agencies. As the 1920s progressed, both organizations would undertake extensive road-building initiatives, which had the double-edged effect of opening the nation's federal lands and forests to the public while reducing the size and number of roadless wild tracts to which many recreationists were attracted.[39]

MacKaye attended the recreation conference as the RPAA's representative. In the meeting's published proceedings, he described the Appalachian Trail as "a movement and not an organization," and he observed that the "regional planning idea seems to be a development of the conservation movement of the Roosevelt-Pinchot days." MacKaye's perspective also represented a counterbalance to the efforts of such organizations as the National Highways Association and the Automobile Association of America. While such boosters urged the development of new roads and promoted "motoring" as a form of outdoor recreation, MacKaye, historian Paul Sutter has noted, reminded the conference that the Appalachian Trail was, "of course, a walking trail or path, and not an automobile road." The gathering also provided an opportunity to foster the network of individuals and groups working on the trail project. During and immediately after the conference, MacKaye huddled with Forest Service officials such as Franklin Reed and other activists to investigate possible trail routes in the South.[40]

❖

Stein's political activities and connections, meantime, had opened up an opportunity for the RPAA, and for MacKaye, in New York State. When Democratic Governor Alfred E. Smith asked Stein to chair a state housing commission, the architect agreed to accept the job on the condition that regional planning be incorporated into the panel's title and mandate. As chairman of the New York State Commission of Housing and Regional Planning, Stein set his colleagues the task of preparing the preliminary outline of a comprehensive regional plan for the Empire State.[41]

Stein recruited MacKaye to do the report's initial fieldwork. During the early summer of 1924, MacKaye traveled across the state, gathering data, meeting officials and experts, speaking to business and civic groups, and surveying New York's physical and social terrain.[42] Officially he reported to George Gove, director of the state bureau that provided support to Stein's commission. But by most accounts Henry Wright oversaw the report's development, and the final product reflected some of Mumford's historical themes and MacKaye's own notions of the "flow" of population, commodities, and natural resources.

The *Report of the Commission of Housing and Regional Planning* was not published until 1926—and then only in a modest printing, as the enthusiasm for

Smith's planning initiative had ebbed. Though not much heeded by New York planners and politicians at the time, the report later became an oft-cited landmark in planning practice and theory. In fact, the New York report was more of a regional survey in the style of Patrick Geddes than a practical guide to planning policy. Stein called it "a series of studies of forces which have shaped the economic history of the State."[43] By this account, New York's era of noncentralized, self-sufficient agricultural settlement, which prevailed until 1840, was followed by three discrete stages of growth and population movement. During "Epoch I," from 1840 to 1880, expansion of agriculture coincided with a "trend toward centralized manufacturing and concentration," propelled by steam-powered machinery and a growing transportation network of canals and railroads. "Epoch II," from 1880 to 1920, saw the demise of self-sufficient agriculture and the dramatic growth of industrial cities. From the grim apotheosis of the urban-industrial age in 1920, the New York planners envisioned a benign "Epoch III," an era of decentralization made possible by the automobile and Giant Power. The report's evocative map of Epoch III depicted highways, greenbelts, forest reserves, cities, and industrial centers arrayed in an appealingly rational and idealized vision of the state's future landscape. The New York report, concludes historian Daniel Schaffer, "stands forth as the clearest and most comprehensive documentation of the RPAA's vision."[44]

The RPAA circle was not then the only source of ambitious ideas about the future of New York State and the metropolitan New York City region. In 1923, the Committee on the Regional Plan and Its Environs, with ample funding from the Russell Sage Foundation and under the leadership of planner Thomas Adams, initiated a comprehensive regional survey and plan of the metropolitan New York region. The committee's efforts culminated in the publication of the ten-volume *Regional Plan of New York and Its Environs*. After the plan's publication in 1931, Mumford engaged in a colloquy with Adams in the pages of the *New Republic*, attacking the Russell Sage planners' acceptance of "the inevitablity of metropolitan development."[45] Mumford charged that the *Regional Plan* represented a betrayal of a genuine regional vision and a capitulation to New York's business and political leaders. When Adams's well-funded committee changed its name to the Regional Plan Association of New York, the confusion of its identity with that of the Regional Planning Association of America was galling enough for Mumford and his colleagues; but the competing visions of the *Regional Plan* and the RPAA embodied "a grave split . . . between the two regional motifs" of the era, writes historian M. Christine Boyer. If the *Regional Plan* represented the views of promotors of metropolitan growth and "apologizers for and rationalizers of the suburban trend," the

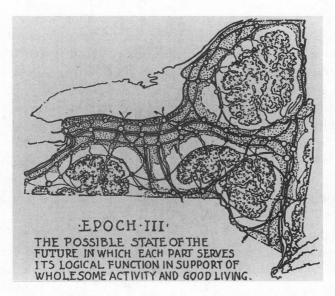

·EPOCH·III·
THE POSSIBLE STATE OF THE
FUTURE IN WHICH EACH PART SERVES
ITS LOGICAL FUNCTION IN SUPPORT OF
WHOLESOME ACTIVITY AND GOOD LIVING.

Clarence Stein chaired the New York state Commission of Housing and Regional Planning during the mid-1920s. MacKaye worked with Henry Wright on the commission's report, which depicted three "epochs" of the state's history and development. The regional planners envisioned a decentralized "Epoch III," in the future, which would exploit such technologies as long-distance electrical transmission and the automobile, develop lowland areas for agriculture and industry, and preserve upland areas for recreation and watershed protection. *Report of the Commission of Housing and Regional Planning. State of New York* (Albany, 1926). (Author's collection.)

RPAA spoke as proponents of "a utopian order" encompassing garden cities, industrial decentralization, and the control of population growth.[46]

The differences between these visions of regional planning did not crystallize until the 1930s. During its mid-1920s heyday, however, the RPAA held to a hopeful vision of organic, decentralized regional development based on what Mumford came to describe as a "neotechnic complex," comprising the technologies of electricity, the automobile, electronic communication, and metal alloys.[47] The RPAA never enjoyed as abundant financial and institutional support as the Regional Plan Association did, but MacKaye, Mumford, and their RPAA colleagues in 1925 found another valuable opportunity to broadcast their own planning notions.

By Mumford's account, it was at MacKaye's instigation that *Survey Graphic* offered to publish an entire issue dedicated to the subject of regional planning and prepared by the RPAA. "In its boldness and comprehensiveness this was as audacious a proposal as that of the Appalachian Trail itself," Mumford later asserted.[48] MacKaye, while in New York that summer working on the state regional planning report, one day visited Robert Bruère at the editor's office. The main topic of their discussion was the RPAA's role as host of the International Town, City, and Regional Planning Conference, scheduled for the following spring. The two came up with the idea for a special *Survey Graphic* issue on regional planning to coincide with the May 1925 New York conference. Mac-Kaye drafted an outline of articles that might be included, and he presented it to Stein and Mumford, along with the request that Mumford serve as the issue's editor.[49]

Mumford accepted the task, but he soon confided to MacKaye that he had some reservations about the RPAA's crash effort to flesh out its regional planning philosophy:

> The damned regional planning number lives up to my worst misgivings. I have a feeling that you and I and maybe Bruère might write a wopping [*sic*] piece, if we could do it after our own fashion: but for the rest of the crowd the idea is still a little unbaked, and in a busy world which gives no one time to think, it is destined to remain so. The regional planning movement exists for the present in the negative state of criticism, criticism of the big city and of "city planning." It is not yet sure enough of itself to offer anything positive: or rather, we are not as a group united on a positive program; we are, in fact, still fumbling around for it.

Wright and Stein, Mumford continued, had confessed to him "that they could plan the physical garden cities, but had nothing to put into them— couldn't visualize them on their social and civic side. This is where you come in Benton, and this is why I hark back again and again to the Appalachian and the New Colonial ideas. . . . We must start a regional movement in America before we can have regional planning."[50] Mumford was determined to instill in his colleagues the tenets of a regionalist doctrine—"a mode of thinking and a method of procedure," not just a professional planning technique.[51]

The *Survey Graphic*'s May 1925 "Regional Planning Number" represented another important intellectual landmark on the path being charted by the RPAA. One of Mumford's two essays, "The Fourth Migration," established a panoramic historical context for the issue's regionalist vision. "Historically, there are two Americas," he asserted, "the America of the settlement and the America of migrations." This "second America," Mumford wrote, comprised three stages: the "clearing of the continent," the "great flow of population . . . from the countryside and from foreign countries into the factory town," and "the flow of men and materials into our financial centers." By Mumford's sanguine account, American culture stood "in the midst of another such tidal movement of population," a "fourth migration," based on such new technologies as the automobile, telephone and radio, and long-distance electrical transmission. Such tools and resources provided the means to create a "stable, well-balanced, settled, cultivated life," which would fulfill the promise of America's founding ideals. Mumford's other article, "Regions—To Live In," was a regionalist manifesto, a challenge to conventional notions of city planning. "The hope of the city lies outside itself," he pronounced. Regional planning "sees people, industry and land as a single unit." It "is an attempt to turn industrial decentralization—the effort to make the industrial mechanism work better—to permanent social uses."[52]

The special issue elaborated these themes in articles by Stein, Ackerman, Bing, Chase, Wright, and Bruère. MacKaye's contribution was titled "The New Exploration: Charting the Industrial Wilderness." The article represented the first appearance in print of his phrase, "new exploration," describing his planning metaphor. The age of terrestrial exploration was over, Mac-Kaye proclaimed. But if mankind had conquered the natural wilderness, it had created in its place a "labyrinth of industry." MacKaye's "new exploration" was devoted to "the untangling of this iron web." Mapping the way out of this "industrial empire" represented, he declared, "the problem of our time."[53]

MacKaye came at the task with the rhetoric and tools of geography and forestry. The modern industrial landscape resembled "a rough hewn organism—a system. Its 'physiology,' in certain ways, resembles that of a river system. It is a flow from source to mouth." Thus he described the "flow" of natural resources and raw materials from their "source" to the "mouths" or markets for finished products, as well as the routes that connected them. "And this," Mac-Kaye concluded, "—the discovery of the most efficient framework for guiding industrial flow—is the job of the new exploration."[54]

MacKaye's article was amply illustrated with his own maps of the Somerset Valley in the Deerfield River watershed of southern Vermont. He adapted to this New England setting a version of the sustainable forestry communities he had proposed for the Pacific Northwest in *Employment and Natural Resources*. The maps depicted the sites of "protection forests," farmland, hydroelectric dams and reservoirs, and new villages, all within the bounds of a single watershed.

MacKaye's graphically illustrated approach to broadscale regional planning

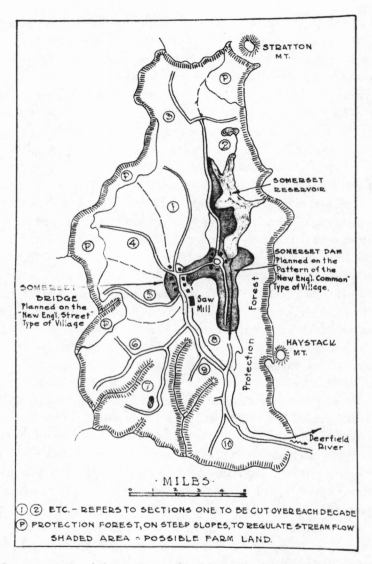

STRATTON MT.

SOMERSET RESERVOIR

SOMERSET DAM
Planned on the
Pattern of the
"New Engl. Common"
Type of Village.

SOMERSET
BRIDGE
Planned on the
"New Engl. Street"
Type of Village

Saw
Mill

Protection Forest

HAYSTACK MT.

Deerfield River

· MILES ·
0 1 2 3 4 5

① ② ETC. - REFERS TO SECTIONS ONE TO BE CUT OVER EACH DECADE
Ⓟ PROTECTION FOREST, ON STEEP SLOPES, TO REGULATE STREAM FLOW
SHADED AREA = POSSIBLE FARM LAND.

The May 1, 1925, regional planning issue of *Survey Graphic* included articles by MacKaye, Lewis Mumford, Clarence Stein, Stuart Chase, Henry Wright, and others in the RPAA circle. MacKaye's article, "The New Exploration: Charting the Industrial Wilderness," showed how southern Vermont's Somerset Valley, along the upper Deerfield River, might be developed for a "forest community" comprising "protection forest" zones, forest units logged on a sustainable rotation, planned residential communities, and a hydroelectric reservoir.

was nothing like any conventional exercise in town or city planning. In fact, he saw his job as complementary to that discipline. "Let us by all means have our city planning," he concluded. "But let us not forget its roots in its vast hinterland."[55] On his own new exploration of regional landscapes, MacKaye was beginning to find his footing. But rhetorically, conceptually, and practically, his was not always an easy trail for others to follow.

In late 1924 and early 1925, MacKaye was also occupied with the next stage in the Appalachian Trail project. He had returned to the Hudson Guild Farm in mid-October 1924 for "an Appalachian revival meeting" that included some of the RPAA core group, such as Mumford, Chase, Whitaker, and Bruère. They hiked, danced, and planned the *Survey* issue.[56] MacKaye then moved on to Harriman and Bear Mountain State Parks, where Major Welch provided him a cabin. For a month, he hiked and scouted trails in and around the parks. (The newly opened Bear Mountain bridge had solved the problem of a Hudson River trail crossing, thereby forging the critical trail link between the New England and Mid-Atlantic states.) He also plotted with Welch and Torrey the next steps in the trail campaign.[57]

Although the RPAA had taken over nominal sponsorship of the trail project from Stein's AIA Committee on Community Planning, only MacKaye, among the RPAA members, felt truly comfortable articulating the trail's significance as an instrument of regional planning. The project's principals well understood that the real expertise, manpower, and enthusiasm to complete the task would necessarily come from the hiking community; and they all agreed that the time had come to pursue more seriously the idea of a centralized Appalachian Trail organization, to be a federation of trail clubs operating the full length of the proposed trail route.

In December 1924, the general council of the National Conference on Outdoor Recreation, which had first convened the previous May, met in Washington, attracting many of the key figures in the trail effort (as well as prominent officials such as Secretary of Commerce Herbert Hoover, who told the conferees that their objective was "to make life less drab").[58] MacKaye, Welch, Chamberlain, Kelsey, and Turner were among the Appalachian Trail activists on hand. Plans were soon being made for a meeting early in 1925 to create a new group to oversee the trail's construction and administration. Welch agreed to preside over an organizational conference. The RPAA approached American Civic Association executive secretary Harlean James to organize an

Appalachian Trail Conference in Washington in March. As editor of the *American Planning and Civic Annual,* James was a respected figure in the fields of planning and public recreation. Motivated as well by a personal interest in the trail project (she had attended the October 1923 Bear Mountain Inn conference), she proved to be the ideal person to organize this critical first Appalachian Trail Conference, which was in fact officially sponsored by the recently formed Federated Societies of Planning and Parks.[59]

When the Appalachian Trail Conference convened at Washington's Hotel Raleigh on March 2, 1925, the impressive and influential array of speakers on the program reflected how powerfully MacKaye's idea had grabbed official and public consciousness since he had offered his proposal less than four years earlier. On the first day of the conference, speaking after Welch and Frederic A. Delano, president of the Federated Societies, MacKaye described the philosophy behind the trail project. "Its ultimate purpose is to conserve, use and enjoy the mountain hinterland which penetrates the populous portion of America from north to south," he declared, according to his own detailed "brief" of the conference. "The Trail (or system of trails) is a means for making this land accessible. The Appalachian Trail is to this Appalachian region what the Pacific Railway was to the Far West,—a means of 'opening up' the country. But a very different kind of 'opening up.' Instead of a railway we want a 'trailway.'"[60]

He went on to explain his vision of the trailway as a "functioning service," comprising a series of camps and stores for shelter and food, a transportation system by train and automobile from neighboring cities, and the footpath itself. But he warned that the "path of the trailway should be as 'pathless' as possible; it should be a minimum path consistent with practicable accessibility." Then he outlined his plan for the project's next stage. As depicted on a map that he had prepared for the conference, the projected trail route would be divided into five regions: New England, New York and New Jersey, Pennsylvania, the central Appalachian states from Maryland through Virginia, and the southern Appalachian states from North Carolina to Georgia. In each of these regions he identified one or two "pivotal sections" where he advised that local scouting crews immediately begin locating and building the trail.[61]

State foresters, park officials, Forest Service district foresters, trail club officers, and local hiking enthusiasts reported on the trail's progress and prospects in their regions. F. E. Matthes of the U.S. Geological Survey discussed the possibility of developing a "nature guide service" in conjunction with the trail, a notion that remained a favorite of MacKaye's. Arthur Comey of the New England Trail Conference, himself a respected landscape architect and planner, gave a talk on "Going Light," which, in an era before the develop-

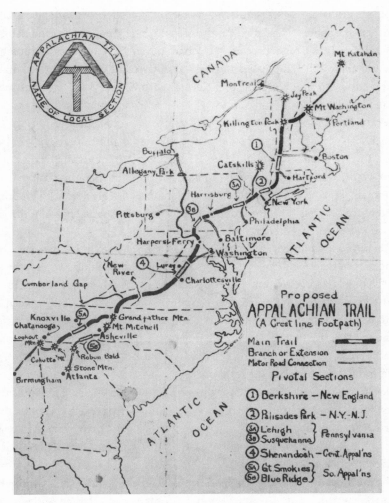

MacKaye prepared a map for the first meeting of the Appalachian Trail Conference in March 1925. Besides depicting the main proposed trail and possible branch trails, it also located five "pivotal sections," on which MacKaye urged trail builders to concentrate their efforts.

ment of high-technology, low-weight camping gear, was a matter of intense interest to long-distance hikers. Clarence Stein, addressing the group in a more philosophical vein, suggested that the recreational development of the Appalachian crestline was necessary to offset the pernicious influence of "Atlantis," the "possible giant city" he depicted evolving along the eastern seaboard.[62]

On the afternoon of the conference's second day, after National Park Service director Stephen Mather spoke, those in attendance voted to establish the Appalachian Trail Conference as "a permanent body." A provisional constitution was approved that created a fifteen-person executive committee. Finally, the executive committee itself was elected, with Major Welch as chairman, Verne Rhoades of the Forest Service as vice-chairman, and Harlean James as secretary. The executive committee was stitched together primarily from the connections MacKaye had made personally while promoting the project. Stein, Torrey, Rhoades, Welch, H. S. Hedges, Paul Fink, Arthur Comey, and Frank Place were among those who had been brought into the project in its earliest stages. And Forest Service chief William B. Greeley and the National Conference on Outdoor Recreation's then chairman Chauncey J. Hamlin added influence and prestige to the ATC's leadership.[63]

MacKaye was named Field Organizer. During discussions that were not made part of the meeting record, plans were laid to raise $5,000 to support his projected fieldwork.[64] Whether by oversight or intention, he was not included on the group's executive committee. His omission from the ATC's governing board provided a harbinger of the ambiguity of his future role and reputation in the conference. Beyond his brief tenure as field organizer, MacKaye would never serve as an officer of the organization that he, probably more than any other individual, was instrumental in creating.

Nevertheless, he declared the 1925 meeting a success. "The Conference was called for the purpose of organizing a body of workers (representative of outdoor living, and of the regions adjacent to the Appalachian Range) to complete the building of the Appalachian Trail," MacKaye recorded. "This purpose was accomplished."[65]

The creation of the Appalachian Trail Conference was an essential step toward the eventual completion of the Appalachian Trail. Already, though, the conceptual scope of MacKaye's original project had narrowed considerably from what he had proposed in his original 1921 article. The leadership the organization chose indicated the nature of that constricted vision. The project was already moving into the hands of well-educated middle-class professionals—lawyers, engineers, educators, scientists—and government officials. The labor unions and settlement houses MacKaye had included in his early depiction of the project were not involved. He had not abandoned his vision of the trail project as a vehicle of social transformation. By his own account of the conference, though, he was no longer advocating "community camps" or "food and farm camps." The trail was taking shape as a recreational project, pure and simple. The challenge it offered to capitalism, urbanism, and industrialism was already subtle at best.[66] But the principle of local groups, federated under the gentle guidance of a modest central organization, working and playing on the terrain they knew and loved, would provide the key to the project's eventual completion and success. In the years ahead, MacKaye's own ideas about the trail would evolve, to emphasize its role as an educational, inspirational wilderness bulwark against the inroads of metropolitan civilization.

❖

When the International Town, City, and Regional Planning conference convened in New York in April 1925, MacKaye could justifiably claim a place in the cause and the profession. Indeed, the Regional Planning Association of America sponsored a weekend gathering at the Hudson Guild Farm for some of the distinguished architects and planners in attendance, including Raymond Unwin and the venerable Ebenezer Howard.[67] For MacKaye, however, the initiation of the Appalachian Trail, the creation of the Appalachian Trail Conference and the Regional Planning Association of America, the preparation of the New York planning report, and the publication of Survey Graphic's "Regional Planning Number" were testaments to the fruitfulness of his endeavors since the death of his wife four years earlier. But once again, despite such accomplishments, he struggled to find support for his "larger work."

"Benton . . . has been living by the skin of his teeth, as usual," Mumford reported to Stein in outlining potential projects for the RPAA, "and yet without his following the numerous scents he opened during the last four years, we'd be far from where we are now in our conception of regional planning."[68]

MacKaye planned a return to his native environs to carry on with his unpredictable vocation as a self-made regional planner. "I guess I can make a living somehow in New England as well as anywhere," he wrote his sister, whose welfare worried him, "and if that is the best place now for those dear to us to get together nothing else need count. . . . I don't want further to *delay* the *real* things in life—I'll delay prospects instead."[69]

The New Exploration

1925–1928

Throughout the mid-1920s, Benton MacKaye was a vagabond in the Northeast, "impracticable, footloose, and a free spirit," as his envious Forest Service mentor Raphael Zon wrote in a letter to him in the spring of 1925.[1] Benton returned more frequently to Shirley Center during those years, for reasons of economic necessity, family obligation, and, not least, because the place was in every sense his home. During the quiet days of the Shirley summer of 1925 he visited with old family friends like the Longleys, Lawtons, Stones, and Wings. The season was an ongoing festival of parties, dances, baseball games, upcountry hikes, and Gilbert and Sullivan sing-alongs.

MacKaye set up a study in the Cottage's "Sky Parlor," an upstairs room on the building's northwest corner overlooking a back field. The little aerie, illuminated with many windows, took shape as his creative headquarters. Though he did not give up his peripatetic ways, he found in the Cottage and his study the setting and solitude that would make the remainder of the 1920s among the most intellectually productive years of his long life. The foremost fruit of this period was the publication in 1928 of *The New Exploration: A Philosophy of Regional Planning*.

It had been MacKaye's ambition, ever since proposing the Appalachian Trail in 1921, to write a more expansive "opus" about his vision for the future of the American landscape. In the months after launching the trail project, MacKaye had conceived the phrase and metaphor—the "new exploration"—to define his perception of the nation's great social challenge. But his early efforts to elaborate the idea had lacked compelling, concrete examples of his perspective and method. His experience with the Appalachian Trail, his association with the RPAA, and the field laboratory of his home state during the mid-1920s provided the substance to illustrate his overarching metaphor. *The New Exploration* was assembled from the articles, manuscripts, and planning projects that had accumulated on his literary workbench throughout the decade following publication of *Employment and Natural Resources* in 1919. With Lewis Mumford's gentle guidance, MacKaye composed a provocative and original work that, at least by Mumford's immodest estimation, deserved "a place on the same shelf that holds Henry Thoreau's *Walden* and George Perkins Marsh's *Man and Nature.*"[2]

The New Exploration has not, of course, achieved a status in the canon of American environmental literature equivalent to those seminal nineteenth-century works. MacKaye's book lacks *Walden*'s literary originality and verve and *Man and Nature*'s magisterial synthesis of history and science. But *The New Exploration* remains a significant, if eccentric, intellectual landmark of the American conservation, planning, and regionalist movements during the early twentieth century.

Before concentrating on his regional-planning book, however, MacKaye for the better part of 1925 and 1926 worked on several projects that reflected a national and global outlook. What he called "A World Atlas of Commodity Flow," for instance, was an elaborate system of maps, flow charts, and diagrams that graphically analyzed global patterns of natural resource distribution, trade, transportation, manufacturing, and population movement. Stuart Chase, then writing for *The Nation,* introduced his friend to the magazine's editor, Freda Kirchwey. MacKaye's three-part series based on the "World Atlas" was titled "The Industrial Exploration," when published in that magazine during the summer of 1927.[3]

"'Commodity flow' is to the industrial wilderness what river flow is to the terrestrial wilderness," he wrote in the first of the articles, elaborating on a worldwide scale the planning analogy he had adapted from geography and forestry.[4] "The commodity stream has three parts: source, destination, and 'flow' between."[5] His "World Atlas," therefore, "would visualize and chart . . . an unseen integrated industrial system, just as the explorer charts for us an unseen river system."[6] MacKaye retained a progressive faith in rational social engineering. His maps and charts could depict the physical and political "entanglements" exploited by capitalists and imperialists. But the same diagrams, he believed, might also reveal the way out of "the labyrinth of world-wide industry."[7]

MacKaye's perspective and graphic method reflected the geographic determinism that prevailed among the turn-of-the-century scientists and scholars under whom he had studied. But he lacked a coherent understanding of existing economic mechanisms and dynamics. While entirely skeptical of capitalism and commercialism, he was no student of Marx. MacKaye's brand of socialist analysis came straight from his brother James's neglected *Economy of*

Happiness, which depicted a rational, steady-state economy, managed by a technocratic elite.

Benton was eager to present his ideas to any audience. He was received enthusiastically by the politically minded and idealistic young people at the National Student Forum camp in Woodstock, New York, where he spent a week during the summer of 1925. But when he arranged a six-week seminar on "The New Exploration" that fall at New York's Civic Club, his audience dwindled from twenty-five the first week to five the final week.[8]

He did not compromise his program to accommodate potential supporters or political allies, however. When Socialist Party leader Norman Thomas asked for guidance about how the "city planning question" might be incorporated in the party's "municipal platform," MacKaye cautioned Thomas to think in broader terms. "My conception of city planning involves all cities at once," he wrote. "It involves a redistribution of the population; not the population of any one city but of the whole United States."[9]

Clarence Stein in early 1926 arranged for MacKaye to prepare a plan for the Russian Reconstruction Farms, Inc., the creation of an American agricultural reformer (and, it was revealed years later, a Communist fellow traveler) named Harold Ware. The Soviet government, under Ware's prodding, designated a substantial tract in southeastern Russia, north of the Caucasus, for a demonstration agricultural project combining collective ownership with American farming techniques.[10] Ware had proposed establishing *artels,* or community working units, of sixty families on parcels of 3,000 acres. MacKaye's ideas for American cooperative farming and forestry colonies appeared to parallel Ware's Russian scheme, but he lacked any firsthand knowledge of the Russian terrain. He traveled to Washington, D.C., where during the late winter and spring he gleaned information from atlases, books, friends, and federal officials. Unlike his American utopias, which derived their vividness and plausibility from his direct knowledge of the landscape, MacKaye's Russian plan was dry, abstract, and uncharacteristically tentative. He proposed expanding Ware's scheme to encompass twenty *artels,* providing for 6,000 people on 60,000 acres. He confessed that the idealized plan he sketched would constitute a drastic upheaval for many already living in the region.[11] Since his contacts with Ludwig Martens in 1920, MacKaye had not pursued further efforts to support the Soviet government. He undertook the Russian Reconstruction Farms project as a practitioner of regional planning, not as a communist partisan. In any event, the dream of a technically modern, collectivized Soviet agriculture would give way within just a few years to the Five-Year Plans, mass starvations, and forced migrations under Stalin's regime.

❖

On April 18, 1926, the fifth anniversary of Betty's death, Benton spent the day in his Washington boarding house room "smoking up and trying to think out plans—for Hazel and self." As he traveled north that spring, stopping along the way for a weekend RPAA gathering at the Hudson Guild Farm, he once again considered how to integrate his efforts and ideas into a book. When he visited Percy in Cornish, New Hampshire, in July, his brother presented him with the finished manuscript of *Epoch,* a massive biography of their father, Steele. Percy's accomplishment—"A Resurrection!" Benton noted in his diary—revealed to him much he had not known about his father; it also inspired him to tackle his own big project.[12]

Benton returned to Shirley late that summer of 1926. The familiar, reposeful setting of Shirley Center helped him come to terms with his almost monastic style of life and his unconventional, unpredictable vocation. "I found myself in an oasis," he later wrote of what he called this "Shirley Renaissance." "Shirley Center, always somewhat insulated from the dust storms of the Sahara of general life, had become even more so. In my own eyes anyhow, she had reverted to type, to the type that I always yearned for, to that old 'Grecian' hue, which, to 'such as me,' constitutes the essence of community culture vs. metropolitan cacophony."[13] From the perspective of the 1950s, Benton invoked his own self-consciously nostalgic effort during the 1920s to preserve Shirley Center, at least symbolically, as a last pastoral bulwark against the inroads of metropolitan civilization.

MacKaye once confessed to Mumford that his "basic source of income . . . consisted in 'not having children.'"[14] But he adopted Shirley Center and its families as surrogates for the family he would never have. During the 1920s, as the automobile opened up the Massachusetts countryside and expanded the commuter's range, the old families of Shirley Center were joined by a cosmopolitan, younger generation of families who were attracted to the community's handsome colonial houses and rural charms. These families provided second homes for him. Reconciled to remaining a widower, he assumed the role of adopted uncle, captain of huckleberry parties, the benevolent pied piper of Shirley Center. On his daily walks around town, he sometimes dropped in unannounced—but just close enough to breakfast or the dinner hour to be invited to stay for a meal. The families of Shirley Center tolerated his mild idiosyncrasies and occasionally insensitive demands, which were more than compensated by his lively wit, raconteurial talent, and interesting friends and acquaintances.

MacKaye frequently visited one such family, the Pifers, who had recently moved into an old home near the Longley homestead. Accustomed since childhood to reading aloud in the evening, he worked through some of his favorite books, such as *The Virginian* and *Huckleberry Finn,* with the three dutiful Pifer children. These obligatory sessions sometimes had the flavor of "command performances," one of the Pifers recalled.[15]

Not far from the MacKaye houses, in a substantial old home on Shirley's common, Howard and Helen Bridgman had recently opened a small, progressive private school for boys. That fall MacKaye led the schoolboys on a series of camping trips and trail-blazing expeditions "upcountry," along a little range of hills to the northwest of Shirley. A trail recently cleared by local hiking enthusiasts extended from Mt. Watatic north for some twenty miles over the state border to the Pack Monadnocks near Peterborough, New Hampshire. The Wapack Trail, in Benton's eyes, was a microcosm of the Appalachian Trail. It included all the elements of the grander path: a rugged walk on a narrow trail through undeveloped mountain terrain; a rustic lodge at the trail's midpoint, providing shelter and food; and a variety of interesting natural features along the way. All of this was within a few hours' range of Boston by train or automobile, enabling city hikers to traverse the trail over a weekend. MacKaye's excursions into the gentle Wapack Range provided inspiration for a flurry of writing about his evolving philosophy of trail building and wilderness preservation. "A Boy's Project for a Balanced Civilization," the title he gave a *Christian Science Monitor* article on the Wapack project, suggested how seriously he took the notion that young people and amateurs could play an important role in transforming both the physical and social landscape. "This development is one not so much of changing the Wapack Range," he wrote at the time, "as of changing ourselves who look upon it."[16]

His promotion of the redemptive potential of such wild landscapes echoed an article Benton had recently read by the eloquent, far-thinking forester Aldo Leopold. In "Wilderness as a Form of Land Use" Leopold provided what some scholars have called "his most sustained, comprehensive statement of the wilderness idea."[17] MacKaye immediately imported Leopold's views and words directly into his own writing on outdoor recreation, trails, and planning. Until this time, he had most often used the term *wilderness* to describe the nation's chaotic urban and industrial terrain, not the undeveloped American "wilderness playgrounds" described by Leopold. "The public interest," Leopold wrote, "demands the careful planning of a system of wilderness areas and the permanent reversal of the ordinary economic process within their borders." In Leopold, MacKaye had discovered a kindred philosopher of the American social and natural landscape, whose ideas resonated closely with his own.[18]

In response to an appreciative note from MacKaye, Leopold wrote, "It naturally gives me much gratification to know, not only of your approval of the wilderness article, but of your understanding of the real issues which it involves."

> . . . It will be particularly interesting to talk over with you the unsolved questions involved in giving the wilderness idea actual expression in the form of a program. From the governmental end I can pretty well visualize the action which is necessary, but the idea really goes a lot further than merely governmental action. It is a point of view, not a piece of land, and I badly need your advice as to how to plant the point of view in places like your Regional Planning Association.[19]

During the 1920s, Leopold and MacKaye were among the leading American conservationists arriving at a new philosophical and ethical perspective on the relationship between humanity and nature. MacKaye provided an intellectual and personal link between Leopold's "wilderness idea" and evolving "land ethic," on the one hand, and the comprehensive, ecologically based regionalist viewpoint of Lewis Mumford and their RPAA colleagues, on the other. MacKaye embodied an expansive, integrated conception of American conservation that set him apart—sometimes to the point of isolation—from many other mainstream conservationists of the decade.

After the RPAA's initial burst of creative activity, some key members were feeling frustrated about the group's progress. "I sometimes wish there were no such thing as regional planning," confessed Stein, "and that I could put all my time into the things which I find most fun—that is to say, drawing. . . . I have no clear idea as to just what the Regional Planning Association is going to do."[20] Mumford likewise reported that the "good old meetings are dead for the present; the spirit of 1924 and 1925 is missing."[21]

MacKaye and Mumford nonetheless pressed ahead. From Shirley, MacKaye reported to Mumford that he and his sister were "living here on a cave man standard, with wood and water and light and every other primal element to be extracted from its source." Since her mother's death, in 1924, Hazel had returned to Shirley Center for summer stays. For a time, she had taught theater at the Brookside Labor College in Katonah, New York. She returned to Shirley Center in September, to join her brother for the winter.[22] The pinched, rustic living conditions of the Grove House and Cottage did not dampen his excitement about the progress and integration of his ideas.

MacKaye explained how he hoped to link the global and "work" aspects of his conception of regional planning, as exemplified in his "World Atlas of Commodity Flow" and his "Industrial Exploration" articles, with the local and "play" dimension represented by the Appalachian Trail and his Wapack Trail excursions.[23] His aim, MacKaye later reported to Mumford, was "to develop a single idea—to meet the metropolitan challenge by the development of the indigenous environment as a synthetic art. (I would pit a functioning art against a perverted industrialism)."[24] He described how his approach to this "synthetic art" had its roots in "the territory around about Shirley Center. This little region embraces the three fundamental environments (as I conceive them) which are necessary for man's full development. These are the primeval, the rural (the 'colonial' in New England), and the urban."[25]

From his immediate surroundings MacKaye arrived at a universal diagnosis of the ills of modern civilization. The three "fundamental" (or "indigenous") environments were already "threatened (and the urban already immersed) by a fourth environment (a diseased environment we might call it) called the 'metropolitan.' . . . The job of . . . repelling the 'invasion' . . . splits itself into three parts: the *preservation* of the primeval; the *restoration* of the colonial; and the *salvaging* (some day) of the true urban. The immediate tasks seem to be the first two just named—the problems of the primeval and the colonial environments."[26]

Mumford sensed that the time had come for his friend to tie together some of the threads of his work. "The work you are doing seems a real crystallization of everything you've been thinking these last six years," he wrote. "Haven't you got the stuff of a book there?" He urged MacKaye to organize and send him his writings, promising to seek a publisher for his friend's efforts.[27] Over the next year-and-a-half, in a continuing stream of letters, manuscripts, and meetings, Mumford coaxed, cajoled, and sometimes funded his friend, as MacKaye completed *The New Exploration*.

Mumford's contribution to *The New Exploration* was an act of self-effacing modesty. As the book's self-styled "editor and sub-rosa collaborator," he added some of his own words and allusions but made no effort to alter the substance of MacKaye's thinking.[28] He recognized the unique integrity of MacKaye's seemingly disorderly work routine and output. When MacKaye sent off outlines, revisions, and thoughts in progress, Mumford replied with encour-

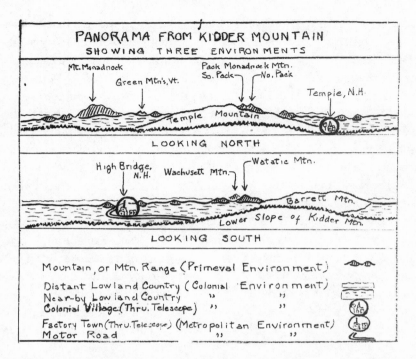

During the late 1920s, MacKaye used the modest Wapack Range, spanning the Massachusetts–New Hampshire border, as an open-air laboratory for his regional planning ideas. His "Panorama from Kidder Mountain, Showing Three Environments," c. 1927, depicts the relationship among "primeval," "colonial" (or rural), and "metropolitan" environments.

agement and gentle attempts to weed out his friend's quirkier ideas and odd neologisms, but he did little to try to steer MacKaye away from his plan to write directly from his own experiences and his own neighborhood. After all, it was in *The Golden Day,* Mumford's own 1926 study of an earlier efflorescence of New England regional culture, that MacKaye had found the final inspiration and confidence to undertake his own book. "Your book has been doing its part in tuning up my thoughts. I am eating it alive," MacKaye wrote to Mumford in the spring of 1927, as his own book at last began to take shape.[29] *The Golden Day* must have provided ample nourishment, because a month later MacKaye reported that he was occupying himself with "just book, book, book! I feel as if I were on the last lap of a six year race."[30]

⁘

Even as he neared the finish of his literary marathon, though, MacKaye was finding and testing new ideas among the thriving network of accomplished planners, architects, and conservationists in his home state. Massachusetts had long been a seedbed of activity and leadership in the cross-fertilizing fields of city planning, landscape architecture, conservation, and outdoor recreation. If Thoreau was, in some sense, the philosopher and progenitor of this regional tradition of landscape awareness, MacKaye was among those who, as Mumford asserted in *The Golden Day,* "might still go out and make over America in the image of" his Concord counterpart.[31]

In the early summer of 1927, Edward T. Hartman, executive secretary of the Massachusetts Federation of Planning Boards, visited MacKaye in Shirley. Hartman, who had heard a vivid talk by MacKaye at that year's New England Trail Conference meeting and had read his 1925 "New Exploration" *Survey Graphic* article, invited him to speak at his organization's October meeting in Greenfield. Boston architect William Roger Greeley chaired the Greenfield conference. Benton's Harvard classmate, writer Walter Prichard Eaton, was also on the program, along with town planner and outdoor activist Arthur Comey. "Historical trip for me," he noted in his diary upon returning from the conference.[32]

MacKaye's talk, "A Cross-Section of Massachusetts: Two Possibilities," favorably impressed Greeley, who would become not only an enthusiastic proponent of MacKaye's ideas but also his professional and financial patron—the Massachusetts counterpart of Clarence Stein. The perspective of the regional planner, MacKaye asserted in Olympian fashion, assumes "the geographic approach, the aeroplane approach, the approach of one on high." Depicting a "cross-section" that traversed Massachusetts from Boston along the northern tier of the state through Shirley, Fitchburg, Greenfield, and the northern Berkshires, he compared the "suburban massings" around Boston with the prospects for the region surrounding Fitchburg. In that small industrial city just west of Shirley, he saw the potential for developing a genuine "regional city," the cultural and economic focus of the small, defined communities surrounding it. These communities would themselves be buffered by belts of open space that would follow the courses of local hills and rivers.[33]

MacKaye was soon spinning out fresh proposals, projects, and ideas for his Massachusetts collaborators. In December, he traveled to Cape Cod at Greeley's behest to address the local chamber of commerce. "The regional planner is the 'environment doctor,'" he said, urging the businessmen to think carefully about the fundamental importance of highway planning. A "main road plan" the length of the Cape, he proposed, should simultaneously take into account the designation of public forests and open spaces, the location of roadside business and services, and the "retention" of existing villages and towns "where they are [and] not as aimless 'stringtowns.'"[34]

MacKaye was also recruited by Harris A. ("Josh") Reynolds, secretary of the Massachusetts Forestry Association, to collaborate with Greeley on a pamphlet dealing with the very swath of Massachusetts he had described in his Greenfield talk. The resulting piece, "Zone the State Highways: The Lesson of the Mohawk Trail," was an exercise in propaganda. Reynolds's organization was promoting legislation to authorize state zoning regulations for roadside commercial development. This modest pamphlet was inspired by the unsightly changes taking place along Route 2, the scenic east-west state highway that ran only a mile north of MacKaye's Shirley home.[35] It was the first of many articles and reports on roads and automobiles that soon led him to his conception of the "townless highway."

With the support of the well-connected Reynolds, MacKaye secured a short-term job with the Governor's Committee on the Needs and Uses of Open Spaces. Chaired by the reform-minded but cautious businessman Charles S. Bird Jr., the panel was charged with inventorying and developing a plan for the state's parks, forests, waterways, beaches, and other outdoor recreational resources.[36] In January, after a meeting with Reynolds and Bird in which he received verbal assurances that he could proceed (and, he thought, would be paid $1,500), MacKaye began drafting recommendations for a state open-space protection program.[37]

The preliminary report distinguished between what he called "industrial" open space, comparable to the national forests, which would supply raw material, and "cultural" open space, equivalent to national parks, for public outdoor recreation. These latter spaces, he suggested, "should lie in lines or belts rather than in scattered fragments: they should form circuits for making outdoor trips from the adjacent cities." The "intertown" environment—the undeveloped roadways between towns that were beginning to be developed in unregulated strip fashion—was, he asserted, a communal resource that should come under the control of the state. A series of maps depicted five different areas for "initial acquisition," from the Berkshires to Cape Cod, each "selected in view of its strategic or pivotal location in a major open way or primary 'levee.'"[38]

When he presented his preliminary findings in April, MacKaye expected the support of his conservationist friends on the board, such as Allen Chamberlain, Arthur Comey, and Laurence B. Fletcher, secretary of the Trustees of Public Reservations. Indeed, after his presentation, the committee voted to authorize his proposed field survey for a Wachusett-Watatic Open Way. But some members gave MacKaye's proposals a frostier reception. Planner Charles W. Eliot II objected to MacKaye's well-rehearsed metaphors describing metropolitan "flows" and the need for open space "levees" to dam the flood of development. Comey politely noted the "intangible character" of the scheme.[39]

More important, Bird's enthusiasm for MacKaye's activities cooled. The committee chairman apparently bridled at the ambitious land-acquisition and regulatory program MacKaye proposed. Bird scaled back the amount of money MacKaye would receive for his efforts. Nonetheless, when he had the time, after putting the finishing touches on the manuscript, maps, and illustrations of *The New Exploration,* MacKaye spent a few weeks in late May and early June surveying and mapping his proposed Wachusett-Watatic Ridgeway reservation.[40]

By August, he had completed a final version of this study, titled "Wilderness Ways," but neither Bird's committee nor Hartman's Federation of Planning Boards was willing to publish his full report.[41] As part of an agreement to settle the payment he had expected for his efforts, however, the journal *Landscape Architecture* published an article based on his findings. "Wilderness Ways" depicted a publicly owned Wachusett-Wapack mountain reservation, less than sixty miles from Boston, linked by an existing forty-three-mile trail following the modest range of hills between Mount Wachusett in Massachusetts and Pack Moradnock Mountain in New Hampshire. "A wilderness way is an open way," he explained, "it is a wilderness area that goes somewhere." MacKaye envisioned a regional counterpart of the Hudson Highlands Interstate Park, "the week-end breathing ground for Boston as now the other is for New York City." He sketched out other wilderness ways throughout Massachusetts, knitting together from the state's existing patchwork of parks, forests, and shoreline a coherent, integrated pattern of recreational open space, "based upon the processes of evolution as these are disclosed for us in outdoor nature."[42]

Of course, Massachusetts politicians and economic interests were not in the habit of implementing public policy on the basis of evolutionary processes. MacKaye's failure to win immediate support for an ambitious open-space program in his home state did not discourage him, though. "It is larger than Massachusetts," he wrote Edward C. M. ("Ned") Richards, a liberal forester who had asked him to help prepare a report on national forest policy for the League for Industrial Democracy. "It has the possibility of growing into a plan for New England; and this indeed is what I have headed for since the beginning."[43] And his Massachusetts efforts provided material for *The New Ex-*

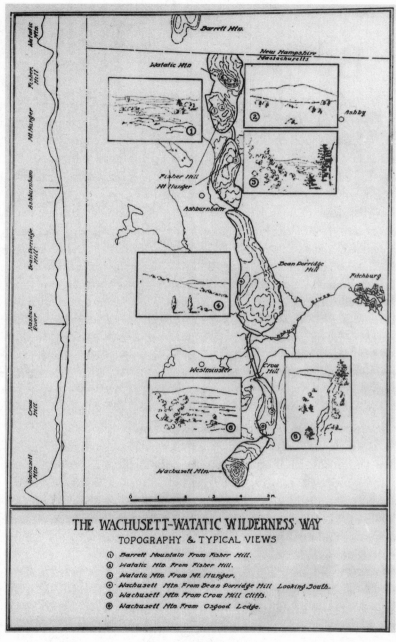

THE WACHUSETT-WATATIC WILDERNESS WAY
TOPOGRAPHY & TYPICAL VIEWS

① Barrett Mountain From Fisher Hill.
② Watatic Mtn. From Fisher Hill.
③ Watatic Mtn. From Mt. Hunger.
④ Wachusett Mtn. From Dean Porridge Hill Looking South.
⑤ Wachusett Mtn. From Crow Hill Cliffs.
⑥ Wachusett Mtn. From Osgood Ledge.

MacKaye's plan for a publicly owned Wachusett-Watatic Wilderness Way, that would surround an existing footpath in north-central Massachusetts, appeared in his article "Wilderness Ways," in the July 1929 issue of *Landscape Architecture*. He developed the plan while working for the Governor's Committee on the Needs and Uses of Open Spaces.

ploration; he incorporated the findings of his open-space report, including maps, almost wholesale into a chapter titled "Controlling the Metropolitan Invasion."

✦

Benton's progress on his "phantom book" during 1927 and 1928 had been hindered by financial and family pressures. Short of cash, he continued to cobble together and send off articles "in hopes of getting some money," as he reported to Percy, ". . . but no one of them is actually landed."[44] For a time, he was hampered by a broken foot, injured while he was pushing a car in the snow.[45] His principal worry, however, was not about his own welfare but that of his sister, Hazel. Since the death of their mother in 1924, Hazel had not found a new social and economic foothold. Unmarried and middle-aged, lacking any income or assets other than her ownership of the Grove House, she was stranded in the early 1920s. The heyday of pageantry, her one arena of professional expertise and success, had passed. The community of militant suffragists and feminists that had once embraced her had dispersed.

Hazel's psychological and emotional well-being took a sharp, irreversible downturn late in 1927. In October, after one severe episode, she left Shirley briefly to recuperate closer to Boston. But by early December, when she and Benton had set up winter quarters in the Grove House, she relapsed into a "sort of delirium."[46] For the better part of the next year, Benton coped with the expense and turmoil created by his sister's troubled condition.

Hazel's maladies, and the cost of her care, created some friction and ill will among Benton and his brothers, who debated mortgaging, renting, or selling the Shirley properties to help pay for Hazel's treatment. Closest to Hazel in age and experience, Benton defended his sister—and his own efforts to help her. He realized that Hazel's symptoms, however vague and difficult to diagnose, were serious and unfeigned. The grim course and outcome of his late wife's similar troubles could never have been far from his mind.

"She is here where I can look after her without having it interfere with my work," he explained to his brothers at one point. "Yet the suggestions pour in on me from 'kindly souls' that Hazel needs a change, that she may be getting into a rut, that if she went somewhere else (to this or that sanitorium) she *might* get better faster. Very beautiful—perhaps she *might*. And perhaps she might *not.* . . . If somebody wants all this then they can take over the whole job and run it—and pay for it."[47] Finally, in November 1928, she began an almost decade-long residence at the Gould Farm, a therapeutic community for the mentally ill, in Great Barrington, Massachusetts, where Benton visited her regularly.[48]

MacKaye pushed *The New Exploration* to completion in the spring of 1928, helped along by Lewis Mumford, who made his first visit to Shirley Center that April. Mumford had another powerful incentive to make the trip. By coincidence, a dark-haired, high-spirited nineteen-year-old poet and pianist named Josephine Strongin was then living at a boarding house in Shirley Center. Strongin was acquainted, even infatuated, with Mumford, whom she had come to know when he resided briefly at the Swiss school she attended. A friend of the Bridgman family, the exotic Strongin helped MacKaye complete work on his book. Now, in the last stages of *The New Exploration*'s protracted gestation, it may have been Strongin's presence in Shirley, as much as Mac-Kaye's need for help, that finally stirred Mumford to make his trip.[49]

It was MacKaye's friend and neighbor, Helen Bridgman, who suggested that he dedicate his book "To Shirley Center, An Indigenous Community."[50] This modest accolade suggested the degree to which he had identified his hometown as the idealized representation of a balanced, harmonious, and authentic community. MacKaye had left the business and practical details of his book's publication to Mumford, Stein, and Henry Wright. The critical and financial success of *The Golden Day* had firmly established Mumford's literary reputation—as well as his clout with his own publisher, Harcourt Brace. When Mumford promised to edit MacKaye's manuscript, Harcourt Brace readily acceded to the desires of their popular young author. Stein and Wright hammered out the contractual details with the publisher. Mumford undertook a final revision of the manuscript and turned it over to Harcourt Brace at the beginning of May.[51]

The New Exploration, when published that fall, was a modest-sized book, comprising 14 chapters, 228 pages, and 25 of MacKaye's distinctive maps and diagrams. There had never been any question that *The New Exploration* would be the book's main title. MacKaye at one point asked Mumford whether his subtitle should include the term "regional engineering." "My attack on things is different from what goes by the name of 'regional planning,'" he admitted, "so how about a different name?" Mumford also recalled that MacKaye had floated Geddes's term, "geotechnics," for a subtitle. But Mumford convinced

MacKaye to describe the work as "A Philosophy of Regional Planning," employing the "more dynamic term that would be more intelligible and useful to both publisher and reader."[52] *The New Exploration* was essentially a series of connected essays and illustrations depicting MacKaye's unique, virtually irreproducible approach to visualizing the rational management and design of the physical environment. Not a textbook or handbook of practical planning techniques, the book was instead an inspirational tract, a work epitomizing the RPAA's conception of regionalism as "romantic-poetic myth and aspiration," in the words of one student of community planning.[53] *The New Exploration* was also a map of MacKaye's active and unconventional mind.

The regional planner's task "is to render actual and evident that which is potential and inevident," MacKaye preached. "Planning is revelation—and all-around revelation."[54] And if planning was revelation, then Benton MacKaye was its self-styled "visualiser," prophet, and evangelist. Others who called themselves community planners grappled with the mundane details of locating and designing roads, buildings, parks, and utilities. A respected contemporary, John Nolen, for example, in his 1927 book *New Towns for Old* presented a series of case studies of recent American community and neighborhood plans.[55] MacKaye, for his part, attempted "to *focus* the people's vision" on an alternative to the forces that were transforming the physical and social landscape of America during the 1920s.[56]

The key metaphor at the heart of MacKaye's *New Exploration* remained his provocative inversion of natural terms and analogies. To dramatize what he saw as the unnatural process of metropolitanization, he beckoned the reader to follow him to the top of the Times Building in Manhattan to observe the city and its surroundings. The prospect from that height, as portrayed through his eyes, was grim and bewildering. He described an industrial "wilderness," strangling in "iron tentacles." "Mankind has cleared the jungle and replaced it by a labyrinth." From his exalted vantage point at the hub of the city, he identified the "streams" and "flows" at work upon the metropolitan landscape. The commuters who rush, morning and evening, in and out of the city by train, subway, and automobile, constituted a "stream of traffic," of "folks." The waterfront, the Hudson River, the railroad lines and freightyards were evidence of the "stream of things," or commodities, that made New York the "mouth" of the continent's interior regions and the "mutinous whirlpool" of international trade.[57]

But upon moving to another observation point, at an environmental extreme from New York City, southern New Hampshire's Mount Monadnock, MacKaye espied a path out of the "wilderness of civilization." From the summit of Monadnock, "the Fujiyama of New England," "the emblem of a unified homeland," "the lofty pivot of an indigenous region and culture," MacKaye visualized through "the eye of the imagination" the metropolitan and indigenous worlds he had observed from the Times Building—"but from reverse position."[58]

If the metropolitan world of New York represented the "conflux or midway" of human and material streams, the "indigenous world" was at once the source of the stream of natural resources and the destination of the metropolitan "invasion." The indigenous world, according to MacKaye, was composed of material resources, like forests and minerals; energy resources, like water power and coal; and "psychologic resources," or "happiness," inherent in the unsullied natural environment. The dynamic interrelationship of these resources, MacKaye asserts, comprised

> the three needs of civilized man, the visualisation of which constitutes the new exploration. They are:
> The conservation of physical natural resources.
> The control of commodity flow.
> The development of environment, or *psychologic* natural resources.[59]

Adapting the sequence of American "migrations" he and Mumford had described in their articles for the 1925 *Survey Graphic* issue, MacKaye described the contemporary stage, the "metropolitan invasion," as a "backflow" of population along the routes of the railways and roadways. This "contact of the indigenous with the metropolitan world," he asserted, "forms the basic problem of regional planning, and one of the big problems of modern civilization."[60]

In the climactic central chapters of his book—"Living vs. Existence," "Environment as a Natural Resource," "Planning and Revelation"—MacKaye's hortatory rhetoric reaches its zenith. As in his 1921 Appalachian Trail article, he recognized that industrial society had not come to grips with the proper, and essential, role of play and leisure. Invoking Thoreau, he identified "the essence of the meaning of *living*: to maintain ourselves on this earth not as a toil and hardship but as a sport and quest." Thus, regional planning was no mere profession; it was a philosophy of living rooted in the doctrine of evolution. "*Living is man's part in evolution,*" MacKaye proclaimed. As an active agent in the process of evolution, humans may pursue objectives beyond "the routine of a merely busy existence. . . . so that finally work and art and recreation and living will all be one."[61]

In describing "environment as a natural resource," "a sort of *common mind*," "the least common denominator of our inner selves," MacKaye departed even further from any conventional notion of regional planning as a merely technical skill or process. Rather, he set down the "indigenous environment," especially in its "primeval" form, as the base line from which humanity should measure its impact on all its physical surroundings. The plea for wilderness preservation at the heart of MacKaye's book was based on his notion that the primeval "is the known quantity from which we came. . . . The less an environment is affected by human hands, the greater the range of human minds it unites. . . . The primeval . . . seems to possess an innate harmony, and this must be why it appeals to us as an innate environment." By this reasoning, he declared, echoing Thoreau, the disruption and desecration of the primeval wilderness represented an assault on the human community as well. And he lauded the "bigness" of Aldo Leopold's idea for the protection of wilderness areas, which MacKaye envisioned as settings for developing "the full *art* of 'living in the open.'"[62]

The chapters "Regional City vs. Metropolis," "Controlling the Metropolitan Invasion," and "Developing the Indigenous Environment" provided concrete case studies and techniques gleaned from observation and experience in his own regional environs. A series of amoeba-like diagrams illustrating the "control of metropolitan flow" in "fictitious" and "actual" localities suggested an almost organic evolution of Ebenezer Howard's more rigid garden city plans.[63]

Likewise, his maps of Massachusetts open spaces and of an Appalachian "Backbone Openway" vividly displayed the book's key practical insight. "The motor ways form the channels of the metropolitan flood, while the open ways (crossing and flanking the motor roads) form 'dams' and 'levees' for controlling the flood," he wrote. "The open way across the motor way would form a barrier not to traffic but to metropolitan development: it would be a 'dam' across the path of the metropolitan flood." MacKaye urged the deployment of trails and open spaces "to hold in check the flow and expansion of a one-sided *civilization,* and advance the growth of an all-sided *culture.*"[64] Some of his contemporary readers may have been unpersuaded by MacKaye's high-flying rhetoric, but his maps depicted concretely how such "dams" and "levees" could be developed on statewide and regional scales. Several generations later, similarly portrayed greenways and urban growth boundaries had become standard planning tools for controlling sprawl.

❖

When *The New Exploration* was published in October of 1928, Clarence Stein and his wife, actress Aline MacMahon, threw a publication party for MacKaye, complete with a square dance, at their spacious New York apartment.

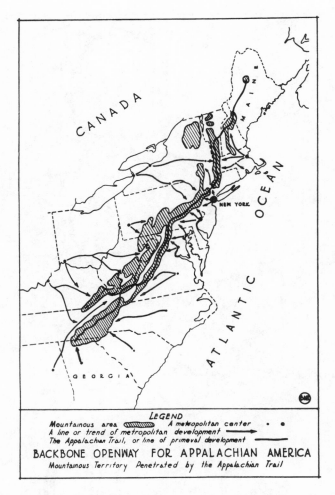

A "Backbone Openway for Appalachian America" depicted the "primeval" environment surrounding the Appalachian Trail as a barrier to the forces of metropolitan development. From *The New Exploration* (1928).

The book's sales were modest, totaling perhaps less than 700 copies of its original edition.[65] But it was reviewed in a wide variety of publications, with a generally favorable response, although one reviewer commented on the book's "occasionally hysterical note" and another "found it exceedingly sad" because of "the comparative hopelessness" of MacKaye's plans to control metropolitan growth.[66] Many of the reviews were written by his friends and colleagues, including, among others, RPAA members Stuart Chase and Alexander Bing,

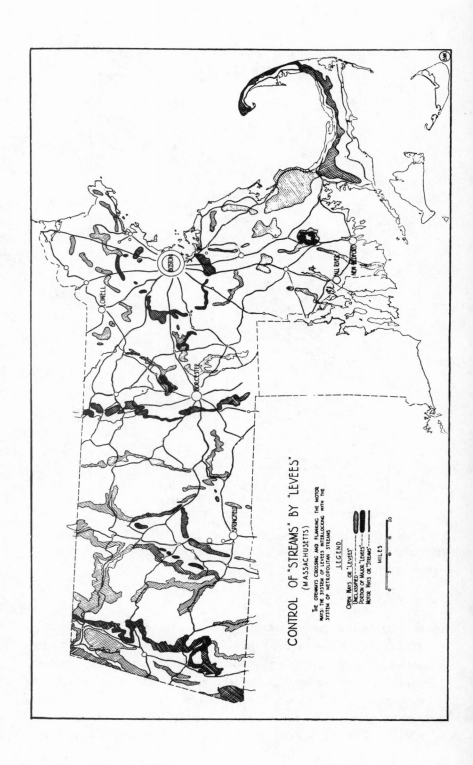

CONTROL OF "STREAMS" BY "LEVEES"
(MASSACHUSETTS)

THE DIAGRAM CROSSING AND PLANNING THE MOTOR
WAYS: THE SYSTEM OF LEVEES INTERLOCKING WITH THE
SYSTEM OF METROPOLITAN STREAMS

LEGEND

OPEN WAYS OR "LEVEES"
UNCLASSIFIED
PORTION OF MAJOR "LEVEES"
MOTOR WAYS OR "STREAMS"

MILES

Hell-Raiser journalists Laurence Todd and Charles Ervin, and New England planning and conservation acquaintances Allen Chamberlain, Arthur Comey, and Austin Hawes.

Such reviewers were, of course, partial to both the book's ideas and the man who conceived them. But MacKaye took particular satisfaction in a review that appeared in the *Journal of Forestry*. Raphael Zon, calling the book "one of the most notable contributions to the philosophy of conservation that has appeared within recent years," traced the source of MacKaye's approach to regional planning to his training as a forester. "Out of the germ of *forest working plans*," Zon wrote, "there emerged a bigger concept of social working plans." MacKaye had in essence adapted the forester's philosophy of sustained-yield timber production on "a comparatively small forest property . . . to entire communities, to entire regions." But the goal MacKaye sought was not just timber, or water power, or recreation. Instead, Zon continued, MacKaye's plans sought to nurture "human happiness and cultural development." He had depicted "the humanization of the philosophy of conservation as applied to community life."[67]

Privately, MacKaye disclosed to Zon that another fellow forester, Austin Hawes, had concluded "that my whole thesis comes down to birth control." In fact, in his book, MacKaye had gingerly discussed population control, a vexing issue that continued to unsettle economists, ecologists, planners, sociologists, and politicians generations after he wrote that "the subject of the birth rate lies as the bottom of every social question known to man." To Zon he confessed, "I must dodge this [subject] or get entangled in a web of sex and syrenic discussion which would lead us far afield."[68]

A "lonely classic," as Lewis Mumford later called it, *The New Exploration* represented the most substantial published work of Benton MacKaye's entire career.[69] While the idiosyncratic book perhaps left some workaday planners and conservationists bemused, appalled, or simply scratching their heads, it opened up provocative new paths of thought that others would continue to follow years later. In 1938, Mumford published his own magisterial regionalist

FACING PAGE

In his 1928 book, *The New Exploration: A Philosophy of Regional Planning*, MacKaye portrayed a series of undeveloped "open ways" or "levees," generally following rivers and mountain ranges, which would dam the "streams" of metropolitan flow originating from Massachusetts cities.

treatise, *The Culture of Cities;* but he never failed to acknowledge that among the RPAA circle it was MacKaye, a decade earlier, "not Whitaker or Stein or myself, who first neatly identified our common purpose" in *The New Exploration.*[70]

In less than a decade, MacKaye had written two significant and original works about the uses of and prospects for the American environment. Much had changed over that period, in his own thinking and on the nation's natural and cultural landscape. In *Employment and Natural Resources,* under the auspices of a federal agency, MacKaye had outlined an ambitious program for the development of planned agricultural and forest products communities on the public domain. Already he was familiar with the most innovative initiatives in the developing fields of community and regional planning. But his 1919 Labor Department report had nothing to say about outdoor recreation, hiking trails, wilderness protection, or the deleterious effects of metropolitan growth.

In *The New Exploration,* by contrast, he was speaking in his own oracular and distinctive voice. The 1928 book, reflecting the success of his conception of the Appalachian Trail, emphasized outdoor recreation on protected lands as a social priority. But the book's perspective also encompassed the entire American landscape, public and private, and envisioned the balanced regional development of primeval, communal, and urban environments. *The New Exploration*'s techniques for controlling the "metropolitan invasion" were responses to the fast-moving technological and social developments of the 1920s, such as the explosive popularity of the automobile, which could not have been accurately foreseen during the previous decade. As a freelance intellectual and activist, trying to fuse his own ideas with those of planners and regionalists such as Stein and Mumford and pioneering conservationists like Leopold, MacKaye had continued his own hopeful exploration of an authentic "indigenous America."[71]

Steele MacKaye, Benton's father, whose resemblance to Edgar Allan Poe was sometimes noted. The playwright, actor, and director "had all the charm of those peculiar men who give no thought to commercialism," commented the inventor Thomas Edison.

Benton, age ten, in 1889.

Sturgis Pray, Benton, and Draper Maury, August 14, 1897, setting forth into the White Mountains from Albany Intervale, New Hampshire. On this trip Benton "first saw the true wilderness."

Benton on a Vermont road, July 1900, during a mountain tramp with his friend Horace Hildreth from the Massachusetts border north to Morrisville.

Benton in his room at 35 Divinity Hall, Harvard College, on his graduation day in June 1900. The room had been occupied consecutively since 1891 by three MacKaye brothers—James, Percy, and Benton.

Benton on assignment with the U.S. Forest Service in the mountains of south-eastern Kentucky, summer 1908.

Benton married Jessie Hardy Stubbs ("Betty") on June 1, 1915, in Chevy Chase, Maryland. Betty (fourth from right) is flanked to her left by Benton, his mother Mary, and his sister Hazel. (The other six members of the wedding party are unidentified.)

"Of course I want to be a good home maker," Betty told a reporter before her wedding to Benton. "The new idea of home making is that women should go out with her home making influence and drive the evil out of the world—in other words, to make the world a better home for the human race." At the time, she was a leader of the Congressional Union, campaiging for a federal amendment ensuring woman suffrage.

The MacKayes in Shirley Center, in front of the Cottage, July 1916. *Left to right: on ground,* Percy, Harold; *in chairs,* James, Hazel, Mary, Benton; *standing,* Marion (Percy's wife), Betty (Jessie Stubbs MacKaye).

Lewis Mumford (*left*) **and MacKaye,** in May 1923, at the Hudson Guild Farm, Netcong, New Jersey. Also present at this early gathering of the Regional Planning Association of American was the Scottish planner Patrick Geddes. (Courtesy Division of Rare and Manuscript Collections, Cornell University Library.)

MacKaye and Myron Avery, probably October 12, 1931, at Bake Oven Knob, Pennsylvania, celebrating the tenth anniversary of the publication of MacKaye's article proposing the Appalachian Trail. Avery had been elected chairman of the Appalachian Trail Conference in June 1931. This photo captures the moment, at least symbolically, when leadership of the Appalachian Trail effort effectively passed from MacKaye to Avery. (Courtesy Appalachian Trail Conference archives.)

MacKaye and Harold C. Anderson along Virginia's Blue Ridge, c. 1931. Anderson joined MacKaye in vigorously opposing "skyline" (i.e., ridgeline) highways on the eastern mountains and in founding the Wilderness Society.

MacKaye led some of the Philosophers Club on a geology expedition to Ewing, Virginia, July 21, 1935. *Left to right:* Olcott Deming, Mable Abercrombie, Fred Beckman, Jim Moorhead, Margaret Broome Howes, Helen Northup, Benton MacKaye, Virginia Moorhead, Allen Twichell, Bernard Frank, John Bamberg, Greta Biddle.

MacKaye on Eagle Rock, Cumberland Mountains, near Jacksboro, Tennessee, September 1935.

MacKaye (*with knapsack*) and (*left to right*) **Hollis Hill, Jim Moorhead, and Virginia Moorhead,** on Brushy Mountain, Great Smoky Mountains, October 1935. (Photo by Harvey Broome. Courtesy The Wilderness Society.)

Left to right: **Bernard Frank, Harvey Broome, Robert Marshall, MacKaye**—four of the eight co-founders of the Wilderness Society—at the Jumpoff, on Mount Chapman, Great Smoky Mountains, January 26, 1936. (Photo by Mable Abercrombie Mansfield.)

The first meeting of "The Woodticks," Halfway House, Mount Monadnock, New Hampshire, May 23, 1937. The informal group of New England foresters and conservationists met almost annually at the mountain until the early 1950s. *Left to right:* Harris A. Reynolds, Karl Woodward, MacKaye, Mrs. Woodward, Allen Chamberlain, Charles Porter, Elmer Fletcher, Lawrence Rathbun, Lee Russell.

MacKaye at Clarence Stein's New York City apartment at 1 West 64th Street, over-looking Central Park, June 18, 1937.

MacKaye on Hunting Hill, near Shirley Center, June 27, 1937. He holds the original journal of his 1893 "expeditions" in the surrounding region.

A sleighing party at Shirley Center, January 16, 1938. *Left to right:* architect and planner Sir Raymond Unwin, Polly Greeley, MacKaye, Roland Greeley, Mrs. William Roger Greeley (*partly hidden*), Lady (Ethel) Unwin, Howard Bridgman, William Roger Greeley, Michael Mayer.

The Wilderness Society Council at its June 1946 meeting at Old Rag Mountain, Shenandoah National Park, Virginia, at which MacKaye presided. *Left to right:* Harvey Broome, MacKaye (president), Aldo Leopold (vice-president), Olaus Murie (director), Irving M. Clark, George Marshall, Laurette S. Collier (assistant secretary), Howard Zahniser (executive secretary), Ernest C. Oberholtzer, Harold C. Anderson, Charles G. Woodbury, Robert F. Griggs, Ernest S. Griffith (treasurer), Bernard Frank. (Photo by Howard Zahniser. Courtesy The Wilderness Society.)

The MacKaye Cottage's "Sky Parlor," October 1958, includes a mounted bittern, a portrait of MacKaye's uncle James Medbery, and MacKaye's map case.

Clarence Stein, age eighty-two, and MacKaye, age eighty-five, at the Cosmos Club, Washington, D.C., June 3, 1964.

MacKaye in Shirley, July 1973, at age ninety-four. (Photo by Norman Miller. Courtesy The Wilderness Society.)

Trailwork and the
"Townless Highway"

1928–1931

T
he *New Exploration* represented a major intellectual milestone in MacKaye's career. But he still had to survive, as Mumford later observed, "by tightening his belt and accepting what the ravens brought him."[1] His friends and colleagues remained on the alert for paying opportunities that matched his unique talents as a self-certified philosopher of regional planning. Immediately upon finishing his book, for instance, MacKaye collaborated with Mumford on an article about regional planning for the *Encyclopedia Britannica*. Their essay succinctly linked English and European regionalist and garden city traditions with their own distinctive variant of the American regional planning movement. The discipline, they concluded, "deals with the ecology of the human community"; it was not to be confused with "metropolitan planning," which "describes plans for city extension over wide metropolitan areas."[2] Benton now worked to influence the ecology of the human community along two complementary "geotechnic" routes: the Appalachian Trail, which was evolving under new leadership, and the "Townless Highway," his vision of a national program for controlling the automobile's growing impact on almost every aspect of American life. There was, of course, an inherent tension between salvaging the wilderness environment of the Appalachian region and developing a national highway system. During the late 1920s and early 1930s, MacKaye strove—and sometimes struggled—to articulate methods for successfully reconciling such competing social, political, and technological trends.

In September 1928, a month before publication of *The New Exploration*, MacKaye began a short stint working in the Hartford office of Connecticut state forester Austin F. Hawes. Hired at the suggestion of Laurence Fletcher, of the Trustees of Public Reservations, he was asked to develop a campaign to educate young people about forest fire prevention.[3] During his eight-month sojourn in his native state, he traveled Connecticut's growing network of public forests and parks "and thereby visited many separate little *regions* of God's own separate making," as he reported to Mumford.[4] His responsibilities kept him in close working contact with his friend Albert Turner, field secretary of the Connecticut State Park and Forest Commission, who had been an active and influential supporter of the Appalachian Trail since its inception. During the autumn of 1928, MacKaye devoted renewed attention to the trail project, then at an important juncture in its brief history. He was a frequent visitor at the Hartford home of the new chairman of the Appalachian Trail Conference, Arthur Perkins. The two men talked about the nuts-and-bolts of the trail project as well as the philosophy behind it.[5]

MacKaye and Perkins had met at the New England Trail Conference in Boston in January 1927. Perkins, a retired judge, was involved in regional trail activities as an active member of the Appalachian Mountain Club. Already filling a vacancy on the Appalachian Trail Conference's executive board, he was eager to devote more of his time and talents to the trail project. By MacKaye's recollection of events, ATC chairman William Welch was equally eager to relinquish his position. In discussions at the meeting, it was agreed that Perkins would take over the job. As MacKaye recalled thirty years later, the 1927 NETC was a "pivotal milestone of the A.T.'s first decade."[6]

Since its creation in 1925, the Appalachian Trail Conference, under the chairmanship of Major Welch and with MacKaye as nominal "field organizer," had provided an organizational skeleton for the trail effort, but the actual work of locating, clearing, and protecting the trail had proceeded slowly. MacKaye had not pursued his undefined task as field organizer in any systematic fashion, partly because sufficient funds had never materialized to support his efforts. In its first years, the ATC had no significant revenue from dues-paying members, its affiliated clubs, or government agencies.

One futile attempt to secure funds for MacKaye's trail work involved Franklin D. Roosevelt, then chairman of New York's Taconic State Park Commission. During 1925 and 1926, MacKaye, Torrey, and Welch campaigned to win Roosevelt's support and the commission's funding for a survey of a trail route east of the Hudson River, up through the Taconic Range to the Massachusetts border. The future president, during meetings with MacKaye in 1925 and in later correspondence, expressed some interest in the project; and MacKaye in January 1926 provided Roosevelt with a detailed memorandum, budget, and map describing the proposed survey. But when MacKaye called Roosevelt's office several months later, a secretary informed him that the project was "off for this year."[7] A decade later, during Roosevelt's presidency, his support for federally constructed scenic mountain parkways would have a profound influence on the prospects for the Appalachian Trail and other American wild-lands.

Some trail work had proceeded since 1925, especially in New England and elsewhere in the Northeast. The indefatigable Raymond Torrey and the New York–New Jersey Trail Conference had pushed ahead with publicity and trail construction in those two states. In Vermont and New Hampshire, the route was substantially in place along existing trails. Progress was slower in Connecticut, Massachusetts, and Maine, although local enthusiasts and Appalachian Mountain Club activists were actively scouting routes. Beginning in 1927 Perkins provided the coordination and impetus the project required to move forward. He recruited trail enthusiasts in the states south of New York and welcomed more clubs, some of them already established, some newly created, into the ATC. The Blue Mountain Club, founded in 1926, and the older Blue Mountain Eagle Climbing Club had begun to link together a route across Pennsylvania. The southern sections of the trail still awaited an organizational framework. The formation of the Potomac Appalachian Trail Club (PATC) in Washington, D.C., in 1927 proved to be another important milestone in the trail's history. The club and its leaders would soon become the guiding forces behind the Appalachian Trail project.[8]

Perkins organized a second Washington meeting of the ATC in May 1928, hosted by the PATC. MacKaye, in Shirley putting the final touches on *The New Exploration* and his Massachusetts open-space report, could not attend. Perkins estimated that 500 miles of the trail route could be traveled on foot, though not all of it was marked. He also set in motion the process of amending the ATC constitution MacKaye had drafted in 1925. The conference's new organizational structure, formally adopted in 1929, created a smaller executive framework reporting to a board of managers. Perkins was formally elected ATC chairman and trail supervisor. But the ATC provided no position or office for MacKaye, the father of the Appalachian Trail.[9] Eking out a living as a free-lance writer and planner, preoccupied with his broadscale regional-planning notions, he had invested little energy in the politics of the ATC or any of its constituent trail clubs.

As a lively address to the 1927 NETC demonstrated, MacKaye had nonetheless assumed the role of Appalachian Trail philosopher and conscience; but the audacious rhetoric of his talk, "Outdoor Culture, The Philosophy of Through Trails," may have left some of his audience wondering just what his wide-ranging allusions and colorful imagery had to do with their down-to-earth outdoor concerns. On an eclectic tour of the cultural horizon, he quoted Lewis Mumford on the differences between "utopias of reconstruction"

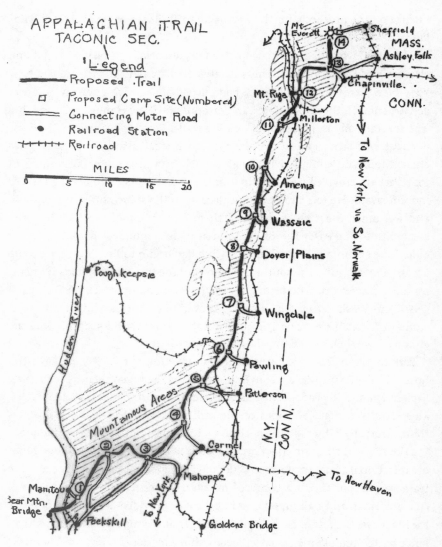

In late 1925 and early 1926, MacKaye met with and wrote to Franklin Delano Roosevelt, then chair of the Taconic State Park Commission, in an unsuccessful attempt to secure backing for a survey of an Appalachian Trail route from the Hudson River to the Connecticut border. The eventual route of the trail verged into Connecticut near Pawling, New York, rather than continuing northward through New York, as MacKaye had proposed.

and "utopias of escape"; and he cited Aldo Leopold on "wilderness areas." MacKaye invoked the celluloid heroism of Douglas Fairbanks, the visionary "pipe dreams" of explorer Magellan and other scientists and inventors, Karel Čapek's grimly futuristic play *R.U.R.*, and the Wapack Trail experiences of the Bridgman School boys.[10]

The crisis of American life, he told the NETC crowd, could be compared to the crisis of Roman civilization, when the Barbarians imposed a "cleansing invasion from the hinterland." MacKaye addressed his audience of peaceful hikers and trailworkers as the Barbarians' "modern counterpart," at the margin of mainstream American culture.

> And now I come straight to the point of the philosophy of through trails. *It is to organize a Barbarian invasion.* It is a counter movement to the Metropolitan invasion. Who are these modern Barbarians? Why, we are—the members of the New England Trail Conference. . . . The crestline should be captured—and no time lost about it.
>
> The Appalachian Range should be placed in public hands and become the site for a Barbarian Utopia.[11]

Though couched in obscure metaphorical language, MacKaye's urgent call to action implied a practical political mission: securing public ownership of the wilderness terrain traversed by the Appalachian Trail. But neither he nor the trail community had yet discovered the political means or influence to achieve that end.

Such inspirational messages to trail club gatherings represented but one element of MacKaye's seamless vocational identity. He carried on his usually uncompensated Appalachian Trail activity simultaneously with his work on regional planning—a field which at least held some promise of providing an income. As he would report to his Harvard classmates in 1930: "Since 1925 I have been working at my chosen job which is also my hobby which is regional planning."[12]

Under the relaxed supervision of Austin Hawes in the state forestry office, MacKaye had considerable freedom to spread his gospel of planning and play to diverse audiences. In January 1929, he accompanied Perkins to Rutland, Vermont, to address the Green Mountain Club with a talk titled "Why the Appalachian Trail?"—the question the new ATC chairman had been asking him during their conversations that winter.[13] In February, at a New England forestry conference in Hartford, he expanded the scope of his Massachusetts "Wilderness Ways" proposal to encompass a network of linear wilderness areas and open-space belts for the entire region. "As shade trees are to the city, so these wilderness ways would be to mechanical civilization generally," he told his fellow foresters. "Forest and city in New England would grow side by side." In the struggle over control of the American environment, MacKaye saw "transportationalists," who were infatuated "with lateral motion and mechanical transport for its own sake," pitted against "evolutionists," gripped with "the inevitable and irresistible impulse of humanity to right the balance of its civilization."[14]

At the International House in New York, he gave a slide presentation on "Regional Planning versus Imperialism: Has Science a Solution for World Conflict?" In March, he spoke once again to the New England Trail Conference, then to the Massachusetts Federation of Women's Clubs. In May, as he completed his Connecticut duties, he traveled to Easton, Pennsylvania, for the third meeting of the Appalachian Trail Conference, where he talked about "The Origin and Conception of the A.T." There he also met firsthand a number of trail and wilderness enthusiasts, such as Harold C. Anderson of the PATC, who would become one of his faithful allies and followers in battles yet to be fought.[15]

"My 50th birthday," Benton tersely noted in his diary on March 6, 1929. Only a few weeks earlier, he had attended funeral services for Sturgis Pray, the friend who had introduced him to the rigorous pleasures of outdoor life more than thirty years before. During the previous year he had lost his oldest brother Hal, the "Gibraltar" among the MacKaye siblings.[16] Benton had also spent a week in the hospital for his chronic digestive troubles, borne the decision and responsibility of institutionalizing his mentally unstable sister, and skirmished with his brother Percy about the use and upkeep of the family's Shirley property. However, when he returned from Connecticut to Shirley Center in late spring of 1929, he set aside his troubles to pursue yet another approach to controlling the "metropolitan invasion."

Benton liked to recall that in 1892 he had heard his "good old 'Cousin Henry MacKaye'" predict that each family would one day have its own "private locomotive."[17] MacKaye had come of age as the automobile took over the nation's roads and streets. During his college days, Benton had occasionally traveled by bicycle or horseback the thirty-odd miles between Shirley Center and Cambridge. When he graduated from Harvard in 1900, there were eight thousand cars on the nation's roads. In 1930, as he introduced his conception of the "townless highway," America's burgeoning road system was traveled by about twenty-seven million cars, trucks, and buses.[18]

During the 1920s, the consequences and control of automobile traffic provided a fundamental challenge for regional and city planners of all methodologies and ideological persuasions.[19] MacKaye had long recognized the essential importance of highway design and location in the process of sound, balanced regional development. In a draft of his 1921 article proposing the Appalachian Trail, for instance, he had suggested the construction of a parallel, valley-level motor road the length of the Appalachians.[20] Indeed, the automobile assumed a relatively benign place in the hopeful regionalism espoused by MacKaye and his RPAA colleagues in the early 1920s. The motorcar, in their vision, was one of the technologies that could promote the development of a carefully designed matrix of decentralized planned communities and balanced, well-defined environments.

But in the latter half of the decade, MacKaye began describing the automobile's influence in sharper, more skeptical terms. In *The New Exploration,* he vividly depicted the automobile as a principal vehicle of the insidious metropolitan invasion, which inexorably followed the roads leading out of the city. "Pygmies have become centaurs," he concluded. "The weakling man, seated in his motor car with hand on wheel and foot on lever, becomes a locomotive running forty miles an hour."[21] Belts of open space and state zoning laws to regulate roadside development were among his specific proposals to control the metropolitan flood and preserve what he envisioned as a government-regulated "intertown" environment.[22] During 1929, he began to fashion a detailed vision of a national highway program, designed both to protect the landscape and to enhance society's cultural development.

Henry Seidel Canby, editor of the *Saturday Review of Literature,* wondering about the implications of the Hoover presidency, had asked Mumford early that year to recommend someone to write about the "cultural possibilities of engineering." The "ideal man," Canby wrote, "would be a first-rate engineer with a breadth of view and understanding of consequences and implications not to be found in any engineers I know."[23] Mumford knew just whom to suggest. MacKaye's article, "Our Iron Civilization," appeared just days after the October stock market crash. Hence his grim and well-rehearsed depiction of the "metropolitan invasion" was imbued with a certain prophetic resonance. But he was also ready to offer America's "chief engineer" a specific policy to respond to the nation's challenges. "Mr. Hoover's power to dyke the flow of metropolitanism throughout the United States lies in the Federal control over government-aided public roads, and the influence, by example, upon state-aided highways," he wrote. "The public motor road is the channel of metropolitan flow, and the regulation of its right-of-way means the regulation of the flow itself."[24]

Other influential conservationists from earlier Republican administrations, such as Gifford Pinchot and Harry Slattery, were corresponding with MacKaye about their efforts to influence Hoover's natural resource policies. From tranquil Shirley Center, heedless of the dramatic events on Wall Street, MacKaye plotted the promotion of his ambitious highway schemes. Indeed, as he wrote Clarence Stein, it had "been suggested that I meet Hoover on this highway plan and I have some leads to him, but I do not want to do this until the right time and as part of some definite policy on all our parts. The headquarters of any campaign should be Washington. A bill should be put in Congress. I have not drawn a bill but could do so on learning the best Congressional approach."[25]

MacKaye perhaps exaggerated his prospects for influencing the president, but he had found a concise, suggestive name for his evolving highway concept: a "townless highway." This idea, he wrote Stein and Mumford in a letter accompanying a memorandum on the subject, "is the complement in a sense of 'An Appalachian Trail.' One follows the primeval crestline or main 'dam' across the metropolitan 'stream'; the other follows the 'stream' itself. . . . There is only one place upon the map for an Appalachian Trail but there are several places on the map for an experimental Townless Highway."[26]

The concept had received an enthusiastic reception at an informal presentation before some Massachusetts planning colleagues, including Greeley, Hartman, Reynolds, and William Leavitt Stoddard. Already he and Reynolds were talking about a route from Boston to Chicago.[27] MacKaye also suggested a route "enclosing the 'superpower zone'" between Boston and Washington. "The point is to make use of the leverage of the Federal-aided public roads to control future policy in laying out State highways," he wrote his friends, "such policy being to avoid the towns and apply the Radburn principle of *cul-de-sac* throughout the countryside."[28]

Mumford expressed concern about "the danger of forming bad little messes of 'towns' on the townless highway itself," but he responded enthusiastically to his friend's latest mental invention. While MacKaye was in New York City in January 1930, Mumford introduced him to *New Republic* editor Bruce Bliven. His article "The Townless Highway" appeared in the magazine in March. "The motor road is a new kind of road, as different from the old-fashioned highway as the railroad was," he wrote; "and it demands, accordingly, a new type of plan." In fact, he continued, "the motor road is a new kind of rail-road." Like the railroad, though, the modern motor road had another impor-tant function and effect besides transportation alone, he contended, namely "the relocation of the population on the map of the United States." This pow-erful new migratory dynamic, moreover, demanded "new communal forms that correspond with its new functions."[29]

MacKaye's townless highway proposal had "four specific objectives." The first sought abolition of "the motor slum, or road town," and the development of "the rural wayside environment." He listed several means to control road-side development: limited roadway access points, zoning, purchase of roadside frontage (in advance when possible, by eminent domain when necessary), landscaping and the planting of shade trees, "strict regulation" of utility lines, and equally "strict control" over the design, location, and management of res-taurants, gas stations, and other roadside services.

A second objective was "the growth of the distinct community, compactly planned and limited in size." The "old New England village" exemplified one such communal form. Radburn, New Jersey, the new "highwayless town" de-signed by Clarence Stein and Henry Wright, represented another. At Radburn, the architects had segregated pedestrians from motor traffic by the careful lay-out and design of footpaths, automobile side lanes, and underpasses. What-ever form they took, these planned communities would be connected to the highway by "side-lane approaches" reserved solely for local traffic.

MacKaye's third major goal, the relief of traffic congestion in larger towns and cities, would be accomplished by building circumferential bypasses. He envisioned new "cul-de-sac towns" such as Radburn, isolated from through traffic with bypasses, as elements of a large-scale program incorporating high-ways and open spaces. "Regional planning with these ends in view will in turn save both the local community and the open wayside environment, and give proper access to the wild places, instead of insidiously wiping out all these pre-cious assets together."[30]

Finally, to increase the safety of motor travel, MacKaye called for the ab-olition of railroad and motor road grade crossings; "safety traps" at highway approaches; restriction of the highway, if possible, to passenger traffic alone (and perhaps the creation of "express highways" for truck traffic); and "dou-ble-tracking" of the roadway into separated one-way roads.

MacKaye envisioned amending federal highway legislation to make road building appropriations to states contingent upon satisfaction of uniform na-tional townless highway standards. A straightforward proposal for a federally funded, centrally planned national highway system, his townless highway scheme would "take the lead in guiding our people, in accordance with some definite policy, into appropriate communities and settings for furthering the cultural growth, and not merely the industrial expansion, of American civilization."[31]

MacKaye spent much of 1930 and 1931 promoting the townless highway program and principles, often using specific geographic examples from his home state. In the pages of the *Boston Globe* he mapped out a two-road parkway—a "super by-pass"—encircling Boston. This "strategic switchboard of New England" would be a circumferential two-mile-wide parkway, about fifteen miles from Boston, with separate northbound and southbound roads flanking a 125,000-acre "belt of public playground"—itself bisected by an "earthy footpath"—from Plum Island in the north to Duxbury Beach in the south. MacKaye predicted that the two roads could be routed to bypass the hearts of sixty-five towns and villages along its route.[32]

Writing in *American City,* he returned to the example of the Mohawk Trail, urging that New York and Massachusetts cooperate in funding the purchase of the roadside frontage along a scenic new section of the road just built over Petersburg Mountain, between Williamstown, Massachusetts, and Troy, New York.[33] In the spring of 1931, MacKaye was recruited by his friend Laurence Fletcher to apply his ideas on wayside culture to the region surrounding Boston. In "Highway Approaches to Boston: A Wayside Situation and What to Do About It," his report for the Trustees of Public Reservations, MacKaye proposed implementing a policy of advance purchase of roadside frontage and road relocation in three specific, heavily traveled Massachusetts locales surrounding Boston.[34] In a *Survey Graphic* article titled "Cement Railroads," he depicted how some of New England's abandoned railroad rights-of-way could be transformed into a regional road network based on townless highway principles.[35]

During the early 1930s, MacKaye prepared a series of planning reports for private conservation organizations in northwestern Connecticut. In those years, some concerned residents and summerfolk from the region's handsome colonial hill towns began to fear that proposed new state highways, and the development they might generate, posed a threat to the rural landscape they cherished. For groups from the towns of Washington and Litchfield, as well as for the Shepaug River Protective Association, MacKaye worked up detailed plans and maps, depicting small-scale applications of his complementary conceptions of wilderness ways and townless highways. His Connecticut re-

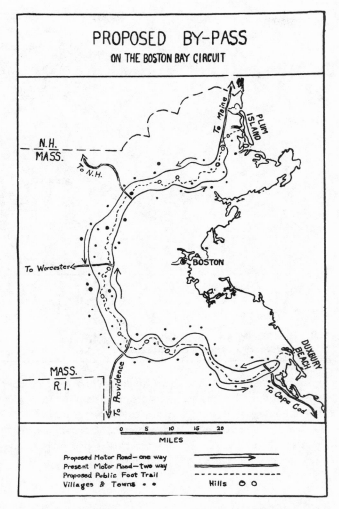

MacKaye's "Super By-Pass for Boston," an article in the *Boston Globe,* October 31, 1930, applied his idea of the "townless highway" to an existing proposal for a Bay Circuit parkway to surround Boston. He called for separate northbound and southbound roads, flanking a 125,000-acre belt of parkland that would encompass a 100-mile continuous trail between Plum Island and Duxbury Beach.

ports were notable as concrete applications of his planning ideas on a local scale. Like the paper exercises of so many other planners, though, his did not solve the essentially political problem of reconciling the different, and sometimes competing, economic and social interests within the communities whose environments he hoped to shape.[36]

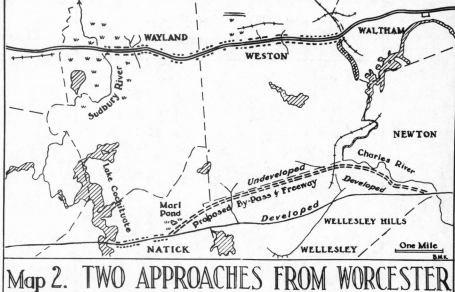

Map 2. TWO APPROACHES FROM WORCESTER

Federal Route 20 - Sudbury River to Waltham,
with 25% of frontage occupied.
Proposed Wellesley By-Pass - Natick to Charles River,
provides another opportunity to keep the wayside clear.

MacKaye's 1931 pamphlet "Highway Approaches to Boston" was published by the Trustees of Public Reservations. The Massachusetts organization promoted "advance purchase" of roadside frontage along newly constructed freeways, to limit what MacKaye called the "wayside fungus" of commercial development. MacKaye mapped several alternative highways to Boston, such as "Two Approaches from Worcester," designed to limit or prevent roadside development.

MacKaye nevertheless persisted in his effort to promote the townless highway on a national scale. He worked up several outlines and draft of a book on the townless highway, but he had no luck peddling the idea to publishers. He also drafted an amendment to the Federal State Aid Highway Act (1916) that included and mandated townless highway standards. Working through some of the old Washington Hell Raisers, he tried to secure the support of Representative Robert Crosser, the vigorous sponsor of several of MacKaye's ill-fated bills calling for a national program of colonization on public lands. The townless highway, MacKaye now wrote to the old congressional crusader, treats "the same old questions under a new name—regional planning—to meet the changes brought about by a new motor power—gasoline."[37] But

Crosser, perhaps recalling his previous lack of success winning votes for Mac-Kaye's ideas, was skeptical. MacKaye's journalist friend Fred Kerby commented of Crosser, "He always wants to be quite sure that he is not sponsoring anything that will in any way conflict with his religion—which of course is Henry Georgeism! I have a private hunch that Crosser is pretty much stumped by this proposal. He is unable to find out 'whose economic interest would be served' by it."[38]

MacKaye's townless highway received its broadest exposure in an August 1931 issue of *Harper's Monthly Magazine.* Coauthored by Mumford, who helped cast his friend's prose into a more popular style, the article emphasized the automobile's potential impact on the prospects for a balanced, healthy community life. "It is only by a deliberate separation of local and through roads, of traffic and residential functions," they asserted, "that the motor road itself can attain its maximum efficiency in the number of vehicles served

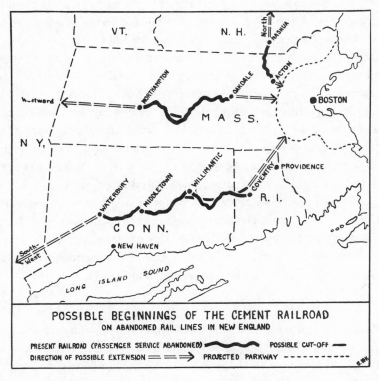

MacKaye, in his November 1932 *Survey* article "Cement Railroads," proposed incorporating some of New England's abandoned railroad rights-of-way into a regional road network based on "townless highway" principles.

at the highest safe speed, and that the community can attain its maximum efficiency as a place for living, recreation, sleep, and the care of the young."[39]

For all his efforts, MacKaye's townless highway received no more than passing attention at the time from policymakers and roadbuilders. Many city planners and highway engineers of more conventional views were already familiar with the technical standards he described, such as limited access, divided one-way roads, and control of development along road frontage. As MacKaye and Mumford acknowledged in their *Harper's* article, some American highways had already been built that successfully incorporated many of these features, especially in New York State itself. The Bronx River Parkway, several Westchester County parkways, and, in the early 1930s, the Taconic State Parkway provided examples of attractively designed, safety-minded, and uncommercialized highways. MacKaye's program went beyond these predecessors in several respects, though. First, he was proposing a truly national, standardized highway network, funded in large part by the federal government. "The Townless Highway would be, like the railway, an institution in itself, a system," he and Mumford wrote. "It would always be a through highway and not a local road. It must follow its own lines of topography."[40] Second, the townless highway was only one element in a comprehensive planning program that incorporated not only roads, but also "highwayless towns," like Radburn, and protected "wilderness ways," such as the Appalachian Trail.

As a planning template for the nation, the townless highway was out of step with both the political disposition of the moment and the thinking of most city and regional planners of the era. A generation later, though, with the launching of the Interstate Highway System in 1956, the federal government did at last initiate a dramatic nationwide highway program based, as MacKaye had proposed, on the federal highway aid formula.[41] But the Interstate Highway System omitted the highwayless towns and the open-space belts that were integral parts of MacKaye's vision. In fact, the Interstate Highway System that eventually evolved became a powerful engine of the metropolitan invasion MacKaye feared, rather than a means of controlling it.

❖

The fourth general meeting of the Appalachian Trail Conference was held in 1930 at Skyland, Virginia, on a stretch of the Blue Ridge that would soon be encompassed by Shenandoah National Park. "Our old project is booming these days," he wrote Stein, upon being invited to address the gathering, "and I've almost lost track of it except when the A.T. Committee sends me word like this." In fact, MacKaye had attended the third meeting of the ATC, in Easton,

Pennsylvania, the year before, delivering a message on the origin and conception of the trail. He corresponded regularly with many trail activists, but his remark to Stein suggests his distance from the details of constructing and maintaining the trail.[42]

MacKaye's talk to the 1930 ATC was aptly titled "Vision and Reality." At the meeting, Harold Allen, a member of the Potomac Appalachian Trail Club, dubbed Benton the "Nestor" of the trail, likening him to the Greek general legendary for his sage counsel. MacKaye assumed the whimsical moniker proudly and with good humor.[43] "Speak softly and carry a big map," he urged the trail folk, at the same time defining his own vocation and method. "The Appalachian Trail is a footway and not a motorway because it is a primeval way and not a metropolitan way," he observed, "because its purpose is to extend the primeval environment and to set the bounds to the metropolitan environment. . . . *To preserve the primeval environment:* this is the point, the whole point, and nothing but the point, of the Appalachian Trail."[44] MacKaye's prescient warning about the automobile's potential to invade the trail's "primeval environment" framed an issue that would soon engulf the Appalachian Trail community in a sometimes fierce debate. The trail enthusiasts would find themselves at odds, not only with the political and economic forces vying for the control and use of the wild Appalachian skyline, but also with one another over the purpose and nature of the Appalachian Trail itself.

Because ATC chairman Perkins had suffered a stroke just before the 1930 meeting of the ATC, his energetic young protégé Myron H. Avery, president of the PATC, stood in as acting chairman. It soon became clear that Perkins's condition would not improve. In 1931 Avery was elected ATC chairman, at the fifth meeting of the conference, in Gatlinburg, Tennessee. He would hold the position for more than twenty years, until his death in 1952.

In his influence on the progress and history of the Appalachian Trail, Avery would eventually rival the project's originator. He and MacKaye probably met first at the March 1929 meeting of the New England Trail Conference, then again at the 1929 and 1930 meetings of the ATC. While the initial dealings between the two men were civil and constructive, their powerful personalities and their differing conceptions of the trail began to clash not long after Avery's ascendancy to the ATC chairmanship.[45]

A Maine native and graduate of Bowdoin College and Harvard Law School, Avery was thirty years old when he assumed command of the Appalachian Trail project. He had come to know Perkins while practicing admiralty law in Hartford. When he moved to Washington to take a position with the U.S. Maritime Commission, Avery immediately helped organize the Potomac Appalachian Trail Club, of which he was elected president. With his election to the ATC chairmanship, the operation of the PATC and the ATC became inextricably linked. "It seemed simply not possible for Myron Avery to be associated with an enterprise without running it," observe trail historians Laura and Guy Waterman. "He was possessed of singular intelligence, energy, discipline, organizing skill, aggressive drive, and personal egotism."[46] "Myron left two trails from Maine to Georgia," according to a PATC acquaintance. "One was hurt feelings and bruised egos. The other was the A.T."[47]

Avery applied his almost obsessive energies and skills to several well-defined objectives in shaping the destiny of the trail. The first, and most important, was the surveying and completion of a continuous trail route, between Maine and Georgia. He believed that the actual existence of an unbroken trail would provide the only guarantee of public use and acceptance of the Appalachian Trail, in both fact and concept. He also insisted that the trail be built, "improved," maintained, and marked according to a uniform system of trail standards. Avery's insistence on such standards—and his assurance that he knew best what they should be—sometimes resulted in clashes with trail builders and clubs who adhered to local or informal trail practices. Avery also believed that broader public use of the trail depended upon the availability of accurate information about trail distances, terrain, water supplies, shelters, and other basic matters. Avery was a familiar figure on trail surveying expeditions, and sometimes the object of amusement, for his routine practice of pushing a measuring wheel as he walked along. In any event, he set the standard for the detailed trail guidebooks that later generations of AT hikers have relied upon.

MacKaye rarely worried about such practical details—unless they represented a threat to his own idealized, if sometimes amorphous, vision of the trail's "primeval" character. Avery, for his part, would later express exasperated disdain for what he regarded as MacKaye's carefree attitude toward the down-to-earth aspects of trailmaking.

The 1931 meeting of the ATC in Gatlinburg, at the western foot of the Great Smoky Mountains, was sponsored by the Knoxville-based Smoky Mountains Hiking Club. The club's young vice-president, Harvey Broome, wrote MacKaye to invite him to attend.[48] Unable to make the long journey, MacKaye prepared a message for Broome to read. "A realm and not a trail marks the full aim of our effort," he wrote.

There were two dimensions to the "policy of 'expansion'" he outlined. The first was broadening the expanse of territory included in the AT, by creating more public forests and parks along the trail route. "Our realm must be broad as well as long." The second dimension of the trailmakers' mission, as Mac-Kaye now saw it, was both psychological and spiritual. He foresaw an expansion "of human understanding . . . of the knowledge (and the feeling) of primeval history. . . . Primeval history is the story of the ages."[49] The rhetoric of political and social reform that had characterized MacKaye's 1921 trail proposal had now evolved into a more oracular, almost pantheistic, vocabulary of self-realization through direct contact with nature.

Such exalted sentiments might have left a Myron Avery cold, but MacKaye's ideas had an electrifying effect on Broome, an ardent hiker, Harvard-trained lawyer, son of East Tennessee, bachelor, and romantic. He and MacKaye soon forged an intense emotional and intellectual bond. MacKaye's words, Broome wrote in a long, unrestrained letter to the Nestor of the Appalachian Trail,

> opened vast new accomplishable possibilities to me. No man can live his best unless he is drawn on by a vision, an idea, grand in conception, and attainable. Without vision we rot. . .
>
> . . . We must strive—healthfully, almost unquestioningly—at some enterprise that penetrates obscurities and discloses unrevealed beauties. There must be movement on—tho in the last analysis we don't know where.
>
> Your conception seems to me to be in line with basic needs of life.[50]

MacKaye, overwhelmed by the emotional enthusiasm of his new disciple, responded in kind. "The thoughts expressed in your letter show to me that the A.T. is here as a spiritual influence . . . The A.T. is an effort to save a piece of earth for a future civilization which will have solved perhaps the problem of poverty and population. By a winning of the wilderness (right now) we play our part (right now) in that coming civilization."[51]

As Myron Avery secured his grasp on the destiny of the Appalachian Trail, MacKaye's blossoming friendship with Harvey Broome signified the changing thrust of his own concerns and activities. He did not abandon his hopes for the possibilities of comprehensive regional planning, even if the townless highway and some of his other planning ideas had not yet succeeded in staunching the "metropolitan invasion." MacKaye and Broome were among a small cadre of conservation leaders who soon undertook the task of "winning of the wilderness (right now)"—not just by means of uplifting rhetoric but also by the organization of other politically pragmatic wilderness preservation activists.

"RP = TH + AT + HT," a Formula
for the New Deal

———

1931–1933

"He lives a very quiet abstemious life: plain living and high thinking; and wastes less of his time on the *means* of living than anyone I know," Lewis Mumford said of his friend in 1932. "He anticipated the present depression long ago, and is completely unaffected by it personally."[1] MacKaye, it seems, had learned to subsist on little more than his own pipe smoke. When in 1930 he had sought medical attention for a condition afflicting his hands, a doctor advised him that the problem was "due to malnutrition—more protein needed and less nicotine."[2] MacKaye's stoic acceptance of such personal deprivations did not allay his growing concern about America's unfolding social and economic difficulties. His endeavors remained something of a hodgepodge. But after a decade of reflection, writing, and activity in his native region, he would gradually be drawn back into the current of national affairs.

✤

For many American intellectuals and activists, regionalism provided one hopeful path toward national rejuvenation. The RPAA circle, preoccupied with the concerns of the metropolitan Northeast, viewed itself as a vanguard of rational, ecologically sound planning. But other folklorists, geographers, sociologists, historians, writers, political scientists, and economists were equally engrossed in regionally inspired pursuits.[3] In the South, for example, the Agrarians, regarded by Mumford as "slightly reactionary," were a literary counterpart of the RPAA.[4] Their 1930 manifesto, *I'll Take My Stand,* included essays by the likes of Robert Penn Warren, Allen Tate, and John Crowe Ransom, who asserted their cultural independence from the forces of northern industrialization. They proposed, as Tate wrote, "a return to the provinces, to the small self-contained centres of life."[5] Clarence Stein also attempted to capture the essence of the regionalist spirit:

Mumford and I have been attempting to find some way of expressing in a simple manner the thing that has been in the back of our heads for so long. We finally concluded that it was much better to call the thing Regionalism than Regional Planning, for after all Planning has come to mean, to most people, something that is put down on paper or in reports rather than something that is constantly developing and changing and relating itself to a new and changing world. . . .

In connection with the meeting we have tentatively defined Regionalism as— How to foster in each region the fullest use of natural resources and economic opportunities so as to improve its social and cultural life.[6]

Stein's thoughts were prompted by a landmark in the era's regionalist movement, a round table held by the University of Virginia's Institute of Public Affairs in July 1931. Stein worked closely with veteran reformer Louis Brownlow, then director of the Public Administration Clearing House, to organize the conference. RPAA members Mumford, MacKaye, Chase, and Wright were all on the program. Geographer Howard W. Odum, sociologist Roderick D. McKenzie, poet John Gould Fletcher, and historian Stringfellow Barr were also among the speakers.[7]

Not the least important of those participating was New York governor Franklin D. Roosevelt, who gave the conference's opening address. Stein had previously appealed to Roosevelt to reinvigorate his state's commitment to regional planning, on the wane since publication of the 1926 report of the Commission on Housing and Regional Planning. Meeting with Roosevelt before the conference, Stein came away with the impression "that he has a very broad point of view in regard to State Planning. In fact, he is already thinking in terms of national planning."[8] Another conference participant, Charles Ascher, later asserted that in off-the-record comments FDR discussed the need for national planning and provided an outline of what would become the Tennessee Valley Authority.[9]

MacKaye, on the fourth day of the conference, participated in a panel called "Cultural Aspects of Regionalism," along with Fletcher and Barr. In his remarks, MacKaye outlined his program for preserving the "basic geographic unit of organic human society[,] . . . the single *town* of definite physical limits and integrity." The "three active American movements" already involved in this quest, he observed, included the crusade to create "public reservations," as exemplified by the Appalachian Trail; the town planning movement; and the "highway by-pass movement."[10]

"Benton certainly looked and spoke like a Yankee after the two Southerners," by Stein's account. "He was good, but a little bit too long. He stuck to the subject of culture most of the time but of course insisted on getting in the townless highway which he connected up by saying it was the means of preventing the metropolitan civilization from flowing through the Shirley Centers."[11] But whether Hudson River squire, southern Agrarian, or New England Yankee, virtually all the self-appointed regionalist delegates who had gathered on the campus designed by Thomas Jefferson shared a vision of a pastoral American landscape undefiled by industrialism. They were confounded by the modern American metropolis.

As MacKaye emphasized to his fellow regionalists the significance of the townless highway, he made no reference to an immediate highway crisis looming on the horizon just to the west of Charlottesville. He had traveled to the conference by way of Washington, where he talked at length with Myron Avery and Harold Anderson about the status of the Appalachian Trail. He then heard in detail about the fast-moving plans to build a scenic highway the length of the recently authorized Shenandoah National Park.[12] Skyline Drive, as the 105-mile road soon came to be called, would bisect the narrow park for its full length. In the eyes of MacKaye and many others in the hiking community, Skyline Drive represented a threat to both the physical environment and the spiritual essence of the Appalachian Trail. The ridgeline road would either obliterate sections of the hiking trail, only recently cleared and marked by the energetic members of the Potomac Appalachian Trail Club, or parallel the trail so closely as to destroy its wilderness character.[13]

When the Southern Appalachian Parks Committee in 1924 recommended establishment of what would become the Shenandoah and Great Smoky Mountain national parks, it also suggested, almost in passing, creating "a possible skyline drive" in the Virginia park.[14] The idea, promoted by William E. Carson, chairman of Virginia's Conservation and Development Commission, won the favor of President Herbert Hoover, who maintained a weekend retreat near the headwaters of the Rapidan River on the eastern slope of the Blue Ridge. In the autumn of 1930, while horseback riding along the ridge, Hoover instructed Park Service director Horace Albright to begin a survey for the scenic mountain highway. In March 1931, the Department of the Interior and the Bureau of Public Roads announced plans to build a 34-mile section of road from Thornton Gap to Swift Run Gap, using recently appropriated drought-relief funds. By mid-July, only months after Hoover's mountain-top revelation and just days after MacKaye had made a visit to the area, parkway construction began.[15]

At Charlottesville, MacKaye discussed the road situation with Harlean James and H. S. Hedges, the original Appalachian Trail scout in Virginia. Hedges chauffeured MacKaye to Skyland, a rustic resort inside the Shenandoah parklands, where they rendezvoused with Avery, Anderson, Harold Allen, and other PATC members. With the official government map in hand, MacKaye inspected the proposed route for the Skyline Drive. The next day, back in Washington, he called on Arno Cammerer, assistant director of the National Park Service, to discuss the road.[16] MacKaye had already written to Cammerer, expressing concern about the federal government's conception of its responsibilities for the eastern mountain terrain. "Does it mean the extension of the policy of motor skyline vs. foot-path skyline in the National Parks?" MacKaye asked.[17] Cammerer's tentative reply provided an early warning to MacKaye that the Park Service, especially after Cammerer assumed its directorship in 1933, could not be relied upon as a guardian of the American wilderness.[18]

At this early stage, the Appalachian Trail's most active leaders were unanimously worried about the likely effects of the road's construction on the integrity of the trail. Myron Avery tried unsuccessfully to convince the Park Service's landscape architects to route the road to the east, off the ridgeline. But by the time of MacKaye's first communications with the Park Service's Cammerer in June and July of 1931, the political momentum behind the "Carson-Hoover-Albright road blitz"—which had been set in motion with hardly any public debate and no congressional hearings or approval—was simply too great to stop.[19]

❖

Later that summer MacKaye returned to New England to work on various projects. He tried to convince his conservationist friends in Massachusetts to promote a linear state park along the attractive, unspoiled Squannacook River, which flowed through Shirley. He prepared a conservation plan for the Shepaug River Protective Association in Connecticut. And he worked on an article he had proposed to Isaiah Bowman of the American Geographical Society about the prospects for nature study along the Appalachian Trail. But the uncertainties and frustrations of his scattershot work routine were beginning to wear. "Trying to get ideas re integration of projects," he noted in his diary in mid-September.[20]

The surveyors' stakes and the maps he had seen at Skyland in July worked on MacKaye's imagination, though. He worried about the grim implications for the Appalachian Trail and the mountain environment of eastern America if the example of the Shenandoah crestline road gained popular and political support. MacKaye reconciled himself to the inevitability, even the desirability, of some form of Appalachian highway. But now, applying some of the principles of the townless highway, he began designing his own alternative to roads along the mountain ridgeline.[21]

Returning to Washington in October, Benton renewed his assault on the Park Service's highest officials. This time, though, he was equipped with an alternative scheme. With Harold Anderson, he had sketched out a "circuit highway" from the Adirondacks to the Great Smokies. This "Appalachian Intermountain Motorway," as he described it, differed from the other interpark and skyline projects and proposals in one fundamental respect: instead of following the mountain ridges, MacKaye's planned highway followed the valleys, sometimes climbing up along the mountain flanks, occasionally traversing mountain passes. He met several times with the Park Service's Albright and Cammerer to promote his alternative highway plan.[22]

But MacKaye had no real constituency, beyond his little band of hikers—who, in any case, were already breaking ranks over the appropriate tactics for coping with the new fervor for scenic roads. For MacKaye, the assault on the integrity of the Appalachian wilderness also represented a cultural and spiritual failure to grasp the underlying philosophy of the Appalachian Trail. He saw that his mission now was to educate his fellow citizens about the timeless and fundamental importance of nature's purpose. Former ATC chairman Arthur Perkins had once asked him, "When we get the Trail, Ben, what are we going to do with it?"[23] MacKaye now sought to provide an answer to Perkins's question.

Through the efforts of an acquaintance, William Carr, a well-known naturalist at the American Museum of Natural History, MacKaye's article, "The Appalachian Trail: A Guide to the Study of Nature," appeared in *Scientific Monthly* in April 1932.[24] More than a year earlier, he had begun writing "New England Primer," modeled on the Calvinist schoolbook of the same title popular in the seventeenth and eighteenth centuries. William Roger Greeley had for a time subsidized MacKaye's work on the primer and had published several installments in a short-lived magazine, *New England*. MacKaye's goal in that work had been to depict for a popular audience the region's landscape and history in the context of geological time and ecological change.[25] MacKaye now tried to approach the complex natural history of the entire Appalachian region in the same old-fashioned pedagogical style. And he also borrowed from the nineteenth-century method and spirit of writers who had been his own guides, such as William Morris Davis, Nathaniel Southgate Shaler, and Thomas Huxley. Cross-sections, panoramas, time charts, illustrations of climatic cycles and food chains—such were the devices he employed to teach "how to read, from the Appalachian Trail, [the] open book of primeveal nature."[26]

MacKaye no longer depicted the trail as primarily a recreational resource or a project in social reform. Rather, the trail environment provided a diverse but well-defined viewpoint from which to investigate the plan of nature and humanity's place in the cosmos. The Appalachian Trail was a vast outdoor classroom for the study of the "earth story" and the inexorable lessons of evolution. "Human society is an offshoot of forest society," he declared. "Each has its history. To know America we must know human history, and to know humanity we must know forest history."[27]

Indeed, the timeless ecology of the forest provided a telling analogy to the economic conditions of 1932. The creatures of the forest, he wrote, constitute "a warring society. . . . Each one is consumer and producer; some 'produce' by preying on others; everybody has a job; the less work the less food; resources in plenty and no markets; no middlemen, no salesman. In unravelling the forest civilization we reveal the contrasts of our own."[28]

Years later, MacKaye described this 1932 essay as "more important" than the 1921 article in which he had first proposed the trail.[29] It is difficult to argue that the later, little-known article has had a greater significance than the earlier manifesto, which launched the trail. But MacKaye's retrospective assessment perhaps reflected his own changing sense of the qualities of human understanding needed to preserve the trail's primeval environment. At the time, he immediately set to work expanding his essay into what he hoped would be book by the same title. To support the educational program described in his article, MacKaye also drafted a proposal for a three-element "Nature Guide Service" for the Appalachian Trail. His plan called for a series of nature trails and nature guide stations, connected to the Appalachian Trail but not directly on its main route; a "travelers' service," comprising shelters, meals, guided nature walks, and transport between cities and the guide stations; and a series of publications describing both the Appalachian region's natural history and basic techniques of nature study.[30] A description and MacKaye's map depicting fourteen "nature guide centers" appeared in an article by William Carr in *Natural History*.[31] Once again, though, MacKaye failed to find a publisher for a book idea.

He had an opportunity to practice his precepts of nature study in the summer of 1932, when the path of a complete solar eclipse crossed northern New England. "For a minute and a half we shall be in quest as citizens of the universe," he observed in the *Christian Science Monitor*, urging hikers to expe-

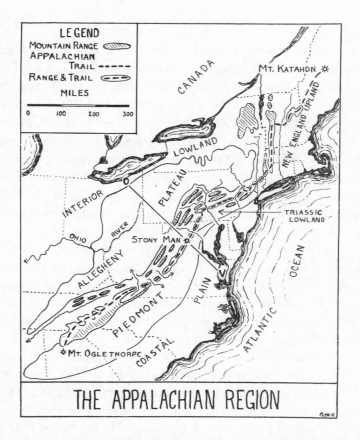

THE APPALACHIAN REGION

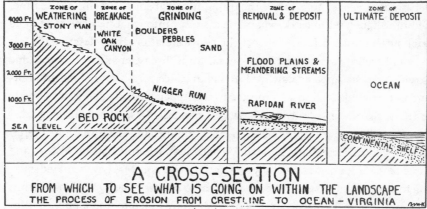

A CROSS-SECTION
FROM WHICH TO SEE WHAT IS GOING ON WITHIN THE LANDSCAPE
THE PROCESS OF EROSION FROM CRESTLINE TO OCEAN – VIRGINIA

MacKaye once described his April 1932 *Scientific Monthly* article, "The Appalachian Trail: A Guide to the Study of Nature," as "more important" than his 1921 article proposing the trail. To accompany it he drew an ecological "cross-section" of the Appalachian Trail at Stony Man Mountain on Virginia's Blue Ridge, to reveal the Appalachian region's "earth story."

rience this "cosmic occasion" when the Appalachian Trail and a "Shadow Trail," the moon's umbra, would coincide.[32] MacKaye joined Horace Hildreth's family for a trip to Limerick, Maine, where on a cloudy day they observed the eclipse from Sawyer Mountain. It was a "world and life time experience," he recorded, a chance to "ride with the stars."[33]

The serenity of Benton's Shirley Center refuge was disrupted during most of 1932 by another family crisis, "a little epic in cacophony," as he wrote Mumford.[34] Busy on projects in New York and Connecticut during much of the winter, he had returned to Shirley in April to find Percy's family in residence. Calamity had caught up with Percy's thirty-two-year-old son Robin. The strapping Harvard man, a once-promising poet and the father of two daughters, had never been quite the same after a serious 1922 accident in which he collided with a truck while riding his bicycle. Robin's deteriorating behavior had undermined his marriage. Unsure what to do for his troubled son, Percy, embroiled in a serious legal dispute with Dartmouth College over a substantial printing bill, fled his New Hampshire home, "lest court papers should be served on him." For a time, Percy installed his troubled son in Shirley.[35] Benton felt his nephew's unsettling presence as an unwelcome, even embarrassing invasion of the peaceful and solitary dominion he had cultivated for himself in Shirley over the previous decade. He spent months living with friends in the nearby town of Littleton, while strenuously urging Percy to find another residence for his son. Not until the end of the year did Robin finally depart, thereby allowing Benton to restake his claim to the MacKaye "Empire."[36]

In the fall of 1932, a lengthy letter from the PATC's Harold Anderson carried harbingers of conflict and controversy about the Appalachian Trail's fate. Anderson reported in detail on the swift pace of events involving the Skyline Drive. He had recently driven the thirty-four completed miles of the road for the first time, along with Avery and the other PATC leaders who had built the trail through the park. He ruefully acknowledged that the road would "be very popular" and that it was "a wonderful piece of engineering." But he added that the road had "raised havoc with our trail. So far as the Shenandoah National Park is concerned, we will have to abandon the idea of a Skyline route for the Appalachian Trail because this route has now been preempted by the new road."[37]

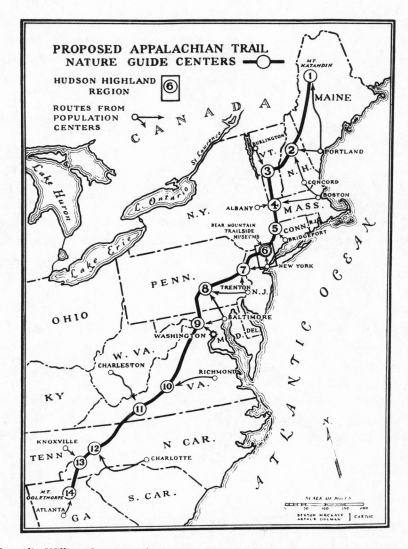

Naturalist William Carr, in a July–August 1933 *Natural History* article about the Appala-chian Trail, promoted MacKaye's map depicting a series of fourteen proposed "nature guide centers" along the trail.

Anderson explained to MacKaye that it would be possible to reroute the trail in an attractive manner on the flank of the ridge, some distance from the road. But Avery saw matters differently. "Myron has not yet grasped the idea of the incompatibility of the road and the trail," Anderson wrote. He urged MacKaye to write Avery to get over to him what he should understand already and that is that the Appalachian Trail should above all things be a wilderness foot-path and that while so far as possible we want to have a skyline trail, it may be necessary in some sections, such as the Shenandoah National Park, to depart from this general idea and that we should as far as possible seek the primitive environment for the trail even if we have to sacrifice the skyline route.

I cannot help feeling that Myron with all his indefatigable energy and enthusiasm has not fully grasped the idea of the Appalachian Trail.

. . . The whole question of road vs. trail is very fundamental. We outdoor folk who love the primitive are accused of selfishness in trying to have preserved inviolate a narrow strip of the little that we have left of the primitive area and preventing the enjoyment (?) by multitudes of the scenery of this area if roads were built therein. It seems to me that it narrows down to the question of whether it is worth while to preserve the primitive.[38]

Anderson had neatly framed the question—"road vs. trail"—that would split the Appalachian Trail community and prod other eminent American conservationists into pursuing more aggressive efforts to preserve the remaining American wilderness.

For the RPAA circle, as for other intellectuals and reformers who felt isolated and ineffectual during the 1920s, the presidential election of 1932 offered fresh prospects for steering the nation down a new path. MacKaye, of course, had a comprehensive public-works program ready at hand. One map depicting his ideas for "Appalachian America" included a highway reaching from northern Vermont to northern Georgia, flanked its entire length by a "belt of public forests" on one side of the road and, in the "Great Valley" on the other, sites for highwayless towns.[39]

MacKaye had in fact reduced his own planning program to a simple conceptual formula, as he had recently related to Stein, "RP = TH + AT + HT": regional planning equals the townless highway plus the Appalachian Trail plus the highwayless town.[40] After the nomination of Roosevelt in July, he took more seriously the prospect of influencing the New York governor and potential president. "How can he run the country unless the R.P.A. tells him how?" he wrote Stein during the campaign. His comment was offered mostly in jest, but the RPAA's contact with FDR at the 1931 University of Virginia roundtable and Stein's prior involvement in the politics of planning in the nominee's home state provided some hope that his ideas might reach Roosevelt directly. The program MacKaye now contemplated called for the building of new communities like Radburn; a moratorium on road building, "until the people can be taught that a road is a highway and not a hog trough"; wilderness protection; and a "build up" of the nation's forests, "gaining thereby the maximum of jobs for the minimum of taxes."[41]

On the brink of the 1932 election, *Survey* editor Paul Kellogg asked MacKaye and others who had been involved in the magazine's 1925 regional planning issue to contribute to another special issue, this one titled "Obsolete Cities: A Challenge to Community Builders." MacKaye's article, "End or Peak of Civilization?" posed a choice for the nation in the apocalyptic precepts set forth by Oswald Spengler and Henry Adams. "The metropolis," he wrote, "is a harbinger of death, . . . the end of the present western mechanical regime." But MacKaye saw the stirrings of hope for restoring the "proper balance" among primeval, communal, and rural environments in "three separate but related lines of action": the conservation movement, the town planning movement, and the highway reform movement.[42]

He was under no illusions about the prospects of a Roosevelt presidency, however. "Well, so far as the depression is concerned if Norman Thomas were to be elected we could go ahead and plan to put together, in plain common sense fashion, machines, materials and men," he wrote Stein. "But the best that might be done under Franklin and Frenzied Finance would seem to be to spend the spendings (while the spendings last) in a way to ration jobs the best and clutter up the earth the least."[43] "Meanwhile," he added on election day eve a few weeks later, "I expect to remain conservative and vote for Norman Thomas."[44]

Shortly after Roosevelt's election, MacKaye drafted yet another version of his national program, in which, as he explained to Stein, "housing, highways, and wilderness areas were placed in one great flying wedge—highways being the 'center rush.'"[45] His memorandum, entitled "A Suggestion Re National Planning," suggested that the Department of Agriculture, home to the Forest Service, would be the appropriate leader of such a program. "All three elements are there: public roads, national forests, community organization. No new administrative machinery is needed; it is a matter of making present agencies focus on a concrete yet comprehensive enterprise."[46] MacKaye did not acknowledge the irony of his newfound faith in the Department of Agriculture's ability to implement a multifaceted planning program, including community organization. Almost two decades earlier, he had left the same agency because of its failure, at least in his eyes, to address such social concerns.

❖

During early 1933, MacKaye stayed at Clarence Stein's apartment overlooking Central Park, keeping his friend company while Stein's actress wife Aline Mac-Mahon was on the West Coast making a film. Nearby Stein's comfortable residence, the worsening depression tightened its grip on other city residents. While MacKaye and Stein gazed out over the city's pleasure ground at twilight, smoking cigars and conversing about the nation's plight, at least one destitute

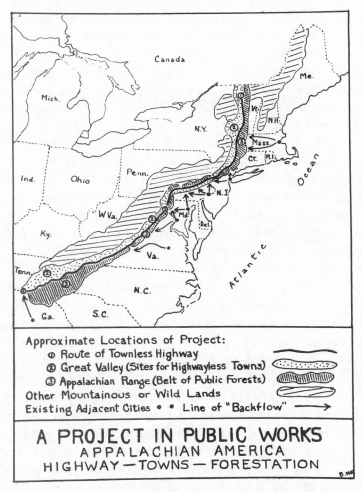

MacKaye's unpublished "A Project in Public Works," prepared about 1932, was a large-scale plan for the Appalachian region, combining the townless highway, sites for "highwayless towns," and a continuous belt of public forests. It described his hopes for the Tennessee Valley Authority. (Author's collection.)

family, by some reports, was living year-round in a cave in the park. MacKaye made plans to go to Washington to investigate his own prospects in the new political era.[47]

MacKaye arrived in Washington in late March, three weeks after Roosevelt's inauguration on the 4th. Those first days of the new administration were exhilarating, hopeful times for MacKaye and other veteran activists of the Progressive Era. He made the rounds of his old friends and associates, who were in the thick of political events and intrigues. And soon he was being introduced to the new generation of reformers and officials. One of the initial legislative initiatives of FDR's eventful first hundred days was a matter of intense personal interest to the new president—and to foresters like MacKaye. The Civilian Conservation Corps (CCC) provided the vehicle for Roosevelt's dream of putting unemployed young men to work in the nation's forests. Indeed, the whirlwind debate over the CCC was the magnet that drew MacKaye to the capital. In the week before passage of the Forestation Act on March 31, MacKaye wrote to Secretary of Labor Frances Perkins, Chief Forester Robert Stuart, and his old Forest Service mentor Raphael Zon. He saw the opportunity to resurrect the plan for forest communities and public employment in the national forests that he had spelled out years before in *Employment and Natural Resources*, uniting "up-to-date forestry with up-to-date town planning."[48]

From Harry Slattery, his one-time roommate and long-time confidant of Gifford Pinchot, who was soon to be named deputy to Interior Secretary Harold Ickes, MacKaye heard "the whole inside story" of the CCC's origins. At the Cosmos Club the day after the Forestation Act passed, Slattery buttonholed Forest Service chief Stuart. "Bob, there's one man you ought to call for in this forestation job and that's Benton MacKaye," Slattery told Stuart. Stuart acknowledged receiving MacKaye's recent letter, but he brushed aside the forest communities scheme with the comment that "there wasn't time now for that."[49]

Undiscouraged by Stuart's abrupt response, MacKaye traipsed from office to office, meeting to meeting, trying to find a niche in the New Deal. "Nothing like it has come to pass since the 'bully' days of T.R.," he reported to Stein. "Things hummed in the Wilson administration but there was butchery in the air. I've seen all three in this Capital city—Teddy, Wilson, and now Franklin: this time combines the imminence of 1913 with the buoyancy of 1908." His old colleagues in the Forest Service and Labor Department told him that "it's fun again to live; they are working like hell and love it."[50]

❖

In another attempt to promote his own prospects, MacKaye turned his attention to a program that had been on the agenda of conservationists and reformers for decades: the public development of the water resources of the Tennessee River valley. In the winter and spring of 1933, under the indefatigable leadership of Nebraska senator George Norris, the Tennessee project was receiving new support. In February, MacKaye, Chase, Stein, and Mumford had planned their own publicity campaign for the Tennessee River proposal. Norris's interest in the project was primarily to promote the public generation and distribution of electrical power. The band of RPAA members took a more ambitious view, envisioning the possibilities for the comprehensive public planning and development of a major river watershed. Mumford composed a letter from the group for Roosevelt's attention. MacKaye set to work writing several articles about the Tennessee valley idea.[51]

MacKaye's old hell-raising ally, Judson King, had long been Norris's principal adviser and propagandist on behalf of the Tennessee scheme. MacKaye turned to King for information and advice as he wrote his articles, which appeared at a timely juncture that spring in *The Nation,* the *New York Times,* and *Survey.*[52] Roosevelt sent a message to Congress on April 10 strongly supporting the Tennessee valley plan. A day later, Norris filed a bill calling for the Tennessee Valley Authority (TVA). On May 18, the president signed the law creating the TVA. The next day MacKaye joined King for a "victory lunch" and plotted ways to secure a position with the new watershed agency.[53] The sweeping legislative authority granted to the TVA represented the realization of the long-time dream of Progressive Era "utilitarian" conservationists, who had advocated the principles of multiple-purpose watershed development and centralized planning by government technocrats. As a veteran of that era and that tradition, MacKaye had articulated those key principles of utilitarian conservation in his articles supporting creation of the TVA.

❖

By late April, MacKaye was stalled in his efforts to secure a position in the new administration. Somewhat to his surprise, though, he was soon receiving the respectful attention of the new leadership of the Indian Service (later known as the Bureau of Indian Affairs). The new Indian commissioner, veteran reformer John Collier, was cut from the same idiosyncratic cloth as the lean, articulate forester who came to his office on May 5. As a youth, Collier had roamed the little-explored mountains of the southern Appalachians, often hiking alone. The experiences inspired his lifelong advocacy of wilderness preservation. As an activist during the twenties in the cause of Indian rights,

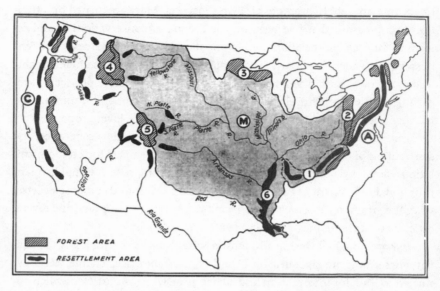

In an April 16, 1933, *New York Times* article promoting creation of the Tennessee Valley Authority, MacKaye illustrated how the region's development would a "first step in a national plan" for "a dozen belts and areas, related by natural geographic features."

he had worked closely with the eminent New York civil rights lawyer Louis Marshall. One of Collier's first acts upon being appointed commissioner was to select Marshall's brilliant, energetic son Robert, a forester and adventurer, to head the Indian Service's forestry division.[54]

On April 24, MacKaye had lunch with Bob Marshall. It was the first meeting of the two men, who both had growing reputations among the small but passionate circle of conservationists and public officials devoted to protecting America's wildlands. Despite their differences in age and temperament, Marshall and MacKaye shared a common education as foresters (both had received Harvard master's degrees), an interest in socialism, and a passion for the wilderness. Their acquaintance now laid the basis for their cooperation as two of the key figures in the creation of the Wilderness Society.

Judson King, an ally of Collier's in previous causes, put in a word with Collier on MacKaye's behalf. But it was one of MacKaye's new acquaintances, Frederick Gutheim, a young staffer at the Brookings Institution, who introduced MacKaye to Collier. MacKaye reported to Stein that he had "had a good long talk with the Great White Father in his office." Collier "seemed to be genuinely interested" in his ideas about forest communities, but suggested they talk them over with several Menominees who were visiting the agency.

That afternoon MacKaye and Gutheim met with two members of the Wisconsin tribe, "a strapping one-legged half-breed named Friedenberg, and a little young Injun named Dodge. Friedenberg was one of the keenest men I ever met; he expressed himself in polished English and showed himself thoroughly versed on every phase of the subject; he was enthusiastic about my ideas."[55]

MacKaye's ingenuous enthusiasm about his first close contact with an American Indian would be tempered considerably in the months ahead. At the invitation of Jay Nash, appointed by Collier as educational director of the Emergency Conservation Work (ECW) program that was part of the Civilian Conservation Corps, MacKaye drafted a memo outlining the first steps for a comprehensive forest community program for Indian reservations. He proposed sample surveys, involving small, "carefully selected" Indian crews, on the Menominee reservation in Wisconsin and the Pine Ridge Sioux reservation in South Dakota. "The Forestation Act provides the chance of enabling the Indian to show the White Man how to do things," he wrote. He depicted the Indian reservation as a "sample plot," providing a laboratory for investigating "sample problems of the region (applicable alike to Indian and White)," solutions which "should exert a widespread influence upon the local region and on the Nation at large."[56]

Nash offered MacKaye a job in the ECW program, although only for two to six months, and sent him to the Rosebud Sioux reservation in South Dakota. By his understanding, MacKaye was to take charge of work camps housing approximately 950 Indians.[57] He left Washington with instructions from Nash and Collier "to use my judgment as to what to do to achieve from this work the most lasting benefit to land and Indian. This means of course giving me a free hand, and I assume nothing will occur to prevent this."[58] MacKaye did not yet grasp, however, that the first priority and purpose of Emergency Conservation Work, like so many other New Deal programs, was to provide jobs quickly to as many people as possible. Building new communities could come later.

When he reached South Dakota, he learned that he would not be the chief of a thousand Indians after all. No sooner had he arrived than he was taken to a Sioux tribal meeting, where he was called upon to address the Indians. Speaking through an interpreter in upbeat generalities, he confessed that he had not yet talked with his local Indian Service superiors about the details of the ECW project for the Rosebud reservation, but "that this was the chance for the Indian to take a big part in a big job." His audience voted to adopt the project.[59] After only a week on the road investigating Sioux reservations with other ECW officials, MacKaye received orders to travel to Gallup, New Mexico. The abrupt, unexpected change of plans indicated the confusion that would characterize the Indian Service's ECW work.

En route by train he composed a long letter to Colonel George Ahern, a veteran federal forester, describing how his first tumultuous experience with Indian culture had unsettled his own idealized preconceptions. He related his conversation with Charles Brooks, a Rosebud Sioux:

> Well, he was talking against the Sioux and I was talking against the Anglo-Saxon (rather we were talking against our respective *institutions*). He was *against* the tribal or communal use of land—the very thing which to my mind should be the big contribution of the Indian to the white (and the human) race. He was *for* the fee simple—the device of my own ancestors, which I consider the Devil's own invention to bedevil the human race. I was for the Sioux idea of land integration; he was for the Anglo-Saxon idea of land disintegration. Can you beat it? . . .
>
> . . . *Land integration vs. land disintegration.* That to my mind is the guts of the Indian question (as far as "Conservation" is concerned). The fight is against disintegration—not only the legal kind (allotment) but the physical kind (erosion).[60]

The Navajo reservation and the desert country of New Mexico and Arizona represented new terrain for MacKaye. But no sooner had he arrived in Albuquerque than he met for the first time a knowledgeable, experienced guide to the southwestern landscape, Aldo Leopold, who was on a short-term assignment with the Forest Service. The two conservationists had been following and corresponding about each other's ideas and writings for almost a decade. Now, during the few days they spent together surveying the eroded washes and arroyos near Albuquerque, MacKaye and Leopold had the chance to take each other's measure.[61]

By MacKaye's account, Leopold drafted the memorandum signed by both for their respective agencies, proposing the creation of two ECW camps on the adjacent Sandia Reservation and the Cibola National Forest. Besides spelling out specific measures to control erosion on Juan Tabo Wash, the memo also raised the possibility that quail-hunting fees might serve as a revenue source for local Indians. They concluded by noting that the Juan Tabo canyon "has outstanding value for recreation of the wilderness type. It seems to be the most suitable area in central New Mexico for restocking with mountain sheep."[62]

Although little apparently came of their proposal, Leopold's collegial gesture cemented MacKaye's respect for his thoughtful and eloquent colleague. "He is one of the few real big men of the country," Benton wrote his sister.

"Not the bigly known men but the real thing."[63] He also reported his impression to Harvey Broome, appropriating a descriptive phrase his Knoxville friend had coined: "Personally I think that 'earth men' are damn scarce [*sic*]—there may be 125 out of the 125,000,000 people in the U.S.A. For I reserve this name (of yours) to the real thing in essence. I've just met one of them out here. Aldo Leopold."[64]

Leopold's just-published book *Game Management* quickly became the wildlife scientist's bible. But that spring Leopold had also reached a watershed in his own thinking about conservation and ecology, which he articulated in a paper titled "The Conservation Ethic." His philosophical and impassioned essay marked the progress of his influential ideas about the "state of *mutual and interdependent cooperation* between human animals, other animals, plants, and soils, which may be disrupted at any moment by the failure of any of them."[65] Leopold and MacKaye shared a common educational and professional background as foresters, but as intellectuals and activists they had each wandered from conventional career paths. Their prolific interests and writings did not always produce coherent and consistent formulas for action. But this first meeting between two important conservationists was another modest personal step toward the development of an organized American wilderness movement. MacKaye and Leopold, on their separate but interconnected paths, had expressed the principal concerns and ideas which motivated that incipient movement: the invasion of America's wildlands by automobiles and commercial development, and the growth of a subtle, holistic ecological perspective.[66]

The quality of MacKaye's experience in the Indian Service swiftly deteriorated. His adjustment to government service after a lapse of more than a decade was difficult. He was stationed at an ECW camp in Leupp, Arizona, in the southwestern corner of the Navajo reservation. The daunting task of organizing from scratch a network of residential camps was made all the more challenging by the clash between the traditions of the Navajo and the expectations of the white ECW administrators. MacKaye's primary duty that summer as an "area supervisor" was the purely clerical chore of trying to keep track of the unpredictable Indians who were enrolled in the ECW program and expected to reside in the government camps.[67]

The frustrations of his job were compensated by the opportunity to attend Navajo and Hopi dances and ceremonies. He also happened to hear a stirring speech Collier gave to a group of Navajo students.[68] But his relationship with his immediate superior, E. H. Hammond, went from bad to worse. His duties were a far cry from the "chance for real planning" he had envisioned and outlined in his early communications with Nash and Collier.[69] "Government work is not what it used to be," he lamented to his brother Percy, "and I'll be glad to go back in my den again. I took the job because I was strictly up against it."[70]

He prepared a memorandum for Nash requesting assignment to a more substantive planning assignment. In late August he fired off a frank letter to Collier, including both his critical assessment of the Emergency Conservation Work program and his plea that he be allowed "to select a small crew of intelligent Navajos (preferably returned students) and make with them an outline regional plan for the Navajo Country."[71] After talking personally with Nash in Gallup, New Mexico, MacKaye thought he might at least be reassigned to publicity duties. He had overestimated his influence with Collier, however, and underestimated the intensity of bureaucratic infighting within the Indian Service, both locally and in Washington. On August 31, he received a telegram from commissioner Collier containing the news of his immediate termination.[72]

❖

Clarence Stein, reporting the news to Mumford, opined, "I fear he is not what is called 'a good soldier,' for which perhaps the world will be ultimately grateful. . . . I hope this does not mean the end of Ben's career as an Indian chief."[73] MacKaye was not being singled out as a malcontent, Collier took pains to explain. Many ECW supervisors at his level were being let go, "a stark necessity," Collier claimed, "due to the economy policies."[74]

MacKaye was eager to return east. "I am anxious to get back to the one project that fundamentally possesses me," he wrote Broome, "to make the A.T. a real access to the secrets of the *earth* sense."[75] He planned a leisurely itinerary through the Southwest and South, stopping with friends and acquaintances in Santa Fe, Oklahoma City, and Florence, Alabama. He traveled by bus, a disagreeably novel mode of transportation for him. The uncomfortable journey, "encased in a motor truck," exacerbated his digestive troubles.[76]

He looked forward to stopping in Knoxville, Tennessee, for two reasons: to prospect for a position with the new Tennessee Valley Authority, headquartered there, and to meet Harvey Broome, with whom he had been corresponding intensively for two years. MacKaye wasted little time promoting his own prospects at the TVA. He made certain a copy of his *New Exploration* got into the hands of Arthur E. Morgan, FDR's choice as TVA chairman. He also met with the head of the agency's forestry division, his long-time acquaintance,

Ned Richards, an outspoken figure in the liberal, activist wing of the forestry profession. Lewis Mumford warned his friend to restrain his expectations about the prospects for the TVA and the New Deal. Morgan, he wrote, was "an egotist," FDR himself "a sort of political Mary Baker Eddy," and Washington "a mess."[77]

The "father" of the Appalachian Trail was fêted as a celebrity by the young and enthusiastic members of the Smoky Mountains Hiking Club, who took him along on several of their hikes. When MacKaye arrived in Knoxville, as Harvey Broome later recounted, he was complaining of abdominal pain. On September 28, after ten days in Knoxville, he met Broome and Ned Richards for lunch, but ate nothing. Later that afternoon, MacKaye called Broome from his YMCA room, asking for immediate help. When Broome and another friend arrived, they found MacKaye "doubled over," pounding the floor in excruciating pain and pleading for help. "His hands were as cold as ice and his face had a ghastly greenish pallor," Broome reported to Stein. "He said, 'I believe there is a stoppage of some kind.'"[78] Without delay, they transported MacKaye to Knoxville's Fort Sanders Hospital, where he was operated on promptly to correct a probably congenital condition diagnosed as Meckel's diverticulum, a potentially fatal blockage of the intestine.[79]

His new friends and acquaintances rallied to his support. MacKaye became a favorite with the hospital's nurses and struck up a friendship with his medical savior, Dr. Eben Alexander, that would survive for decades. Harvey Broome visited daily during the weeks of MacKaye's hospitalization and kept in touch with MacKaye's worried family and acquaintances in the East. Clarence Stein provided assurance that his friend's medical expenses would be covered.[80] Fate had selected Knoxville, Tennessee, as the next major way station on Benton MacKaye's life journey.

✤ 13 ✤

The Tennessee Valley Authority

1934–1936

"Well, I've landed the T.V.A. at last!" MacKaye would report to Stein early in the spring of 1934.[1] During his early years with the Forest Service, he had been a disciple of the leading American advocates of comprehensive river basin planning, such as Gifford Pinchot, W. J. McGee, Frederick Newell, Raphael Zon, and Marshall Leighton. The Tennessee Valley Authority would offer the opportunity—he hoped—to put such ideas to work on the spacious scale of his own vision. His tenure with the TVA would last from April 1934 until July 1936, a period he came to call "The Two Damndest Years." During these formative years of the controversial new agency, MacKaye would play a modest but meaningful role in the rise—and the demise—of the regional planning idea within the TVA.[2] This interval would also encompass increasingly bitter differences between MacKaye and the ATC's Myron Avery, as well as creation of the Wilderness Society.

MacKaye convalesced from his surgery during the autumn of 1933 in Knoxville and returned to New England during the winter of 1933–1934. From Hanover, New Hampshire, where his brother James had been teaching philosophy at Dartmouth College, he campaigned by correspondence for a position at the TVA. A gentlemanly competition for his services soon developed, pitting the agency's Division of Forestry, under Ned Richards, against the Division of Land Planning and Housing, headed by Earle Sumner Draper. Richards had received an enthusiastic recommendation from Bob Marshall, written the day MacKaye was released from the Indian Service. "I have seldom felt so strongly that any man had just the right qualities for a job tha[n] I have that Benton MacKaye is the ideal man to head the forest recreational work in the Tennessee Valley," Marshall wrote, adding that "any other liberal forester in the entire United States" would testify to MacKaye's talents."[3]

Marshall, Richards, and MacKaye in fact represented forestry's most liberal and activist faction. In 1928, Richards had approached MacKaye about collaborating on a proposed national forestry survey and program, sponsored by the League for Industrial Democracy (LID).[4] But it had been Marshall, not Richards, who in a 1930 LID report, *The Social Management of American Forests,* had advocated an expansive program of public ownership and control of the nation's forests. "It appears incontrovertible that government ownership has been distinctly more conducive to social welfare than has private ownership," Marshall bluntly asserted. "With such a clear record, the rational conclusion seems inescapable that commercial privately owned timberlands should be socialized."[5] Marshall also pursued the campaign for public forest ownership in his fervid 1933 tract, *The People's Forests.*

With his own TVA prospects in mind, MacKaye recapitulated for Richards the sweeping program for Tennessee Valley development he had spelled out in his several articles early in 1933. "Tennessee—Seed of a National Plan," appearing in *Survey Graphic,* had suggested how a Tennessee Valley regional planning experiment, if successful, might provide a model for reshaping American society. His map of the region depicted sites for dams, reservoirs, and powerlines, as well as expansive tracts of potential public forests along the region's mountain chains. But MacKaye's thumbnail sketch of the valley was only a fragment of his plan for the entire Appalachian domain.

"These three developments of townless highway, of highwayless town, of forest wilderness, I would, as public works dictator, place side by side in one long belt connecting the Appalachian valleys. The heart of this triple project is the highway," he had written.[6] But the TVA law failed to take into account "the highway as a means of population distribution," he asserted in a February 1934 manuscript, "Regional Planning and The Tennessee Valley," which he provided to Richards. "The neglect of this function of the highway marks the one signal omission in the Roosevelt school of planning."[7]

Richards was predisposed to endorse MacKaye's bold objectives and programs, but the nominal head of the TVA's regional planning efforts, Earle Draper, became interested in what MacKaye could bring to his own division. A landscape architect and a former member of the planning firm headed by John Nolen Sr., Draper had earned his professional reputation designing industrial communities in the South, but he had little experience with truly regional projects.[8]

At the TVA, Draper proved to be an efficient and shrewd administrator. Carefully navigating his division through the TVA's first hectic years, he oversaw the design of housing and communities for relocated families and TVA construction workers, the planning of new roads and parkways, and, not least important, efforts to coordinate the agency's varied programs and projects in a comprehensive plan. In late March of 1934, Draper won approval from the TVA's three-member board of directors for a budget and additional staff to "prepare a comprehensive regional plan of the entire Tennessee Valley Basin."[9] The title of Draper's proposal—"Project R.P. 1: Tennessee Valley Section of the National Plan"—reflected President Roosevelt's (and MacKaye's) vision of the TVA as a national laboratory of regional planning. "If we are successful here," Roosevelt had prophesied in his April 10, 1933, message promoting the TVA law, "we can march on, step by step, in a like development of other great natural territorial units within our boundaries."[10]

When Draper presented his Project R.P. 1 proposal to the TVA directors, he admitted that the "type of personnel required for this work will not be readily found."[11] But immediately after the TVA board approved the project, Draper sent a telegram to MacKaye offering a position in his division, with the title of "Regional Planner."[12] Stein and Mumford had provided the TVA an enthusiastic yet revealing recommendation of their friend. Praising MacKaye's intelligence, integrity, and diligence, they described him as someone who had "given more time and thought to Regional Planning, both in theory and practice, than possibly anyone else in the country." They also gently disclosed some of his quirks of character. "MacKaye is a normal and well-adjusted personality, but for the fact that he has made no attempt to repair the breach in his personal life occasioned by his wife's death," they wrote. "Except in the environment of big cities, he functions at full efficiency."[13] When MacKaye accepted Draper's offer, he joined the wave of employees who tripled the TVA's payroll from 3,000 at the end of 1933 to 9,000 by April 1934.[14]

The TVA's seemingly expansive regional planning powers were encompassed in Sections 22 and 23 of the TVA Act. MacKaye would repeatedly proclaim that the two sections constituted "the Magna Carta of American big-scale regional planning."[15] Section 22 authorized the president, "by such means or methods as he may deem proper," to prepare "surveys" and "general plans" for the "Tennessee basin and adjoining territory . . . for the general purpose of fostering an orderly and proper physical, economic, and social development" of the region. Section 22 also granted the president authority to cooperate with the states, and their "subdivisions or agencies . . . or with cooperatives or other organizations . . . to make such studies, experiments, or demonstrations as may be necessary and suitable to that end."[16] Section 23 authorized the president to recommend legislation to effect the purposes and findings spelled out in Section 22. The scope of such legislative proposals was not restricted to the maximization of flood control, navigation, and electric power; the promotion of reforestation, "the proper use of marginal lands," and "the economic and social well-being of the people" of the Tennessee River basin were also identified as appropriate legislative objectives.[17]

Arthur E. Morgan, the TVA's first chairman, later took credit for drafting Sections 22 and 23, but his contribution apparently consisted only of the last-minute addition of a few provisions and phrases. According to two TVA staffers who later traced the sections' origins, it had been MacKaye's young friend Frederick Gutheim, along with John Nolen Jr., working behind the scenes with Senator George Norris in the spring of 1933, who slipped the politically potent sections into the law almost unnoticed as it made its brisk way through Congress.[18]

The genesis of the river basin planning idea could be found in the Progressive conservation movement and in the decades of congressional debate over the federal government's role in developing Muscle Shoals, on the Tennessee River in Alabama. There was also a legacy of planning in FDR's home state of New York which had been substantially influenced by Clarence Stein and the RPAA circle, including MacKaye. Gutheim and Nolen had attended the 1931 University of Virginia regionalism conference, where they heard Roosevelt's off-the-record remarks on regional planning and met some of the RPAA members. "There is no record of any direct connection between the Conference and the emergence of the regional planning idea in the Tennessee Valley Authority Act," concluded the TVA's in-house historians, "but certainly it provided fertile soil for the transplanting of this idea from New York southward." MacKaye, of course, personified the various strands of the regional planning idea that converged in the TVA Act's planning sections.[19]

From the outset, the scope of the TVA's authority was a matter of intense regional and national controversy. The most public and dramatic dispute engulfing the TVA was its running battle with the region's private utility companies. The power industry resorted to law suits, congressional lobbying, and an extensive public relations campaign to challenge the competitive threat allegedly represented by the TVA. The TVA's board was soon divided as well—not only over how to respond to the private utility challenge but also over the appropriate scope of the agency's regional planning activities. The debate within the agency mirrored the New Deal–era's highly charged national discourse about economic, social, and land-use planning.

The board's three members were led by Arthur E. Morgan, who came to the chairmanship of the TVA with a national reputation as both a hydraulic engineer and an educator, but a streak of righteousness, a suspicion of conventional politics, and an idealism "tinged with mysticism" soon led him into conflict with his more pragmatic and politically canny colleagues. David Lilienthal, the young protégé of such Progressive Era figures as Louis Brandeis and Felix Frankfurter, had fearlessly battled with private utilities while chairing the Wisconsin Public Utilities Commission. At the TVA, he dedicated his efforts to establishing the agency as a successful public electrical utility. Harcourt A. Morgan, an agricultural entomologist who had served as president of the University of Tennessee, soon began carving out a powerful bureaucratic dominion within the TVA's agricultural programs.[20]

In August 1933, Harcourt Morgan won Lilienthal's support for a plan to divide administrative responsibilities three ways; Arthur Morgan had no choice but to acquiesce. Harcourt Morgan took charge of the agency's agricultural programs, including fertilizer production and "rural life planning." Lilienthal oversaw the power program, land acquisition, and the legal department. And Arthur Morgan supervised engineering and construction programs, forestry, and the coordination of social and economic planning.[21]

Draper's Project R.P. 1 hence fell under Arthur Morgan's authority. MacKaye reported to Tracy Augur, head of the Regional Planning Section in Draper's division. Augur organized the staff of Project R.P. 1 to study separately several different fields: architecture and housing, parks and recreation, highways and parkways, general land use, planning legislation, colonization, public relations, and cartography. He assigned MacKaye responsibility for overseeing general land use, but he also suggested to Draper that perhaps their new planner "should be left free for more general land use planning and general determination of policy."[22] For his part, MacKaye understood his role to be "the task of formulating a method of procedure for carrying out the general program of R.P.-1."[23]

MacKaye developed a colorful metaphor, harking back to his boyhood love of railroads, for his work at the TVA. "I was given the little chore of laying out a regional plan for the whole Tennessee Valley," he reported to Stuart Chase, after he had been with the agency for a year, "—a plan for a plan; this constitutes my job; I have the best job in the TVA and the best job of my life; Arthur Morgan runs the locomotive while little me resides in the caboose; each end commands the total view; the caboose can study the landscape but the poor old in-gine must watch the danger signals."[24] MacKaye's upbeat dispatch to Chase, whose popular articles and books promoted the TVA's accomplishments and prospects, contained no clue of the bureaucratic crack-up for which both he and Morgan were headed.

The philosophical differences among the directors were soon reflected at lower levels of the agency. Richards and the Forestry Division, along with some of the planners, landscape architects, and geographers in the Land Planning and Housing Division, overseen by Draper, have been identified by some TVA historians as "superidealists," who supported Arthur Morgan's more utopian intentions. Harcourt Morgan's subordinates, the "agriculturists," with stronger ties to local farming and economic interests, hewed to more conservative land-use and economic development policies. Among the superidealists who served in the TVA in those early years, none was more supremely high-minded than Benton MacKaye. Indeed, the philosophical and political tenets of his approach to regional planning soon placed him at odds with even some of his colleagues in Draper's division.[25]

By training and tradition, foresters like MacKaye and Richards were accustomed to thinking and working on regional terms and according to watershed boundaries. Some of the landscape architects in Draper's division, by contrast, had dealt primarily with discrete sites and projects, which did not necessarily coincide with broader natural or political divisions. Jim Moorhead, a young TVA landscape architect friendly with MacKaye, later described how "the basic differences" between MacKaye and Draper were grounded in their professional backgrounds. The landscape architects, whose numbers prevailed in the division, "were trained in the Olmsted and English view of nature, appreciating its beauty and preserving scenic values. Our training was largely in developing land for estates and parks," Moorhead recalled. MacKaye's forestry experience, on the other hand, provided "a broader palette of thinking than the Landscape Architect's concern for physical beauty."[26]

Land-use, economic, and social planning nevertheless came to be viewed with suspicion or even contempt in some important quarters of the TVA. A pragmatist like Lilienthal described planning as fundamentally incompatible with what he called "grassroots democracy." "There is something about planning that is attractive to that type of person who has a yen to order the lives of other people," he wrote. "It has an attraction for persons of a vague and diffuse kind of mind given to grandiose pictures not of this world. Planning is a subject that attracts those who are in a hurry but are rather hazy as to where they want to go."[27] Some TVA skeptics, however, regarded Lilienthal's assault on broadscale planning as a rationalization of his own ambitions to control the TVA's burueacratic domain.

In his own planning rhetoric, MacKaye attempted to emphasize the need for flexibility and adaptability. Regional planning "is government by vision, not dictation," he wrote just before taking up his TVA duties. "It would picture opportunities, not formulate commandments; its eyes are on the map and not the person."[28] Yet MacKaye in fact brought his own preconceived, schematic vision to the Tennessee Valley, and he quickly began laying out this vision in a steady stream of reports and memorandums to his new colleagues. Earle Draper soon detected that MacKaye's ideas and approach embodied certain potentially unsettling implications. "Inasmuch as it looks as if most land in the country will be brought under some sort of control within the next generation," Draper observed in response to one of MacKaye's earliest memos, "I am wondering if your comments carried to their logical conclusion would not mean Government ownership or socialization of all land. . . . Complete ownership of all land by the government and leasing for a specific use might bring us benefits of a certain sort but probably also attendant disadvantages."[29] Draper's premonition identified a latent ideological fault line within the agency concerning the appropriate balance between public and private control of land and natural resources.

By midsummer, MacKaye was ready to present the outline of his Tennessee Valley planning program to Chairman Morgan. He practiced his presentation one night before Harvey Broome, Ned Richards, and another like-minded TVA forester, Bernard Frank. With his maps and charts, "Benton laid his mind open to us," as Broome recorded the scene. "Almost intolerant of details, it is the vision that counts to him. He said privately, 'I have done my best. If they don't see it or receive it, I don't give a damn.' It was great—a sort of historic occasion—the first rehearsal of the first plan for the first region of the country definitely to come under planning."[30]

MacKaye's presentation in Arthur Morgan's office fulfilled his most pessimistic expectations. Prepared with thirteen complex charts, he was allowed less than ten minutes to make his case during a half-hour meeting. Morgan's terse response to MacKaye's abbreviated presentation perhaps implied its own answer. "The question," Morgan said of MacKaye's array of charts, "is whether they are worth the paper they're written on."[31]

The extensive progress report MacKaye presented to Morgan a month later amplified the theory behind his maze of flow charts and diagrams. His TVA planning toolkit resurrected the method of charting industrial processes and "flows" that he had developed nearly a decade earlier for his "World Atlas of

Commodity Flow." Indeed, his evolving report comprised an intellectual grab-bag of past plans and projects, integrating ideas that he had previously worked on separately. Wilderness belts, hiking trails, working forests, new communi-ties, townless highways, hydroelectric dams and powerlines, upstream flood control reservoirs—all would eventually be incorporated into what MacKaye called his TVA "Opus One."[32]

He addressed directly the relocation of some of the region's population which would inevitably result from the TVA's varied activities. Unpredictable new population patterns, MacKaye insisted, could exacerbate the very prob-lems the TVA was intended to help solve—unless the agency identified pre-cisely an optimal population density for various "environments":

> What is the decent limit to the congestion of an urban population? What health-ful and desirable limits, in terms of persons per square mile, go with a truly rural setting? Are there bounds beyond which large cities should not grow? What pro-portionate rural areas are required to give adequate opportunity for the more se-cluded manner of living? What areas should be preserved as primeval refuges?
>
> Some agreement on such questions must be made before we can base on rea-son and not on dogma, any policy of population distribution.[33]

The "method of procedure" outline in MacKaye's "Opus One" consisted of a series of compelling and fundamental questions, but it offered few specific ideas for quick translation into action. Arthur Morgan urged his subordinates to develop proposals that provided concrete evidence of progress and activity. He had already pressed Draper, MacKaye, and the other planners to narrow the focus of their efforts to the Norris "sub-region," named in honor of the TVA's political champion.[34] The TVA's various programs and problems were developing rapidly in the area surrounding the site of Norris Dam, north of Knoxville. The Norris Basin, two historians of the area have observed, "pro-vided TVA's first laboratory in regional planning and one of the most impor-tant," its people "the raw materials of TVA's planning experiments."[35] While work proceeded on the dam itself, Draper's division applied certain garden city precepts to the creation of a new community, Norris, nearby. Draper's en-gineers and landscape architects were at the same time designing a similarly named freeway from Knoxville to the town and dam. Meanwhile, the agency also confronted the sensitive task of relocating families whose farms and homes would be inundated by Norris Reservoir.

MacKaye, working with Richards and others in the forestry division, devel-oped a plan for the Norris Forest Working Circle. Based on the concept famil-iar to foresters, and which he had earlier proposed in *Employment and Natural*

Resources, the forest working circle first suggested by MacKaye called for the management of 120,000 acres on a genuinely sustained-yield basis. His social blueprint for cohesive and stable communities envisioned a continuous, long-term supply of forest products, permanently supporting the loggers, millwork-ers, and their families—3,000 persons or more, he estimated—who would live in the area. The foresters' eventual proposal covered only half the land sug-gested in MacKaye's original scheme. Even so, their bid to create new lumber-sustained communities on public land in the Tennessee Valley made no more headway than had MacKaye's similar World War I–era proposals for the Pa-cific Northwest. MacKaye incorporated the Norris Forest Working Circle into "Opus One."[36] But outside political forces, as well as the TVA's internal ideo-logical and personal disputes, soon constrained the organization's tentative attempts to fashion an extensive regional planning program. In an hour-long meeting, Arthur Morgan instructed MacKaye to continue work on the Norris project.[37] Draper was gauging the director's cool reaction to his regional planner's "Opus." Though relations between MacKaye and Draper remained civil and professional, MacKaye's already vague responsibilities within the di-vision became increasingly marginal.

MacKaye's daily Knoxville routine, he liked to say, was bounded by "an alpha-betical triangle—the *YMCA* where he slept, the *S & W* Cafeteria where he ate, and the *TVA* where he worked."[38] His TVA colleagues responded variously to his unique personality and odd work habits. One officemate recalled that MacKaye "would drift in about ten in the morning, nod a good morning and sit staring at the wall until he dozed off."[39] If some of his skeptical co-workers wondered just how—or if—MacKaye completed any work, part of the explana-tion was his habit of working at his TVA office alone and late into the night. He was an outright inspiration to some of the younger men in the office, how-ever. Robert Howes, fresh from the landscape architecture program at the Uni-versity of Massachusetts, recalled that MacKaye's "cubbyhole of an office" on the sixth floor of the Arnstein Building in downtown Knoxville, a room "forever redolent of tobacco from his ever-present pipe," was always open "to any of us who elected to drop by for an exchange of ideas, comment or opin-ion—for a 'pow-wow,' as Benton used to say."[40]

On Benton MacKaye's ninetieth birthday, Earle Draper recounted how "about once a week" his former colleague "would come down to a staff meet-ing with fresh diagrams showing flows of water, men and goods in a region.

He would pep up the others with his fresh ideas and imagination and then disappear upstairs again."[41] After MacKaye's death, though, Draper recalled his contribution to the department in a somewhat more ambiguous light. "I can't put my fingers on anything definite that he accomplished in the planning, but I know he had an effect," he commented. "You might say he was a contributing influence to the thinking of people in my division."[42]

With the advent of the TVA, the small city of Knoxville became a bustling regional center. "This place is more stimulating than Washington," MacKaye excitedly reported to Stein. "And why not? It is a little United States Government packed into four small office buildings on two narrow streets. One is completely and perpetually surrounded by experts and specialists on every subject under the sun and all talking at once."[43]

Knoxville was also a mecca for journalists, politicians, foreign dignitaries, and reformers who came to observe firsthand the progress of the TVA's efforts. Stuart Chase, Judson King, Oscar Ameringer, Horace Albright, David Cushman Coyle, Harry Laidler, Robert Marshall, Clarence Stein, Sir Raymond Unwin, Charles Whitaker, and Henry Wright were among those who made a point of calling on MacKaye.[44] Norris Dam, rapidly taking shape on a round-the-clock work schedule, was a favorite destination for visitors' pilgrimages. MacKaye enjoyed explaining to his friends the feats of civil engineering on display at the dramatically illuminated dam site, which swarmed with workers and machinery. "More than once," he recounted, "people whom I've casually known in the whirl of New York or Washington come here and linger in the buildings, and squat in the cafeterias, rumble out to Norris, and then return to talk long into the night."[45]

If MacKaye became frustrated by the resistance and controversy his own efforts faced within the TVA, he reveled in the uncompromising devotion offered by a group of younger friends, who soon dubbed themselves "The Philosophers' Club." The name originated from a conversation overheard between two busybodies in a Knoxville restaurant. "Do you know that pipesmoking man over there?" one was heard to say. "I understand that his name is Benton MacKaye and that he belongs to some sort of a philosophers' club." Comprising some of MacKaye's friends in the Smoky Mountains Hiking Club, as well as young TVA staffers and their spouses, the self-styled philosophers kept up a busy schedule of the social activities that MacKaye loved: square-dancing, led by his calls and accompanied by his harmonica; high-minded readings and discussions—often from his own works or those of his brother James; and weekend hikes in the Smokies. Harvey Broome and his sister Margaret were among the members. When MacKaye at one of their dances introduced her to his young TVA colleague Bob Howes, he sparked a romance that led to their marriage within a year.[46] The philosophers' gatherings became more formal and regular in September 1935, after MacKaye moved from his spartan digs at the YMCA to a more spacious and comfortable apartment at 21 Maplehurst Park, in a pleasant riverside neighborhood. Indeed, by MacKaye's parsimonious standards, his TVA salary of $4,500 was princely compensation, representing the steadiest paycheck he had received in almost fifteen years, the most substantial of his lifetime thus far.

Broome, as compulsive a recordkeeper as MacKaye, calculated that during one year the hiking philosophers spent forty-five weekends in the mountains.[47] MacKaye, years older than his companions, his physical vigor diminished in the aftermath of his surgery, sometimes stayed at camp or napped beside the trail while others tramped on ahead. One hike up Brushy Mountain included MacKaye, Bob Marshall, and Stuart Chase. Throughout the day, other members of the party were amused by the one-word exclamations of these three distinguished climbers—"Bully" (MacKaye), "Swell" (Marshall), "Marvelous" (Chase).[48]

"They are a gorgeous lot," MacKaye wrote a long-time New England friend about his young disciples. "They dance and they tramp and they sing—and through it all they *think*. Indeed they do just about the opposite of the average mundane jackass that one meets in the present 'whirling dervish' civilization."[49]

Among the core group of seventeen "philosophers," MacKaye soon singled out one for special attention. In May 1934, at a Southern Appalachian Trail Conference held in Highlands, North Carolina, and attended by more than 150 members of four southern trail clubs, MacKaye had been lionized as "Creator of the Trail, only a little removed from the earlier Creator."[50] At that event he had met an attractive, personable twenty-one-year-old member of the recently created Georgia Appalachian Trail Club, Mable Abercrombie. The young woman had arrived in Knoxville later that summer seeking a job at the TVA. Abercrombie failed her stenography test, but MacKaye immediately took her under his wing, arranging for her to take the exam again and finding her a job in the forestry division.[51] For the duration of MacKaye's stay in Knoxville, Abercrombie was his regular companion.

Their relationship had a formal, slightly comic aspect. As Abercrombie later recalled, she became his Pygmalion, a "project" in personality building. MacKaye referred to her as "the child" and prepared for her a course of reading suitable to the education of a young "pantheist." Abercrombie was rather startled by some of MacKaye's eccentricities, such as his ignorance of music (besides country dance tunes and Gilbert and Sullivan ditties) and his habit of leaving a restaurant in the middle of a meal if a radio was heard. She remained cheerfully immune to most of her courtly admirer's earnest intellectual intentions. She may have been too young and ingenuous, though, to appreciate the anguish obscured by the three-decade gap in their ages and MacKaye's old-fashioned reserve. Once, inadvertently, he let slip that he had been married. When Abercrombie pressed him on the subject, MacKaye immediately shifted their conversation to other matters.[52]

His protective relationship with Abercrombie also involved matchmaking. During one of Bob Marshall's whirlwind visits to Knoxville, MacKaye introduced his two young mountain-climbing friends. The bachelor Marshall, who courted a variety of spirited women, later invited Abercrombie to visit him in Washington, and the two spent part of a summer hiking in Montana.[53]

When MacKaye left Knoxville in the summer of 1936, his further communications with Abercrombie cooled. Fearful of the anticommunist climate that overtook the TVA in the late 1930s, she returned to her native Georgia and married.[54] Until the last decade of his long life, when he enjoyed the company and care of his neighbor Lucy Johnson, MacKaye perhaps never again came so near to enjoying close female companionship as during this innocent two-year friendship with Mable Abercrombie.

At the end of 1934, Benton MacKaye took a leave of absence to return to the Northeast, where family matters demanded attention: his brother James was hospitalized, and his sister Hazel expected his annual Christmas visit to the Gould Farm. Moreover, Bob Marshall had summoned the Wilderness Society organizers to Washington, to flesh out a program and platform for their fledgling group. A few days before Christmas, with seven other members of the Philosophers' Club who had come to see him off, MacKaye danced a quadrille on the platform of Knoxville's Southern Railway station. He did not then know how emotionally intense the coming six-week sojourn would prove to be.

Benton spent several weeks in the Boston area, to see James through gall bladder surgery. The doctors predicted recovery. Benton headed to Washington, where, a week later, just days after helping draft the Wilderness Society's founding charter, he received word that his beloved brother had succumbed to pneumonia.[55] He quickly returned to Cambridge for his brother's memorial service, at Harvard's Memorial Church, and visited his sister at Gould Farm. Percy, then living in Florida, scolded his surviving brother for keeping him in the dark about James's medical problems. Then, after cooling down, he urged Benton to write a memoir of James. That summer, back East for vacation, Benton was joined by Horace Hildreth, and they scattered James's ashes from the summit of Mount Watatic, northwest of Shirley Center.

James's death set Benton to rereading his brother's writings, including several unpublished manuscripts. James MacKaye's austere, arcane philosophy may have left some of his colleagues on the Dartmouth faculty dubious; but a dedicated band of student disciples had gathered around him during those final serene years of his life, yielding to the force of personality that had once caused John Reed to remark that Jamie "talked like Confucius and looked like Apollo."[56]

No one had fallen more passionately under James MacKaye's influence than his younger brother Benton. "His mind and mine came closer to coinciding than any other that I knew," Benton later observed.[57] It had been James who had most clearly articulated the "religion of evolution" that became Benton's personal creed. James's doctrine, as Benton explained in a memorial tribute, citing a 1909 lecture by his brother, suggested "not that man is a low caricature of a pre-existing divinity, but that he is a dim prophecy of a divinity that is to be; not that the universe proceeded from the deity, but that it is proceeding to him. . . . The purpose which [science] suspects in the universe is one of improvement—a striving after something infinitely better than has ever been before."[58]

Benton returned to Knoxville determined to infuse his own work at the TVA with the spirit of his brother's ideas and of the Wilderness Society's creation. "We have many maps upon our walls but no single map upon the public mind," he attested to Draper, who had asked for his latest thoughts on TVA "land planning matters."[59] MacKaye's efforts at the TVA revolved increasingly around the issues of outdoor recreation and wilderness preservation. The previous September, only a few days after composing a draft statement of principles for the nascent Wilderness Society, he had prepared a memorandum proposing that a wildlife management and study program be incorporated into the TVA's regional planning efforts.[60] Also in the autumn of 1934 he, Richards, and Bernard Frank had pressed Draper and Augur to create a TVA-managed wilderness area in the Linville Gorge of North Carolina, as an experiment in "environmental development."[61]

In fact, recreational development quickly became a significant element of the TVA's overall program. Recreation, including its potential economic benefits, provided an important justification for the public development of reservoirs and their new shorelines. MacKaye again found himself at odds with his TVA colleagues, however, as they tried to establish a common vocabulary to describe their efforts and objectives. One of his brother's obsessions had been the fundamental need for the precise definition of terms. Benton recounted these difficulties to one of James's former students: "We spend, I truly believe, more than half our time trying to find out what on earth the other means by the words which they . . . use."[62]

MacKaye struggled to make his colleagues understand the central importance of what he called "habitability," a concept and standard by which he believed all TVA programs, including recreation, should be measured. "Habitability is human ecology," he said. "Like every ecology it demands its *balances*."[63] MacKaye now used the term *habitability* to describe his conviction that planning, by whatever description, must simultaneously embrace urban, rural, and wilderness settings, and he urged the agency to conduct "habitability surveys" throughout the region.[64] "Another name for . . . habitability," he insisted, "is *environment*."[65] For MacKaye, it went without saying that the influence of a wholesome physical environment would enhance the lives of the valley's inhabitants. Outdoor recreation, parks, and wilderness protection could therefore be justified on their own terms, regardless of their potential economic benefits. But the pursuit of habitability as a TVA objective was, if anything, even more remote and exasperating to his superiors than his lofty depictions of regional planning.

In one lengthy and impassioned memorandum, "Primeval Environment as Natural Resource," MacKaye quoted liberally from his own *New Exploration* to make his case that the TVA could properly undertake "wilderness planning."[66] Just a few weeks before submitting it, he had met with Harvey Broome, Bernard Frank, and Robert Howes (three of the four were founding members of the Wilderness Society) to discuss the prospects of promoting a wilderness program within the TVA.[67] MacKaye's memorandum could only have unsettled his cautious TVA colleagues, however. Reflecting the wilderness proponents' growing obsession with the precise classification and description of outdoor recreational resources, he described the difference between the "outing area" and the "wilderness area." Development of the former creates "the opportunity for scenery," he observed; the second represents "the type needed for solitude." The distinction between the two, he continued, was that "while the outing area may be a mixture of urban and primeval influences, the wilderness area is the exclusion of the city."[68] And the outing area, he enthusiastically predicted, would inevitably "spread discontent." Citizens exposed to modest and accessible parks and recreational resources would clamor for the protection of larger and more numerous tracts of wild land. "Wilderness planning," he charged, "in terms of habitability, seems not to have been carried on, by us or others, on any scale or understanding commensurate with its importance."[69]

"I won't say that you have convinced me that primeval environment is a natural resource," Draper coolly responded, but he granted that the "majority of people will agree that a certain acreage of primeval land forms properly protected is desirable." The "burden of proof is on you to show that they are needed," he firmly instructed MacKaye.[70]

Late in 1935, after MacKaye was named by Draper to the three-member Resort and Recreation Committee, he immediately lobbied to change the group's name to the Scenic Resources Committee, arguing that the new label would more clearly reflect the tenets of habitability.[71] Tracy Augur, the committee's chairman, accepted MacKaye's suggestion but took issue with his reasoning. The TVA "must be concerned with the scenic resources of the Valley in the same way that it is concerned with its agricultural or forest cover or mineral or water power resources," he responded, "namely as a means of sustaining livelihood."[72] MacKaye, abandoning bureaucratic discretion, reaffirmed his own conviction that the TVA's "*direct* goal is public use as distinguished from private profit."[73]

MacKaye and his colleagues were near the point of talking past one another. His patience was running thin with the endless succession of meetings and memorandums that sometimes appeared to be the sum and substance of the TVA's regional-planning efforts. In a mock memorandum to one colleague, prepared after a particularly futile staff meeting, he quoted both graphically and verbally from Lewis Carroll's nonsense epic, *The Hunting of the Snark.* MacKaye's one-page memo included an empty rectangle depicting "A map they could all understand. . . . A perfect and absolute blank."[74]

❖

In late 1935 and early 1936, as the impetus behind Project R.P. 1 waned, MacKaye continued to promote his ideas on highway policy, recreation, and wilderness protection. His March 1936 draft of a national scenic resources law was a pathbreaking proposal for genuine national wilderness-protection legislation. Though MacKaye did not say so in the TVA memorandum accompanying his proposal, the scenic resources bill grew directly from the ideas he shared with his Wilderness Society comrades.[75] From long and bitter government experience, he and other battle-worn wilderness advocates such as Marshall, Aldo Leopold, and Robert Sterling Yard well understood the ambiguous, arbitrary nature of bureaucratic designations of public lands, especially those identified as "wilderness" or "primeval."

In 1934, Marshall had tried without success to convince his chief, Secretary of the Interior Harold Ickes, to establish a wilderness planning board. He also pestered Ickes to create a recreation committee under the National Resources Board, to coordinate official nomenclature for the classification of federal lands devoted to recreation, wildlife, and wilderness. "When one talks about a Wilderness Area," Marshall wrote to MacKaye, "it should mean the same in a National Park, a National Forest, or an Indian Reservation."[76]

The 1916 law establishing the National Park Service, MacKaye pointed out in the memorandum accompanying his scenic resources bill, commanded the agency "to provide for the enjoyment" of the parks so as to "leave them unimpaired for the enjoyment of future generations." For MacKaye, though, the law's fatal flaw was its failure to define *enjoyment*. So the purpose of his bill, he wrote, was "to fill the gap . . . between the vague utopia of 'enjoyment' and the varying notions of officials. 'Policies' would not (as now) be left to said officials but would be written down as law by Congress itself."[77]

MacKaye's bill called for a scenic resources board, consisting of the secretaries of agriculture and the interior and a third member appointed by the president for life. The board would be empowered to classify all federal lands for scenic, recreational, and wilderness purposes. The heart of MacKaye's 1936 proposal was a detailed four-class "scheme of classification." The first class, "Outing Areas," he divided into nine subcategories, ranging from "Intensive Use Areas," designed for "a specially large number of visitors per acre-day," to varieties of "Wayside Sites" along highways for picnicking, camping, or scenic observation. A second class, the "Wayside Zone," designated "a belt of rural or semi-primitive territory" bordering a highway. The third, the "Wilderness Zone," envisioned a belt of "primitive or semi-primitive territory" running along a natural feature such as a mountain chain (as the Appalachian Trail did), "isolated largely from mechanical sounds and sights." MacKaye divided the fourth class, the "Wilderness Area," into types: "Extensive," "Restricted," "Superlative," and "Primeval."[78]

Many of the apparently minute differences among the various proposed categories of federal recreational and wilderness areas were based on the presence or absence of roads and other improvements to accommodate automobiles. MacKaye's legislative blueprint attempted to control the federal government's growing role as the promoter of intensive recreational development on public lands, whether in the form of new man-made lakes in the Tennessee Valley or scenic mountain parkways along the Appalachian range.

MacKaye envisioned a regional approach to the designation and administration of federal recreational, scenic, and wilderness lands. The Scenic Resources Board would be empowered to classify lands within each of six census regions over a five-year period. Furthermore, he proposed that application of the law's land-classification definitions take into account differing conditions in each region, rather than being applied according to a rigid and uniform national standard. In MacKaye's scheme, the Scenic Resources Board would publish regulations for each region, to establish the maximum number of visitors per acre per day for outing areas and visitors per square mile for wilderness areas.[79] In other words, he was calling for density standards—and limits—for the use of federal recreational and wilderness lands, a notion that few public officials had yet dared whisper.

MacKaye's provocative 1936 scenic resources proposal never saw the legislative light of day. He sent a copy to Robert Sterling Yard, the secretary of the Wilderness Society. Bob Marshall's brother George later speculated that Mac-Kaye "may have been the first" to propose a national wilderness-protection law.[80] The suggestion stirred no response at all within the TVA, however—other than, perhaps, to speed his departure from the agency. MacKaye's draft of a scenic resources law proved to be one of his last TVA tasks.

Regardless of his fate at the TVA, MacKaye's influence within the agency was being felt in subtle yet significant ways. Draper had detailed several staffers, including Bob Howes, to develop a scenic resources inventory of the region. Some of the young landscape architects and planners leaned heavily on MacKaye for guidance and ideas. Their substantial report, *The Scenic Resources of the Tennessee Valley*, published in 1938, incorporated a version of the detailed and discriminating classification of outdoor recreational areas that had become the fixation of MacKaye, Broome, Marshall, Leopold, and other like-minded wilderness advocates. More than thirty years later, Robert Howes, near the end of his own long TVA career, claimed that the scenic resources inventory represented "perhaps Benton's most permanent record" at the TVA.[81]

❖

In early April 1936, Augur and Draper reported to MacKaye that Arthur Morgan was favorably impressed with his proposal to develop a "Wayside Handbook." But just as he prepared to focus his energies on an environmental guidebook for motorists, MacKaye fell victim to a TVA organizational shakeup. On April 16, Draper informed him that the TVA no longer required his services, effective at the end of June.[82] When Roosevelt, over Morgan's fierce objections, a few weeks later extended Lilienthal's appointment to the board for another term, the die was cast for the agency's future. The days of the TVA's "superidealists," such as MacKaye and the Forestry Division's Ned Richards, were already numbered. The ongoing feud among the TVA's directors would climax in 1938, when Roosevelt dismissed Arthur Morgan and Congress conducted an intensive investigation of the agency. The mechanism on which MacKaye had pinned his ambitious hopes for the TVA—the preparation of "surveys and plans" and the recommendation of legislation under the provisions of Sections 22 and 23 of the TVA Act—faded as a rationale for the agency's activities.[83]

MacKaye accepted his termination with equanimity—even a measure of relief. "I am told, and I believe in all sincerity, that my philosophy of planning has been a 'real contribution' but that its further pursuit just won't fit the program for the next fiscal year," he told Lewis Mumford. "Well, the point could be made that said contribution should be clinched and not left waving (and such point has been made quite independently of me). But *as* for me I'm ready and itching to do the clinching on my own outside, and so I've said 'amen.'"[84]

Draper may have come to regard MacKaye's powerful influence over some of their younger colleagues as a threat to his own authority. Shortly after MacKaye left the TVA in 1936, the young Jim Moorhead reported to MacKaye that the division was in turmoil and that he and some of his co-workers were "termed by Mr. Draper to have been too much affected by your ideas and that in my case this amounted to 'idolatry' and that I had developed an 'immature cynicism.'"[85]

In his perceptive account of MacKaye's years at the TVA, historian Daniel Schaffer concludes, "The fact that he stimulated so many people shows how richly his ideas were textured. The fact that he ultimately failed to have his ideas implemented illustrates how alien his planning concepts were to mainstream agency thinking."[86] MacKaye's immediate failure to imbue the TVA with his "alien" planning ideas exemplified, on a small and personal scale, the collapse of other grand social and economic planning proposals of the New Deal era. Despite his rocky TVA experience, however, MacKaye never abandoned his faith in the valley authority idea as a cornerstone of comprehensive regional planning. In the early 1920s, after he had failed to win government support for his planning and colonization ideas, he had successfully initiated the Appalachian Trail. During the mid-1930s, partly as a consequence of another frustrating stint in the federal government, he joined other activists and recreation-minded citizens in organizing the growing movement to protect wilderness lands throughout America.

✤ 14 ✤

The Wilderness Society

1934–1936

As an apostle of the utilitarian tradition that equated conservation with multiple-purpose river valley development and forest management, MacKaye tried to convince his TVA colleagues to adopt principles of regional planning that emphasized wilderness preservation as the equivalent of other land and resource uses. But he expressed few strong reservations about the scale of environmental disruption and change that the TVA's massive projects created along the region's river channels. The preservation of "rivers for their own values," in fact, remained a neglected purpose of the American conservation and environmental movements for decades to come.[1] However, MacKaye, along with other wide-ranging land-use thinkers, like Aldo Leopold and Bob Marshall, became a leader of the small but growing number of American conservationists who during the New Deal era addressed head-on the federal government's paradoxical role as both the developer and the protector of the nation's wildlands and natural resources. His central role in an intense struggle for the organizational and intellectual control of the Appalachian Trail idea provided a catalyst for the remarkable group of articulate wildland preservationists who united in January 1935 to create the Wilderness Society.

Scenic mountain highways like Skyline Drive, the location of which MacKaye had protested, had proven their popularity with ever-growing legions of motorists, local boosters, and the many workers employed to build the roads. The skyline highway idea also had an important new enthusiast in President Roosevelt himself. His polio-stricken legs useless to carry him, Roosevelt provided a poignant symbol for those who advocated greater public access to the rugged mountain crestline. "How would I get in?" the president replied plaintively, when conservationist Irving Brant once urged him to prevent further road-building in the proposed Olympic National Park.[2]

When Roosevelt visited CCC camps in Shenandoah National Park in the summer of 1933, reporters heard him muse about the potential for a mountain highway from New York to Georgia, passing through the Shenandoah and Great Smoky Mountains parks. In November of that year, Secretary of the Interior Harold Ickes announced that FDR had approved a scenic, ridgeline highway connecting the two national parks. The Public Works Administration, which Ickes then headed, quickly authorized an initial $4 million allotment for the Blue Ridge Parkway, which would eventually stretch almost five hundred miles.[3]

The immediate public appeal of Skyline Drive, the Blue Ridge Parkway, and the federal government's program of public works, expanding during Roosevelt's first term, soon inspired proposals for numerous eastern mountain parkways. William Carson, a booster of Skyline Drive, suggested a parallel road on the western side of the Shenandoah Valley. In the Great Smokies, plans were proceeding for a so-called skyway that would run the length of that park. One Park Service project plotted a Green Mountain parkway the length of Vermont. Another proposal called for a road across the Presidential Range of New Hampshire's White Mountains, New England's highest, most hallowed hiking terrain. Plans for mountain parkways in northern Georgia and eastern Pennsylvania also appeared.[4]

For trail and wilderness enthusiasts, the potential consequences of all these road proposals were nightmarish. Many of the schemes would impinge directly on the Appalachian Trail and other popular hiking paths. For MacKaye, they epitomized the "metropolitan invasion" he had prophesied. In March 1934, just before taking up his TVA duties, he wrote an article, "Flankline vs. Skyline," for *Appalachia,* the journal of the Appalachian Mountain Club. His plea became a manifesto for the small but vocal band of skyline highway opponents. The article again promoted the virtues of what he called "flankline" parkway routes, which, he asserted, could serve the recreational and scenic needs of the motoring majority as well as ridgetop routes could. And he addressed directly the accusation of elitism and selfishness often directed at preservationists. Like Robert Marshall in his 1930 article "The Problem of the Wilderness," MacKaye argued that the American tradition of protection for minority rights gave hikers and wilderness lovers a legitimate claim to a modest portion of wild mountain landscape.[5]

MacKaye conceded some scenic highways would inevitably be constructed. At issue, then, was their location and route. The "skyline" alternative, he charged, "cuts the wilderness in two: . . . it violates the wilderness solitude not merely here and there but throughout its whole length—its violation tends always toward the maximum."[6]

The influence of MacKaye's article reached beyond *Appalachia*'s modest readership. A few months later, Harvey Broome promoted the "flankline" idea in the pages of the liberal *Nation*.[7] And Horace Albright, after reading the article in *Appalachia,* assured MacKaye that he had taken up its message with his recent successor as director of the Park Service, Arno Cammerer. "I had a long talk with Cam in Washington about these problems of preserving the wilderness, and I found him very sympathetic," Albright wrote. "We can be absolutely certain that the old policy established when I was Director, of holding the eastern half of the Great Smokies as a wilderness area, keeping out all roads, will be maintained."[8]

The idea that a Great Smokies skyline highway might slice through only half the park provided little solace for wilderness enthusiasts. From Knoxville, MacKaye orchestrated a campaign of resolutions, correspondence, and propaganda intended to persuade the Appalachian Trail Conference to take a strong, unequivocal stand against skyline roads. As the parkway fever burned, however, it also widened the latent philosophical and personal rifts in the eastern trail community.

The Appalachian Trail Conference, since its creation in 1925, had been slow to establish a strong independent identity. As a federation of independent amateur organizations stretching from New England to Georgia, the group received varying forms and degrees of support from its constituent clubs. In fact, loyalty to the ATC as an organization—and to its chairman Myron Avery—seemed to vary in proportion to a member club's distance from the conference's headquarters at the Potomac Appalachian Trail Club in Washington. The hard-driving Avery, who was intent on completing and linking the AT's unfinished sections, could expect support from his own PATC and from neighboring clubs, such as the recently formed Natural Bridge Appalachian Trail Club of southwestern Virginia; but MacKaye still had many faithful followers in the northeastern clubs, such as the Appalachian Mountain Club, the Green Mountain Club, and the New York–New Jersey Trail Conference, and in such southern groups as the Smoky Mountains Hiking Club and the Georgia Appalachian Trail Club. Even so, most members of all of these clubs were interested primarily in recreational and social activities; probably only a few were involved in, or even knew much about, the intense behind-the-scenes disputes that were flaring up among leading trail activists like MacKaye, Avery, and their closest supporters.

❖

MacKaye could not attend the Sixth Appalachian Trail Conference, held in Rutland, Vermont, in late June 1934, but he sent the gathering a written message, "Expression of Sentiment," regarding skyline roads. He hoped that the meeting would adopt the position that the conference "stands unequivocally opposed to the skyline or crestline type of highway as suggested for the Mountain Ranges of New England and for the Park-to-Park highway in the Southern Appalachians; and, in lieu thereof, suggests that whatever highways or parkways are built near the Appalachian Ranges be located along the lower flanks and levels."[9]

MacKaye's proposed resolution directly rebuked Avery's policy of cooperation and compromise with government agencies. The two men's differences had also surfaced a month earlier at the Southern Appalachian Trail Conference. "Our main problem," Avery said of the Appalachian Trail, in a message sent to the gathering, "is to actually create it. Then we may discuss how to use it." MacKaye, speaking personally to the southern hiking enthusiasts, expressed the belief that Avery had matters backwards. "A wilderness is like a secret: the best way to keep it is to *keep* it. Keep the wilderness *wild*. Do not manicure it. To a manicured civilization the wilderness (real wilderness) comes as something new—as a new path in the public mind." His allusion to "manicured civilization," as some of the audience may have recognized, revealed MacKaye's growing skepticism about Avery's methods for "improving" the trail, such as grading it and adding signs. As it happened, though, the Georgia conference adopted a brief resolution urging "that the crestline or skyline in general be preserved exclusively for primitive recreation via foot paths where desirable."[10]

Things went differently a month later at the Vermont meeting of the full ATC. MacKaye's skyline highway resolution appeared at an inopportune moment for the host Green Mountain Club. The Vermont club had opposed the original skyline route of the proposed Green Mountain Parkway, which would have obliterated stretches of the Long Trail. But the Vermont road proposal had substantial support in the economically hard-pressed state; and when government planners indicated that they were considering a low-level flankline route of the sort proposed by MacKaye, GMC members were reluctant to be associated with a resolution that, as one observer related to MacKaye, "would be interpreted as opposition to *all* mountain parkways."[11]

MacKaye's resolution, offered by Ruth Gillette Hardy of the Appalachian Mountain Club, was the subject of lengthy and unfocused floor debate. According to her account, Major Welch, presiding over the meeting, indicated that the conference could not adopt such resolutions. Avery, she observed, "tried to be neutral, which is equivalent, under such circumstances, to opposition." Hardy withdrew the resolution, "after it was clear that all the views had been aired."[12]

But MacKaye's efforts had forced many outdoors activists to examine more carefully their positions on the skyline road controversy. Raymond Torrey reported in his *New York Evening Post* column that the debate promised "to make the matter a vital one for outdoor organizations, in the immediate future, if the Federal plans as announced are carried forward, although opposition already aroused may tend to modify them in recognition of the protest formulated by Benton MacKaye."[13]

MacKaye's resolution, and its equivocal reception, had opened up a number of festering wounds within the ATC. Avery had angered and alienated some clubs and individuals, including long-time MacKaye friends and allies in New York and Pennsylvania, in his obsessive and sometimes insensitive determination to get the trail completed. Battle lines were soon being drawn within the eastern trail-club community. Avery tried to reconcile the considerable anti-road sentiment in the ATC with his own, probably quite accurate, sense of political realities. On behalf of the ATC, he urged the Park Service's Cammerer to support reconstruction, at federal expense, of any stretches of the Appalachian Trail destroyed by extension of Skyline Drive southward. Asserting the incompatibility of hiking trails and motor roads, he further requested that "the footpath or Trail be as far removed from the Skyline Highway as practicable."[14]

Avery's effort at compromise with federal authorities also revealed that the trail community did not speak with a single voice. MacKaye vigorously disagreed with Avery's assertion that "those interested in the Trail have preferred to merely ask that what was theirs be restored when destroyed." In fact, he wrote Cammerer, there were many who were concerned "far more for the preservation of the wilderness itself than for the restoration of a particular trail route."[15] "You represent the ultra conservative in conservation," Cammerer responded, defending the Park Service's wilderness policies:

> The fact of the matter is, old man, you and I must remember that about the only safeguard we who love the wilderness have for keeping a lot of it in primitive condition is the national parks because in the national forests they must harvest their crop in time and on the private lands in time the forest would be denuded. A parkway, with a right of way of some 200' to 1000', at least assures for all time the retention of wilderness conditions in that width, and I daresay if you and I were living fifty or one hundred years from now we would find these parkways in most places the only wilderness beauty spots left.[16]

MacKaye took delight in being branded an ultra conservative. "I've been called so often a 'red-eyed radical' that the change is quite refreshing," he replied. But he was astounded by Cammerer's assertion that a 1,000-foot-wide strip of parkland with a highway down the middle could constitute anything resembling genuine wilderness. "This appears offhand like a thoroughly amazing statement," he chided the Park Service director, "especially from the Guardian of American-owned wildernesses."[17] This exchange distilled certain deep-seated differences between the Forest Service and the Park Service about the meaning and management of wilderness on federal lands. Cammerer could not imagine that the Forest Service would ever refrain from intensive timber harvesting in national forests. MacKaye was appalled that the Park Service's leader could not comprehend how parkway development undermined the very resources he was charged to protect. And such institutional differences and suspicions would soon sharpen: four of the Wilderness Society's eight founders were Forest Service veterans.[18]

Avery took sharp exception to what he viewed as MacKaye's meddling with his duties and policies as ATC chairman, complaining, "your letter [to Cammerer] doesn't help what we are trying to accomplish."[19] By now, though, the clearly expressed positions of Cammerer and Avery had convinced MacKaye and his like-minded friends and allies that more aggressive tactics were required in the contest for the fate of the wild mountain skyline.

Harold Anderson, the Washington accountant and PATC activist who had been MacKaye's early and impassioned informant about the progress of Skyline Drive, was also becoming exasperated with the ATC's equivocation under Avery's leadership. "My idea is to play for bigger stakes," he wrote Guy Frizzell, president of Knoxville's Smoky Mountains Hiking Club, in early August 1934. A committee "of four or five persons of some prominence in hiking circles," he proposed, should first try "to spike the idea" of a Blue Ridge skyline highway. Failing that, the committee should seek government support and permanent recognition for the Appalachian Trail, perhaps moving the trail from the Blue Ridge to the Alleghenies, if that were the only hope "to preserve the primeval environment. Then perhaps out of the movement would evolve a real federation of hiking clubs which is now so sadly needed."[20]

Anderson sent a copy of this letter to MacKaye, who coincidentally received a telegram from Bob Marshall containing word that he was arriving in Knoxville on August 11. "By gum," MacKaye told Harvey Broome, "we'll put up to him this proposal of Anderson's."[21] In fact, as MacKaye and Broome learned when they met Marshall at the Andrew Johnson Hotel on the night of the 11th, their friend had been quietly dispatched to Tennessee by Interior Secretary Ickes to investigate possible routes for the southern stretches of the proposed Blue Ridge Parkway. Marshall's party included Park Service landscape architects Stanley W. Abbott and Thomas Vint. Over dinner that night, MacKaye and Broome made their case to the influential federal planners for a "low-level, valley routing" for the road rather than a crestline route.[22]

The next day, MacKaye and Broome joined the surveying party, which was headed for Asheville, by way of Gatlinburg. Marshall accompanied Broome, while MacKaye rode with the rest of the group. At Newfound Gap, where work was already well under way on a road to the summit of 6,643-foot Clingmans Dome, Marshall and Broome took a brisk hike up Clingmans, exchanging thoughts about the road's effect on the surrounding forest. When the hikers returned, MacKaye joined them in Broome's roadster. Along the way, they discussed in earnest Anderson's proposal for a new group to fight the skyline roads. "Bob was enthusiastic," Broome recalled.[23]

But Anderson's idea and the prodding of MacKaye and Broome convinced Marshall to pursue an even more ambitious goal than simply a coalition of eastern hikers. "His thought," Broome reported to Anderson, was "to take up other projects than the preservation of the Trail; the protection and preservation of the wilderness wherever it might occur." Anderson, in fact, was already familiar with Marshall's 1930 essay, "The Problem of the Wilderness," in which he had called "for the organization of spirited people who will fight for the freedom of the wilderness." Now, in one swift stroke, Marshall had transformed Anderson's idea for a regional anti–skyline drive committee into a plan for a national organization "uniting the most influential and aggressive wilderness defenders of the country."[24] Thus was initiated the first concrete step in the fast-moving chain of events that soon led to the creation of the Wilderness Society.

By mid-September Marshall and Anderson had met in Washington to map out the next steps in their plan to form a wilderness preservation group. They began compiling lists of possible members, and, at Marshall's request, Anderson drafted a statement of principles, to be sent as an invitation to the organizing members.[25] Anderson sent his draft to MacKaye, suggesting that he revise it in anticipation of Marshall's return to Knoxville in late October for a convention of the American Forestry Association (AFA). MacKaye was under intense pressure at TVA during these months to produce his "Opus One," but he set to work writing "Invitation to Help Organize a Group to Preserve the American Wilderness." The preamble to his statement was a plea: "help us try to integrate the increasing sentiment which we believe exists in this country for holding its wild areas *sound-proof* as well as *sight-proof* from our increasingly mechanized urban life." And, in one of his eight points, he boldly declared, "the time has come, with the brutalizing pressure of a spreading metropolitan civilization, to recognize this wilderness environment as a serious human need rather than a luxury and a plaything."[26]

"I have not had much experience in organizing societies," MacKaye wrote Anderson, when he sent off his draft, "but enough to know the value of reducing the job to its very simplest terms. And our motto should be Jefferson's paraphrased—the less organization the better." Anderson had proposed "The Wilderness League" as the group's name, Broome "The American Wilderness Association." Either suited him, said MacKaye; "the only suggestion I have as to a name is to keep it short."[27]

On October 19, during the AFA convention in Knoxville, MacKaye, Marshall, and Broome squeezed into a car belonging to TVA forester Bernard Frank and joined a motorcade headed to inspect a CCC camp near LaFollette, north of Knoxville. En route, near Coal Creek, Frank (who was accompanied by his wife Miriam) pulled his car off the highway. On a bank beside the road, as their forester colleagues sped by, the four men arrayed themselves on the ground around Marshall, who worked from a copy of MacKaye's draft statement of principles. "One by one we took up matters of definition, philosophy, scope of work, name of organization, how we should launch the project, the names of persons who should sign the statement and those to whom it should be sent," Broome recalled.[28] By the time they were finished drafting their manifesto for saving America's primeval environment, they had agreed on a name for their fledgling organization: The Wilderness Society.[29]

The roadside brainstorming session was not the only landmark in the wilderness preservation cause that occurred that October day. Marshall was slated to address the AFA convention that evening, ostensibly to describe some of his adventures in the Brooks Range and Koyukuk region of Alaska, which were chronicled in his popular book, *Arctic Village* (1933). Inspired by his discussions earlier that day, though, he launched into an impassioned plea on behalf of the wilderness—making pointed reference to the Clingmans Dome "skyway" he had inspected in August with Harvey Broome. "*Great speech* by Bob Marshall," MacKaye noted in his diary. "*Most* brilliant affair yet."[30]

But some government officials were not so pleased with Marshall's remarks. Arno Cammerer, during a speech at Newfound Gap the next day, took vigorous exception to what he called Marshall's "improper and ungracious attack." As this public dispute between two highly placed government officials was played out in the newspapers, MacKaye came to Marshall's defense in a letter to his old friend Harry Slattery, assistant secretary at the Department of the Interior. "What [Marshall] said constituted a discerning, constructive criticism, which under friendly human 'chemistry' would be welcomed." Cammerer's testy response, he went on, reflected "a tension in the local Park atmosphere which has arisen from the growing opposition to the skyline roads."[31]

Marshall reported that Ickes gave him "a bawling out" for his role in the dispute with Cammerer, who, Marshall added, "apparently got a worse bawling out." Marshall's wealth, self-confidence, and sense of mission insulated him from fear that his blunt public words would have any detrimental consequences for his career. Indeed, he felt sure that he was winning Ickes' support for a more aggressive official wilderness policy. "At any rate," Marshall concluded, "I did not mind being bawled out, nor did I mind apologizing to him for the good of the cause."[32]

On the day after Marshall's controversial remarks and the roadside conference, Mable Abercrombie typed up the revised Wilderness Society "invitation," which went out over the signatures of MacKaye, Marshall, Broome, and Frank.[33] The invitation was sent to six others, including Harold Anderson, Aldo Leopold, John Collier, John C. Merriam, Ernest Oberholtzer, and Robert Sterling Yard. By coincidence, Merriam, president of Washington's Carnegie Institution, and Yard, former president of the National Parks Association, had earlier that year discussed forming a group they called "Save the Primitive League." Yard eagerly accepted an invitation to launch the Wilderness Society, but Merriam declined, pleading other commitments, as did Indian Commissioner Collier. Leopold, then teaching wildlife management at the University of Wisconsin, accepted the invitation. Oberholtzer, a Harvard-trained landscape architect long involved in the fight to protect the Quetico-Superior lake country along the Minnesota-Ontario border, also agreed to join the organizing committee.[34]

In the space of a few years in the 1920s and 1930s, these wilderness-protection advocates had concluded that it was not just the traditional villains—loggers, miners, ranchers, speculators, and assorted other capitalists and entrepreneurs—who were undermining both the wilderness idea and the actual wilderness environment. Just as worrisome was the threat posed by individuals, private organizations, public officials, and government agencies whom the wilderness crusaders had counted among their allies. At the root of "the problem of the wilderness," as Marshall called it, were matters of perception and language. The debate over the coexistence of the Appalachian Trail and the eastern mountain parkways revealed that many Americans were virtually speaking different languages when they used the term *wilderness*.

The founders of the Wilderness Society believed from the outset that they were fighting literally a war of words. The group's key members, veterans of harsh experience in government agencies, were obsessed with questions of terminology, as applied to wilderness and other classifications of public lands. "The Tower of Babel, New Style" is how Yard once described the confused wilderness nomenclature used by different federal agencies.[35] Aldo Leopold in 1921 had offered the memorable definition of wilderness as "a continuous stretch of country preserved in its natural state, open to lawful hunting and fishing, big enough to absorb a two weeks' pack trip, and kept devoid of roads, artificial trails, cottages, and other works of man." Marshall in his 1930 essay, "The Problem of the Wilderness," used the term "to denote a region which contains no permanent inhabitants, possesses no possibility of conveyance by mechanical means and is sufficiently spacious that a person in crossing it must have the experience of sleeping out." Yard recommended that wilderness be defined "as land not entered or developed by roads," a description he called "simple, accurate and clear." The American wilderness, whatever else it represented to these champions of the primeval, obviously was a place without automobiles and other machines.[36]

The Wilderness Society's founders considered not just rear-guard actions to slow the pace of wilderness despoliation; some envisioned a positive program to expand public land protected from development. Broome, for his part, suggested, "We should look about and by a program of withdrawal and protection add other areas which might become wilderness in character after the passage of a few decades." Marshall, writing the chapter on recreation for the 1933 "National Plan for American Forestry," the so-called Copeland Report, had recommended setting aside almost twenty million acres of public land for "primeval" and "wilderness" areas, barred from "mechanized development."[37]

"The time has come, I believe, to hit hard for what we want," MacKaye wrote Yard. "The subject is a complex one and in urgent need of being clarified. A large body of sentiment is with us, I believe, but thus far it has been largely inarticulate."[38]

❖

On January 20 and 21, 1935, during his winter sojourn, MacKaye met with Marshall, Broome, Anderson, and Yard at the Cosmos Club in Washington to hammer out a more detailed statement of the Wilderness Society's principles, purposes, and operations. The four-page printed document produced immediately after the two-day meeting was a genuinely collaborative work, signed by the eight-member organizing committee (including Leopold, Oberholtzer, and Frank, who did not attend); but its principal conceptualizers and authors appear to have been Marshall and MacKaye. Its opening paragraphs, "Reasons for a Wilderness Society," straightforwardly stated the group's intent: "We have constituted ourselves a committee to organize an aggressive society for the preservation of the wilderness. We desire to integrate the growing sentiment which exists in this country for holding its wild areas sound-proof as well as sight-proof from our increasingly mechanized life."[39]

The eleven-point "Platform of the Wilderness Society" consisted of essentially the same statement that had evolved out of the October 19 roadside revision of MacKaye's draft. A section titled "Types of Wilderness" grew from the outline of definitions prepared earlier by MacKaye and Broome. The five varieties of wildlands described were distinguished by size, purpose, and scenic and natural attributes; they were: extensive wilderness areas, primeval areas, superlatively scenic areas, restricted wild areas, and wilderness zones. A list of "Common Types of Wilderness Invasion"—including everything from Mac-Kaye's nemesis, the radio, to power lines and cattle fences—illustrated the challenge of defining the appropriate human place in the wilderness.

The organizers also emphasized their intent to become a genuinely national organization. Other preservation-minded groups, such as the Sierra Club, the AMC, and the Save-the-Redwoods League, customarily focused on regional concerns. And the existing national conservation groups, such as the National Audubon Society and the Izaak Walton League, concentrated on protecting particular natural resources, such as birds, fish, wildlife, and their habitats. The Wilderness Society founders proposed to work throughout the nation for the preservation of the wilderness environment itself. "The principle that an invasion of one area which should be preserved is in effect an invasion of all will be held paramount," the organizers asserted, "so that members of the Society on the West Coast will be enlisting support for a battle in the Appalachian Mountains, while members on the Atlantic Seaboard may be working to defeat a road through some primitive Sierra canyon which they may never expect to see."[40]

The eight-man organizing committee, united by a common love and understanding of a wilderness ideal, also reflected a diversity of regional backgrounds and professional expertise. The group comprised four foresters (MacKaye, Marshall, Leopold, and Frank), a lawyer (Broome), an accountant (Anderson), a landscape architect (Oberholtzer), and a writer and publicist (Yard). A year earlier, MacKaye had suggested to Broome that some *"earth men could start a Carboniferous Cult."* The founding membership and principles of the Wilderness Society came close to fulfilling his whimsical prophecy. Indeed, the editor of the *Journal of Forestry* soon branded the society's 1935 platform as the manifesto of a new "cult of the wilderness."[41]

The first meeting's minutes recorded an anonymous $1,000 contribution—which was actually from Marshall—as well as the election of Marshall as president and of Yard as secretary and treasurer. Marshall quickly relinquished his position, however, on the firm advice of Ickes, his Interior Department chief. Aldo Leopold, when offered the society's presidency a few months later, turned it down, on the grounds of his distance from Washington and the pressure of his other responsibilities. Consequently, Yard ran the group, in close consultation with an executive committee composed of himself, Marshall, and Anderson, from his home at 1840 Mintwood Place in Washington.[42]

From the outset, the Wilderness Society's organizers vigorously debated the eligibility qualifications for the group's leadership and membership. "We are not looking for large number, but only for those people who believe wholeheartedly in the necessity of preserving our remnant wilderness," their founding statement proclaimed. "Above all we do not want in our ranks people whose first instinct is to look for compromise." Marshall, when he had written to Anderson to report on the outcome of the October 19 roadside meeting, had set the tone: "We want no straddlers, for in the past they have surrendered too much good wilderness and primeval areas which should never have been lost."[43]

Indeed, from MacKaye's point of view, the society's standards had to be high enough and its selection process rigorous enough to exclude someone like Myron Avery. "My own doctrine of organization is that any body of people coming together for a purpose (whatever it may be) should consist of persons wholly wedded to said purpose and should consist of nobody else," he wrote Marshall. "If the purpose is Cannibalism (preference for Ham a la Capitalist) then nobody but a Cannibal should be admitted." Marshall explained to an exasperated Yard the source of MacKaye's cranky persistence on the subject of membership. "Benton is really a grand fellow but very eccentric," he wrote, "and he has been hurt so badly by the traitorism of his own Appalachian Conference that he naturally is worried about any other organization going bad."[44]

✛

MacKaye's anxiety about qualifications for Wilderness Society membership grew in proportion to the increasing tension within the Appalachian Trail Conference. Efforts to ameliorate the effects of skyline roads on the Appalachian Trail were meeting with mixed results—and equally mixed responses from ATC members. Ickes, after public hearings on the politically charged issue of whether to route the southern stretches of the Blue Ridge Parkway through Tennessee or North Carolina, in November 1934 had opted for the North Carolina route. A blow to Tennessee politicians and boosters who had anticipated the economic benefits of a scenic highway through their state, Ickes' decision represented at least a partial reprieve for the Appalachian Trail, in the mind of enthusiasts. A North Carolina route for the portion of the parkway south of Virginia put the roadway on ridges to the east of the trail, generally avoiding relocations and disruptions of the trail like those created by Skyline Drive in the Shenandoah National Park.[45]

At the same time, though, Ickes had raised the possibility of a "great national scenic highway" in the eastern United States, linked to Skyline Drive and Blue Ridge Parkway. In fact, Ickes sketched out two such roads, one starting in New Hampshire and following "the first definite line of mountains west of the Atlantic Seaboard all the way to Georgia . . . referred to generally as the Blue Ridge." Another north-south highway would follow a route farther west, following "roughly the main chain of the Appalachian system," from western New York south to Tennessee.[46]

Marshall, from inside the Interior Department, worked incessantly to restrain the government's road-building plans, firing off memos to Ickes concerning the proposed Green Mountain and Great Smoky parkways. Gradually, Ickes' public statements and actions began to reflect a more pro-wilderness attitude. "I think we ought to keep as much wilderness area in this country of ours as we can," he declared to a group of CCC workers in February 1935. "I do not happen to favor the scarring of a wonderful mountain side just so what we can say we have a skyline drive." "The motorist doesn't need encouragement," he told a group of state park officials a few months later, "but the walker does."[47]

As the 1935 meeting of the Appalachian Trail Conference approached, MacKaye, who could not attend, once again attempted to secure a resolution opposing skyline highways. But Avery, after the experience at the Vermont meeting the year before, had carefully lined up his forces to ensure control of the seventh meeting of the conference. This gathering, held at Skyland in Shenandoah National Park, adjacent to the parallel routes of the Appalachian Trail and Skyline Drive, crystallized the growing personal differences between

MacKaye and Avery. MacKaye's message to the conference reiterated his conception of the Appalachian Trail's transcendent purpose. "The physical path is no end in itself; it is a means of sojourning in the primeval or wilderness environment whose preservation and nurture is your particular care," he observed. "The Appalachian Trail as originally conceived is not merely a footpath *through* the wilderness but a footpath *of* the wilderness."[48]

Inveighing against the "crassitudes of civilization," such as billboards, radios, automobile horns, highways, and overengineered graded trails, he asserted that "the Appalachian Trail is a wilderness trail or it is nothing." The policy of negotiating relocations when and where the trail was impinged upon by motorway projects consumed energy, he urged (pointedly citing the efforts of the Wilderness Society), that might better be "devoted to controlling causes." Rebuking what he saw as Avery's compromised conception of the trail, MacKaye stated his own position bluntly: "In any given case, on any particular stretch from Maine to Georgia, let the Appalachian Trail be *real* or else be absent."[49]

But Avery and the majority of those attending the meeting were not ready to sacrifice the possibility of a continuous trail for the principle of undefiled wilderness. When Anderson and three members of the New York–New Jersey Trail Conference, Raymond Torrey, Frederick Scheutz, and Frank Place, offered a revision of MacKaye's anti–skyline drive resolution, Avery offered, and the meeting approved, another alternative. Instead of adopting a blanket policy of opposing skyline highways in principle, the ATC now determined that "each project should be considered on its particular merits" and that federal agencies should be urged to rebuild the trail, "where interfered with by such highways."[50]

Avery also engineered amendments to the conference's constitution and nominations to its board of managers that placed him and his followers in the Potomac Appalachian Trail Club in secure control of the organization. Indeed, half of those in attendance were members of the PATC, of which Avery was president. The members passed amendments that granted voting power on the basis of trail mileage maintained by individual clubs—and the Washington-based club maintained by far the longest stretch of the trail, as it traversed Virginia.[51]

At the same meeting, MacKaye's allies, such as Harold Anderson and Raymond Torrey, resigned from or lost positions of significant influence within the organization. Torrey concluded, "The Conference, as now dominated, can not be regarded as a conservation organization, in respect to such dangerous projects as the skyline highways, built and proposed." Torrey contemplated that as a result of Avery's heavy-handed tactics, the future of the ATC as a federation of clubs was threatened. Indeed, he considered proposing that the New York–New Jersey Trail Conference withhold its contribution to the ATC and "resume entire autonomy. . . . We are likely to be outnumbered, in any future Appalachian Trail Conference, by Avery's ingeniously managed bloc, and the Board of Managers, which he may not quite control, though he now does very nearly, is a non-entity as he never consults it, between general meetings. He and Miss James are going to run what survives of the main Conference." The ATC held together as a federation of clubs, but Torrey's prediction about the extent of Avery's control proved to be substantially accurate.[52]

This dispute over the control of a tiny organization of amateur enthusiasts could easily be dismissed, MacKaye acknowledged to his friend Stuart Chase, as "a humorous squabble among a bunch of picknickers." But its consequences for the future of the American public landscape, he believed, were more profound. "This clash of Trail vs. Highway on the mountain tops is something bigger than it seems," he reported to Chase. "It is an early skirmish, perhaps the first significant skirmish, in the retention of a humanly balanced world. This is the world that the Wilderness Society was formed to fight for."[53]

Recounting to Chase the news of the meeting, he said, "I'll not go into the whole story, but the gist of it was that the Conference forsook the Trail and the whole wilderness concept and went over bag, baggage, and pantaloons to the skyline highway interests. The National Park Service sold out to them about three years ago. This has made a schism in all but one or two of the A.T. clubs from Maine to Georgia—the celluloid outfit going one way and the real folks (the 'left wingers') going the other. The latter are looking to me to lead them not out of but *into* the wilderness."[54] MacKaye may not have regarded the "left wingers" among his trail disciples as crypto-socialists, as he had when he originally conceived the Appalachian Trail as "a retreat from profit." By 1935, however, he had explicitly incorporated wilderness preservation into his comprehensive social vision of "a humanly balanced world."

In the months after the Skyland conference, relations between MacKaye and Avery swiftly deteriorated. Though his control of the Appalachian Trail Conference seemed secure, Avery was angered by the blunt opinions Torrey vented in his *New York Evening Post* column. At one point he threatened to sue the paper for libel and sought to prevent Torrey from publishing any more about the matter.[55] When Avery sent MacKaye a "Statement of Facts" countering Torrey's reports, MacKaye directly explained his differences with the man who had appropriated his place as the leading figure in the Appalachian Trail effort:

The purpose of the A.T. Conference, we all agree, is to preserve the primeval or wilderness environment. One means to this end, and only one, is a "connected trail." To preserve the wilderness the trail must of course be a *wilderness* trail (as to sounds, sights, and tread). There are hundreds of *non* wilderness ways well connected from Maine to Georgia; to make one more is pointless. Here then is the first issue between us. You are for a *connected* trail—whether or not wilderness; I am for a *wilderness* trail—whether or not connected.

. . . Your ideas on the skyline road seem not to be consistent. Your ideas on the trail itself seem wholly consistent and I do not agree with them: you put *connected* trail first while I put *wilderness* trail first. You have put great zest and energy into a connected trail from Maine to Georgia, and I have praised you for it, privately to you as well as publicly. But this very zest for a *means* (a connected way) has dimmed apparently your vision of the *end* (a wilderness way). . . . *Wilderness, not continuity, is the vital point.*[56]

Avery's acerbic reply distinguished between the knowledge gleaned from his seven years of trail work and what he termed the "pure theory" on which MacKaye based his opinions. "There are limits to our amateur abilities," Avery wrote. "It takes little effort to criticize and much to accomplish." Avery's detailed letter included many valid points about the ambiguous character of the eastern "wilderness," the practical difficulties of trail location and maintenance, and the political battles over skyline roads. But he could not refrain from personal barbs. "You speak of 'such time as you can give to the A.T.,'" he wrote. "I know of no one who should have more. You have a leisurely employment, under no pressure or responsibilities." He went on, with more than a trace of self-martyrdom, to describe his own prodigious efforts on behalf of the trail. He charged that in 1927, the year that Perkins and he became closely involved in the trail effort, the "project was all but dead," following MacKaye's tenure as field manager:

It is very pleasant to sit quietly at home and talk of primeval wilderness, and to think of a Trail that will make and maintain itself. But to bring such a Trail into being requires hard work, hours of labor under broiling suns and pouring rains, camping out in all kinds of weather, as well as almost incessant "office work" in connection with guidebooks, maps, markers, publicity, and a thousand and one other details. It is, don't you think, significant that the majority of those who are loudest in their demands and in their abuse of workers, have covered but little of the Trail and have done little physical labor on it. . . .

. . . Why those who differ in their views cannot also attempt to accomplish something rather than dissipate their energies against those who do not go along with them is unanswerable.[57]

MacKaye did not dispute—indeed, had often praised—the labor, time, and skills Avery and other trailworkers had devoted to building and maintaining the trail. He would not attempt to recount for Avery the several years of intense, meagerly compensated activity he had devoted to the project, or his efforts, more recently, to organize the Wilderness Society, which he recognized as an essential complement to the Appalachian Trail idea. Nor did he explain the constraints of his personal and professional life, such as his tenuous health, his age, his financial responsibilities for his sister, Hazel, or his full-time TVA duties. He took a month to mull over Avery's six-page screed before replying that he would not "answer your statements one by one":

> For sometime past I have noticed in you a growing, self-righteous, overbearing attitude and bullying manner of expression. Your statements to me now—of assumption, distortion, and accusation—constitute a piece of insolence which confirms my former observations, as well as various reports of your conduct which have come to me from individual club members both in the North and South. In your present frame of mind, therefore, I feel that further words are futile.[58]

After one more curt exchange between the two proud, stubborn men, Mac-Kaye and Avery permanently ceased any direct communication. Avery's control of the ATC was now complete. MacKaye had little practical influence over the organization he had been instrumental in founding. Thereafter, he devoted his time and effort as a conservation activist principally to the Wilderness Society.

A trail club leader had once proposed to use MacKaye's likeness on AT signs. MacKaye reminded him that the trail was "larger than any single personality."[59] In fact, the contributions of both MacKaye and Avery were indispensable to the trail's creation, perpetuation, and success. What has appealed to the public imagination are the paired attributes of wilderness *and* continuity that have always characterized the Appalachian Trail. MacKaye's ideas and efforts defined the first attribute, Avery's the second. But the two men who represented each characteristic were personally irreconcilable.[60]

❖

Within only a few weeks of his final correspondence with Avery, MacKaye drafted his proposed scenic resources law at the TVA. While the proposal failed to gain any official support, it was one indication that wilderness advocates were finding their voice in articulating coherent principles of wilderness protection. The wilderness idea had come far during the 1920s and 1930s. "The fight to save the wilderness has grown during the past ten years from the personal hobby of a few fanatics to an important, nation-wide movement," Marshall could accurately claim in 1936.[61]

In the first issue of the Wilderness Society's magazine, *The Living Wilderness,* in September 1935, Aldo Leopold offered a succinct answer to another question: "Why the Wilderness Society?" Besides resisting the encroachments on wildlands generated by the "pressure of public spending for work relief," the society, Leopold observed, would promote the value of wilderness as a setting for recreation and its "scientific values . . . still scantily appreciated, even by members of the Society." But Leopold and his colleagues were principally attempting to lead the public toward a new understanding of their relationship to the natural world. "The Wilderness Society is, philosophically, a disclaimer of the biotic arrogance of *homo americanus,*" Leopold wrote. "It is one of the focal points of a new attitude—an intelligent humility toward man's place in nature."[62] The Wilderness Society and its first leaders were redefining the terms of the debate over the future and meaning of the American wilderness. And they had begun to nurture a growing constituency of citizens who supported the preservation—and the expansion—of the nation's system of public wildlands.

Benton MacKaye's impassioned message to the 1935 Appalachian Trail Conference also appeared in the inaugural issue of *Living Wilderness,* under the title "Why the Appalachian Trail?" His words to the ATC turned out to be a bittersweet valediction: "Hence these words of caution sent to you good people of the Conference. With them go my deep appreciation for your devoted years of effort in building an idea—and an equal faith in your courage to maintain it."[63] Another formal message from MacKaye would not be heard by the Appalachian Trail Conference for more than a decade.

✢ 15 ✢

"Watershed Democracy"

1936–1945

During the decade after leaving the Tennessee Valley Authority, Mac-Kaye would return to work for the federal government twice, serving stints with the Forest Service and the Rural Electrification Administration. During these years, he continued to promote an approach to planning—what he called "watershed democracy"—that attempted to reconcile the centralized powers of federal authority with the grassroots aspirations of citizens and local communities.

As he proceeded eastward in the summer of 1936, the homes of friends and relatives marked way points along his route back to Shirley Center. At Amenia, New York, where Lewis Mumford and his wife, Sophia, had recently settled year-round into their rural Dutchess County home, MacKaye read the manuscript-in-progress of Mumford's *The Culture of Cities* (1938), the expansive regionalist treatise that complemented his own *New Exploration*. He also visited with Stuart Chase, in Fairfield County, Connecticut. Chase shared with his friend the proofs of his own conservationist manifesto, *Rich Land, Poor Land* (1936), which vividly depicted the paradoxical American physical and social landscape of the Depression and New Deal years. "If our old R.P. Ass'n would only resurrect and assemble itself," MacKaye promptly reported to Mumford, "we'd plant the pompous pansy planters in their pretty proper places!! *Think this over.*"[1]

The ecological principles underlying the Regional Planning Association of America's regionalist ideology were encompassed in the trio of books by MacKaye, Mumford, and Chase. But MacKaye's acerbic characterization of landscape architects ("pompous pansy planters"), such as those with whom he had recently worked at the TVA, signified how quickly the RPAA's hopeful visions for broadscale regional, land-use, social, and economic planning had faded after the first flush of New Deal optimism. Clarence Stein, during 1935, had worked briefly as a consultant for the Resettlement Administration's greenbelt towns program. The RA, under the leadership of the ambitious New

Dealer Rexford Guy Tugwell, had originally envisioned new "satellite towns" throughout the country, designed according to garden city principles. But only three such communities were built: Greenbelt, Maryland; Greendale, Wisconsin; and Greenhills, Ohio. Henry Wright, Stein's collaborator in designing such landmark projects as Sunnyside Gardens, in the New York City borough of Queens, and Radburn, in New Jersey, died in 1936. Mumford and Chase nurtured their own growing literary and journalistic reputations. During the 1930s, the leading figures of the RPAA had remained on close terms, personally and intellectually, but their collaborative efforts under the association's loose organizational framework had essentially ceased.[2]

That fall and winter in Shirley Center, MacKaye wrote a postmortem of his TVA experience, titled "Magna Charta: An Interpretation of the T.V.A. Planning Law." His effort was inspired by the ongoing public and political controversy about the TVA. Congress was debating proposals to create powerful multiple-purpose authorities—"little TVA's"—for other American river basins. MacKaye's latest TVA "opus" reworked familiar themes.[3] A job-hunting visit to Washington in the spring of 1937 somewhat tempered his ever-hopeful outlook regarding the agency's political prospects, but its principal weakness, he remained convinced, was its organizational structure, not the purposes for which it had been created. As he reported to Howes:

> I am more than ever of the notion that what I call the "Planning Law" of the TVA (viz. Sections 22 and 23) should have a *single head,* that said head should have some *single objective,* that said objective (whether "good" or "bad") should be *definite,* and that the workers on the job should be led to *see* it as something definite. "Definite" does not mean something rigid and dictatorial; it means (in my mind) something pliable and democratic.[4]

He had identified a fundamental question facing all American planning advocates: Could planning take a form that was "pliable and democratic" or must it always evolve into "something rigid and dictatorial"? Well-rooted American traditions of individualism, capitalism, property rights, and states' rights generated ample resistance to the planning idea—as did the era's ominous totalitarian examples of national socialism in Germany, fascism in Italy, and communism in the Soviet Union. MacKaye abandoned work on his TVA "Magna Charta," but it is a measure of his abiding passion for the TVA's potential—and perhaps of his own political naïveté—that he made a strenuous but unsuccessful effort in 1938 to secure a position on the staff of the congressional committee then conducting extensive hearings into the TVA's activities.[5]

In January 1937, MacKaye attended the annual meeting of the Massachusetts Forestry Association in Boston, where he talked with many of his long-time New England forestry and conservation acquaintances, including Allen Chamberlain. The two recalled that they had first met thirty years earlier in Fitzwilliam, New Hampshire. MacKaye suggested that in the spring they make a trip to Mount Monadnock to celebrate both the anniversary of their 1907 hike up the mountain and the recent publication of Chamberlain's book, *The Annals of the Grand Monadnock* (1936). Some of their other colleagues at the Forestry Association meeting asked to join the expedition. So, on a weekend in late May, eight men rendezvoused at Monadnock's Halfway House for the first meeting of "The Woodticks." MacKaye, Chamberlain, Harris Reynolds, Elmer Fletcher, Karl Woodward, Charles H. Porter, Lee Russell, and Lawrence Rathbun showed up. They climbed the mountain with Chamberlain's book in hand, visiting Thoreau's campsites of 1858 and 1860. That 1937 jaunt set the pattern for the Woodticks' future gatherings. They would stay at either the Halfway House or the nearby Wapack Lodge. Some climbed Monadnock, but their paramount purpose was to share memories, experiences, and ideas concerning the many regional conservation issues in which they all remained intensely involved. This modest fraternity of friends and colleagues represented a hall of fame of distinguished New England foresters and conservationists from the first half of the century. The Woodticks, whose membership changed from year to year, met on a late May weekend (with interruptions during the war years) until the early 1950s, by which time many of the key members had either died or were too frail for mountain climbing.[6]

A month after the Woodticks' inaugural gathering, MacKaye had the gratifying opportunity to offer the Cottage as the setting for the marriage of Harvey Broome and Anne Pursel. Pursel had appeared in Shirley Center on bicycle the previous November, riding out from Cambridge to introduce herself as a long-time friend of Broome's. She was also on hand for a surprise party a few months later, at the Cambridge home of Horace Hildreth, when MacKaye was presented with a large and elaborate scrapbook from the Philosophers' Club. The album's maps, songs, charts, and more than 200 photographs humorously documented the "Two Damndest Years" he had spent among his Knoxville disciples. MacKaye usually balanced the sentimental pleasures of nostalgia with a self-deprecating sense of humor, but in this instance he was overwhelmed with emotion. The event was a "life time occasion," he noted. "It grows in significance," he reflected as he perused the volume later. "It's a Christmas present, a book, a scrapbook," he wrote Stein. "But it's more than this sounds. It's a classic, a masterpiece—the first ever of its kind." Pursel was equally moved by MacKaye's response. "Nothing so heart-rending," she reported to Broome, "as seeing an old person receiving affectionate tributes from youth."[7]

Shortly thereafter, Pursel announced to MacKaye that her friendship with Broome was in fact a full-fledged romance. Furthermore, the couple wished to be married at his Shirley Center home. On June 29, 1937, a small group of friends and family gathered at the Cottage as Broome and Pursel exchanged vows. The unwavering affection and respect offered by the Broomes was another notable indication that MacKaye's life as a childless widower was enriched by many profound friendships. In these durable relationships, he often provided as much devotion and support as he received.[8]

A few weeks after the wedding, Benton celebrated with his next door neighbors Harry and Lucy Johnson the 50th anniversary of his first day in Shirley Center as a boy of eight.[9] Such festivities compensated for the more somber family responsibilities which now preoccupied him. In the spring of 1937, he had received word that his sister, Hazel, had suffered a "nerves attack" of such severity that the Gould Farm was unable to accommodate her needs.[10] Unemployed, scrambling for a new opportunity or position, Benton searched for an institution or living situation that would meet Hazel's new requirements—and which he could afford. Hazel resided briefly at other sanitariums, one in Vermont and one in Massachusetts (the latter of which she tried to escape). In September, Benton moved her once again, this time to the Hartford Retreat. "It was a matter of emergency," he reported to one friend. "This move has helped her (she is better) but it has also nearly 'broke' me."[11] By early December, when it had become clear that Hazel's condition would not improve, he installed his sister at a private hospital in Green Farms, Connecticut. Benton received financial help from Mumford, Stein, and his nephew William, Harold's son; and he pursued every other available recourse to meet the substantially higher cost of her care. The Connecticut sanitarium would prove to be Hazel MacKaye's last place of residence. The financial and emotional demands imposed by his sister's condition forced Benton to take a pragmatic approach to his own employment prospects. He needed a paying job—and he could not be too particular about what it was. "Being 'jobless,'" he explained to Harvey Broome, "I'm working night and day."[12]

During the summer of 1937, as he worked on his own ideas about a trans-Massachusetts nature trail, MacKaye learned from Laurence Fletcher, of the Trustees of Public Reservations of Massachusetts, about efforts to resurrect a proposal for a one hundred–mile beltway of parks, trails, and nature trails surrounding Boston. The idea had originally been promoted by the private statewide land conservation organization in the mid-1920s. During his work for the state's Committee on Needs and Uses of Open Spaces, in 1928, MacKaye himself had been briefly associated with the proposal, which he had further elaborated to include a townless highway in his 1930 proposal for a circumferential "super by-pass" around Boston. As Fletcher had a decade earlier, he turned for financial support to the conservation-minded businessman Charles S. Bird Jr., chairman of the Trustees of Public Reservations. With the promise of a $300 fee, MacKaye set to work on what would prove to be a handsomely produced sixteen-page publication illustrating the updated project.[13]

In the maps, panoramic sketches, and text he prepared for "The Bay Circuit," MacKaye integrated, on the rapidly changing landscape of his home state, some of his ideas about roads, nature trails, open space protection, and planning. His depiction of the beltway encircling Boston, from Plum Island in the north to Duxbury Beach in the south, included existing and potential open spaces, recreation areas, and canoe routes. The "Bay Circuit Motor Route" paralleled the belt beyond the existing Route 128. His text described the evolutionary "Drama of Nature" on display along the proposed Bay Circuit. The challenge before Massachusetts residents in the twentieth century, according to his familiar formula, was to achieve a balance between primeval, communal, and rural environments. "We have devastated the forest, metropolitan-ized the village, and motor-slummed the wayside," he concluded. "What can be done? How to conserve our countryside? There are many ways of DOING but first comes the WILL TO DO."[14]

Characteristically, MacKaye viewed the Massachusetts scheme in more universal terms. "'Bay Circuit' applies only to Boston," he explained to Stein, "but its counterpart would apply to any *other* metropolitan dumpheap."[15] The Bay Circuit as a distinct official project, connecting many separate municipalities, made little more headway in the late 1930s than it had a decade before; it would be revived on a smaller scale in the 1980s and 1990s.[16]

PANORAMA OF THE BAY CIRCUIT

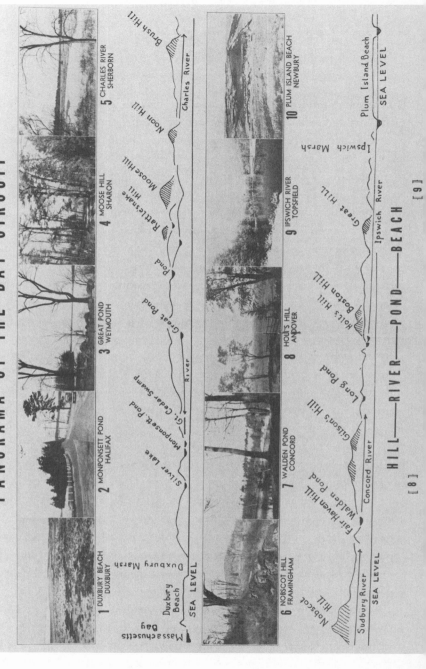

1 DUXBURY BEACH DUXBURY   2 MONPONSETT POND HALIFAX   3 GREAT POND WEYMOUTH   4 MOOSE HILL SHARON   5 CHARLES RIVER SHERBORN

Massachusetts Bay — Duxbury Beach — Duxbury Marsh — SEA LEVEL — Silver Lake — Monponsett Pond — Gt. Cedar Swamp — River — Great Pond — Pond — Rattlesnake Hill — Moose Hill — Noon Hill — Charles River — Brush Hill

[8]

6 NOBSCOT HILL FRAMINGHAM   7 WALDEN POND CONCORD   8 HOLT'S HILL ANDOVER   9 IPSWICH RIVER TOPSFIELD   10 PLUM ISLAND BEACH NEWBURY

Nobscot Hill — Sudbury River — SEA LEVEL — Fair-Haven Hill — Walden Pond — Gilson's Hill — Concord River — Long Pond — Holt's Hill — Boston Hill — Hall's Hill — Great Hill — Ipswich River — Ipswich Marsh — Plum Island Beach — SEA LEVEL

HILL——RIVER——POND——BEACH

[9]

After completing work on the Bay Circuit publication at the beginning of 1938, MacKaye spent months prospecting continuously, and with little success, for paying projects and employment. For a time he tried to promote the notion of a "camp train." Modeled after the "snow trains" that flourished as the sport of alpine skiing gained popularity, the camp train service he envisioned would connect New England hikers by rail to the region's mountains and forests. Despite his appeals to the Boston and Maine Railroad, social agencies and youth groups, and the reform-minded businessman Lincoln Filene, MacKaye's latest recreational notion found no economic or institutional support.[17]

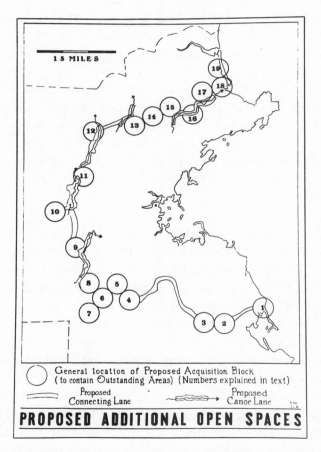

PROPOSED ADDITIONAL OPEN SPACES

Facing page: In 1937, MacKaye prepared a report for the Massachusetts Trustees of Public Reservations, "The Bay Circuit," which promoted surrounding Boston with a belt of parklands connected by trails, "canoe lanes," and a motor route. *Above:* One of his maps depicted the new open spaces to be created along the route. (Author's collection.)

He and Clarence Stein also tried to breathe some new life into the RPAA. In January, Benton enjoyably hosted an informal "Planning Pow-wow" in Shirley. Sir Raymond Unwin was the magnet; Stein and a small group of Mac-Kaye's Massachusetts planning colleagues gathered for a festive weekend at the Empire. Intense talks about New England sites for "satellite cities" and other favorite planning themes were interrupted by walks and sleigh rides around what Benton proudly described to Mumford as "America's *most* indigenous village (and *best* example of the 'communal environment')."[18]

Several weeks later in New York, some of the same group, joined by Mumford, reconvened an informal RPAA meeting, at which the discussion touched on "decentralization, Federated City, and the contradictions of capitalism."[19] In April, MacKaye traveled to Washington in search of a job. Bob Marshall, then heading the Forest Service's Division of Recreation and Lands, hinted at the prospect of a position, but not for a year or more hence.[20] Benton had no better luck when he knocked on doors at the Park Service, the National Resources Committee, the Department of Agriculture, the Bonneville project, the staff of the TVA congressional investigation, the Brookings Institution, and the Congress of Industrial Organizations. By July, just as his depleted finances presented him with the grim prospect of moving his sister once again, he finally secured a job. He would return to the federal agency where he had started his government career: the United States Forest Service.[21]

Bernard Frank, MacKaye's former TVA colleague and fellow Wilderness Society founder, was among the many government officials charged with implementing the expansive new federal flood control and river development powers provided in the Flood Control Act of 1936 and its 1938 amendment. Frank, who had recently transferred to the Forest Service's Division of Forest Influences, found a position for his friend. MacKaye, self-described "cross between a forester and a regional planner," took up his Forest Service duties in September 1938, assigned at first to the Northeastern Forest Experiment Station in New Haven.[22]

The 1936 Flood Control Act, the congressional response to the floods that ravaged the country that year, "set a policy of dam building that clinched the fate of hundreds of rivers," in the words of one environmental historian.[23] Despite Roosevelt's declaration that the bill was "thoroughly unsound," the law passed overwhelmingly.[24] For planners and foresters of MacKaye's liberal persuasion, the flood control acts of 1936 and 1938 fell short of the unified river basin planning approach exemplified by the TVA. In fact, the 1936 law, described by another scholar as "ill-conceived and wretchedly drafted," codified the longstanding tension between two competing schools of flood control—and between competing federal agencies.[25] The Corps of Engineers was authorized to construct flood control works on rivers throughout the country. The agency had traditionally called for downstream dikes and large storage dams; but hydroelectric power, watershed planning, and other land-use policies were not high priorities for the Corps. The Department of Agriculture, principally through the Forest Service, was charged in the law with conducting watershed investigations to measure and propose means of controlling run-off and soil erosion.[26] The role seemed to offer a potent opportunity to apply scientific forestry's tenets of upstream watershed management by protection of headstream forests and construction of more but smaller headwater dams. Thus the Forest Service's Division of Forest Influences, headed by agency veteran E. N. Munns, was poised to assume important new duties.

Foresters and the Forest Service had long used the phrase "forest influences" to describe the relationship of forest lands to other natural processes and resources, such as climate, precipitation, and water run-off. For Forest Service veterans like MacKaye, as for somewhat younger acolytes like Munns and Frank, the water-retaining power of forested lands—and the corollary necessity of comprehensive watershed management—was preached as unimpeachable scientific dogma. "There can be no rational *water management without forest management*," Raphael Zon had once asserted.[27] MacKaye was well prepared to promote the Forest Service's long-held point of view. Frank and Munns soon approved work on his proposal for a "Primer on Flood Control," using as a case study the modest-sized watershed of southern New Hampshire's Contoocook River, a tributary of the Merrimack.[28] In 1938, New England was reeling from another year of serious flooding and a severe hurricane that had mown down a broad swath of forests across the heart of the region. As MacKaye traveled the familiar New England terrain with Frank and other officials, from both the Forest Service and the Federal Power Commission (the 1938 flood control bill had strengthened the commission's role in selecting hydroelectric sites), he contemplated the social and civic questions provoked by the numerous dam and reservoir projects planned for the region. That winter, as an auditor and reporter, he sat in on a series of public hearings held by the various federal agencies in communities in Massachusetts, Connecticut, Vermont, and New Hampshire within the watersheds of the Connecticut and Merrimack rivers.

"This flood control is a wondrous game—a wilderness of conflicting interests and opinions," MacKaye reported to Harvey Broome in the midst of his northcountry travels. "We need control of the flood of ideas upon the subject. But it's great.—The job takes me all over hell's kitchen. I'm once more the tin can on the end of the dog."[29] As had so often been the case during his government career, his assigned responsibilities were somewhat vague, but they suited his interests and skills. The landscape, history, and traditions of MacKaye's native New England created what he described in one lengthy memorandum as a "jumble" of conflicts. One issue pitted proponents of public power, including MacKaye's pamphleteering friend Judson King, against a New England "power trust" composed of the region's private utilities and their powerful political supporters. Another protracted conflict among the states sharing the Connecticut and Merrimack river watersheds revolved around the question of cooperation—with each other and with the federal government—to develop an effective regional flood control and river development program. Proponents of TVA-like river authorities for New England made little progress. But the states also struggled to find common ground in exercising their rights to create interstate flood-control compacts.[30]

For MacKaye, the federal government's activities in New England's river valleys were linked to fundamental issues of democratic government. "It is at bottom a clash of upstream community interests with downstream community interests," he wrote Munns.[31] He explored this conflict in a series of published articles based on his on-the-ground investigations. For flood control workers such as himself, he confessed, "it would be a boon if our country were a *United Watersheds of America* rather than United States."[32] But even the most ecologically rational policy prescriptions, he admitted, had to acknowledge certain existing political boundaries and realities. For *Survey Graphic* he described in upbeat terms the experience of the small town of Hill, New Hampshire, on the Pemigewasset River (a major tributary of the Merrimack). Hill stood at Franklin Falls, the site selected by the Corps of Engineers for a substantial flood-control dam and reservoir, designed to protect such downstream industrial cities as Manchester, Nashua, Lowell, and Lawrence. The inundation of Hill by the project was certain, but the town's 350 residents were determined not to abandon each other as an ongoing community. MacKaye attended a March 1940 town meeting where Hill's voters appropriated $50,000 to supplement state and federal funds for the move to a new townsite on higher ground. Both the process and the outcome were inspiring for MacKaye, who saw them as examples of the successful reconciliation of local, regional, and national interests. At Hill, "upstream interests conceded to downstream interests," he reported. "[Hill] and its sister towns may have started a new kind of American migration, a civic movement from sites and areas of less habitable value to those of greater habitable value."[33]

Nearby on the Contoocook River, he had watched another scenario play out, to a different but equally uplifting result. A Concord, New Hampshire, hearing drew almost 700 concerned citizens to hear about the proposed River-hill reservoir, which would have flooded out several small downstream Contoocook communities. MacKaye spent several days in Keene, Peterborough, Rindge, and Fitzwilliam, among activists representing the watershed's upstream communities, who felt that the Riverhill plan did little to control floods or provide electrical power for their part of the region.[34] The public outcry and the offer of alternative proposals by the Federal Power Commission persuaded the Corps of Engineers to abandon the Riverhill project. Instead, the Corps designed a series of smaller, less disruptive upstream dams, hydroelectric facilities, reservoirs, and temporary storage areas. Here, by MacKaye's account, downstream communities (the "Big Reservoir Method") gave way to upstream interests (the "Small Reservoir Method").[35]

But even as he extolled such exercises in grassroots democracy—which were, he later suggested, reflections of a "social 'watershed consciousness'" indigenous to New England—MacKaye asserted an overriding ecological determinism.[36] "Any downstream town, or any portion of a town, that insists on squatting on the lower flood plain level partakes of amphibian character and defies the laws of fluvial physics," he explained. "If not doomed to extinction it is doomed, despite the irrational efforts of man, to the eternal recurrent irritation of becoming a civic refugee and a drain upon the body politic."[37]

With the support of Munns, MacKaye during 1939 and 1940 fashioned a traveling seminar on flood control, planning, and "watershed government."[38] Using the landscape of New England as an open-air classroom, he worked with planners Carl Feiss of Columbia University and Frederick Adams of MIT to investigate flood-control options for the Connecticut River and Merrimack River watersheds. Accompanying MacKaye, Adams and some of his students visited the lower Merrimack cities of Haverhill, Lawrence, and Lowell, studying the prospects of flood plain "evacuation"—removal and resettlement of residents—as one of the elements of watershed planning. Sir Raymond Unwin was on campus at Columbia at the time; the eminent British planner joined other participants in what Feiss remembered as "MacKaye's March," an excursion up the lower valley of the Connecticut River in the raw days of December 1939.[39]

The 1938 amendment to the Flood Control Act authorized the Army Corps of Engineers to expend funds on the "evacuation" of flood plains and the "rehabilitation of the persons so evacuated" elsewhere, if such measures were demonstrably less expensive than building dikes and dams to protect the same areas.[40] The reform-minded planners and students looked hopefully to this provision as the source of funds and opportunities to establish dispersed new communities of the sort promoted by Clarence Stein. But MacKaye revealed his anti-urban bias by offering an invidious distinction between the "upstream community" and the "downstream slum." The first represented "a true *community*," in its "normal and healthful form." The second constituted "a *slum*—an area of social decay," MacKaye observed in an article for a February 1940 issue of *Survey Graphic* focused on housing. "Every slum cleared . . . requires a community created."[41]

The planners and students were addressing the "problem of distributing people as well as distributing water," he wrote in "Regional Planning and Ecology," an article for *Ecological Monographs* which originated from a paper for a symposium of the Ecological Society of America. "Regional planning is ecology," he asserted. "It is *human ecology*; its concern is the relation of the human organism to its environment. The region is the unit of environment. Planning is the charting of activity therein affecting the good of the human organism; its object is the application or putting into practice of the optimum relation between the human and the region. Regional planning in short is applied human ecology."[42]

For a time, Munns encouraged MacKaye's philosophical speculations on the relationship between humans and natural resources. By February 1940, MacKaye had completed a lengthy confidential report for his chief, summarizing his investigations of the Merrimack River valley. He pulled no ideological punches. The Army Corps of Engineers, he asserted, pursued a "commercial" interpretation of the 1936 Flood Control Act, to protect downstream millowners. The Forest Service, he urged, should promote a "social" interpretation of the law, which would protect upstream interests. The Engineers' plan, he confided, "approaches the subject from the standpoint of the Merrimack 'waterway'—and not from that of the Merrimack 'watershed.' Its origins spring from the pioneering needs of river navigation, rather than from the modern needs of regional conservation."[43] Munns may have agreed with his aide's trenchant political analysis, but the report offered few practical resolutions to the conflicts it documented.

❖

At Dartmouth College, where MacKaye had befriended many students and scholars during his brother James's tenure, he briefly participated in an experimental program called "Local Problems and Institutions."[44] A byproduct of his Dartmouth experience was his involvement with Camp William James. The short-lived social experiment was a pet project of President Roosevelt, his wife Eleanor, and certain prominent literary Vermonters like columnist Dorothy Thompson and novelist Dorothy Canfield Fisher. An enthusiastic and idealistic group of students from Dartmouth and Harvard, working with several Dartmouth professors, hoped to introduce a new element to the CCC program, at the site of a recently abandoned CCC camp in Sharon, Vermont. These visionaries saw a program that would go beyond providing construction work for unemployed men. The out-of-work farmers, who were the usual CCC recruits, and college students would work together on various community programs, especially land-improvement projects. In honor of the Harvard philosopher who had proposed such community endeavors as "a moral equivalent of war," they named the place Camp William James.[45]

The young college men, some of whom had accompanied MacKaye on his traveling flood-control seminars and heard his inspirational discourses on "watershed democracy," approached him to serve as the camp's "commander." MacKaye briefly considered the possibility, urged on by Bernard Frank. "You have a habit of striking sparks in young people," his friend wrote him. MacKaye contemplated a program "to train a set of men to cope with our coming land problems both in detail and in scope." But he turned down the position, which was filled by another Agriculture Department staffer, E. G. Amos.

The early enthusiasm and high-minded hopes that had marked the initiation of Camp William James were soon dashed. Attacked by some in Congress as a program for the "overprivileged" and for being somehow suggestive of Nazi forced-labor camps, the project was shut down by CCC administrator James J. McEntee within only a few months. Some of the young creators of Camp William James relocated to the nearby town of Tunbridge to pursue the program as a private project.[46] Finally, the onset of direct American involvement in World War II demanded that young men of all classes and backgrounds fight not merely the moral equivalent of war but the real thing.

During World War I, MacKaye had been an unapologetic opponent of American participation; but as he followed the calamitous news of Axis advances during the late 1930s and early 1940s, he came to support a strong American role in their opposition. In substantial measure, he followed the lead of Lewis

Mumford. Breaking ranks early with American antiwar liberals, Mumford became a vigorous proponent of American resistance to fascism. On MacKaye's sixtieth birthday, he had discussed with some of his Washington friends Mumford's passionate book, *Men Must Act* (1939), which urged Americans to resist the appeal of isolationism. And later that year, only days after war in Europe at last exploded, with Hitler's blitzkrieg of Poland on September 1, MacKaye heard Mumford personally make his case for immediate American involvement in the war. The effects of the European conflict also cut close to home. He had been receiving urgent dispatches from his brother Percy, then living in Europe; and he helped Percy's family secure funds and make arrangements for a retreat back to America.[47]

Like many on the noncommunist American left, MacKaye had remained hopeful about the social and political experiment represented by the Soviet Union. By the late 1930s, though, some of his friends, including Isaac McBride and Mumford, had opened his eyes to the true nature of Stalin's brutal regime.[48] While he didn't abandon his own social idealism, he acted pragmatically in exercising his political rights. In the 1936 presidential election, MacKaye's vote had been one of two cast in his small Shirley precinct for the Socialist Party's Norman Thomas. In 1940, however, the political exigencies of the impending American war role swung his vote to Roosevelt.[49]

As in his previous stints as a federal employee, MacKaye's Forest Service duties were somewhat nebulous and erratic. Munns sometimes assigned him the task of reviewing the flood control preliminary examinations drafted by agency personnel for national forests and watersheds throughout the East and South. But Munns also permitted him to pursue projects of his own devising. In May of 1940, MacKaye started considering how the Forest Service's flood control work could be linked to the impending defense effort. "Every calamity should be an opportunity," he told Mumford.[50] During the next year, he churned out a steady stream of reports, memoranda, and articles that adapted his ideas on conservation districts and "watershed government" to wartime priorities. In his continental design, separate, locally managed watersheds would be the basic units or cells of an organic national system. The material demands of the war could thus be reconciled with the creation of stable communities and the fulfillment of other conservation purposes.

"'Defense Time' Conservation," an article he wrote for *Planners' Journal*, was a plea for decentralization in the name of patriotism. In preparing to resist the spread of totalitarian dictatorship, MacKaye urged, Americans could rely upon their grassroots democratic traditions. He outlined a national program of "defense by 'scattering' . . . of industry, of folks, of government." A watershed might thus become "a sphere or unit of government." National defense production needs, MacKaye argued, revealed the logical necessity of decentralizing population. Seaboard cities, where industrial production and population were concentrated, remained vulnerable both to enemy attack and to flooding of their congested flood plains. Decentralized production, in inland and rural areas, would be safer and more resilient. "Then population would follow industry." Indeed, the effort to reconcile the needs of wartime production, sound conservation, and democracy "merely emphasize once more that population lies at the bottom of conservation," he concluded.[51] The proposal struck a chord with the gloomy Mumford, who foresaw "that the bulk of our planning and living, in future, will have to be in terms of a mainly rural environment," as he reported to his friend. "Your notion of collective control at the grass roots is a fundamental one."[52]

MacKaye mentally scouted the nation's watersheds, looking for an opportunity to develop a working model of his idea. In a memorandum for Munns, he used the watershed of the Big Sandy River, encompassing a border region of Kentucky, Virginia, and West Virginia, to illustrate the potential location of munitions plants, housing units, and hydropower sites, as well as the flood plains from which people would be evacuated.[53] When Munns instead assigned him, in late 1940, to work with Vermont officials in drafting a state flood control and hydroelectric power plan, MacKaye used the exercise to explore these same notions. He worked closely with Paul F. Douglas, a likeminded state senator who chaired the Vermont study committee. MacKaye's chapter for the committee's published report was a standard summary of his watershed planning notions.[54] But he used the Vermont experience—especially his investigations of the Winooski River watershed in the northwestern part of the state—as the basis for a more emphatic demonstration of "'defense time' conservation." By March, he had delivered to Munns an elaborately illustrated report.

"The notion," he explained to Munns, "is for every watershed to take quick stock of its latent war supplies and set up a scheme to *tap* but *not exhaust* them."[55] The report's acronymic title, "A.R.M.," sought to invoke the spirit of President Roosevelt's own increasingly urgent rhetoric of preparedness. It stood for "Attacking power. Resistance capacity. Morale, or the 'will-to-do.'"[56] Again, for MacKaye, the essence of democracy could be nurtured watershed by watershed. "To make America one big arsenal of democracy we must make of each cell and segment of America a little arsenal of democracy," he summarized. "Within each cell there lies, active or dormant, the virus of self-government."[57]

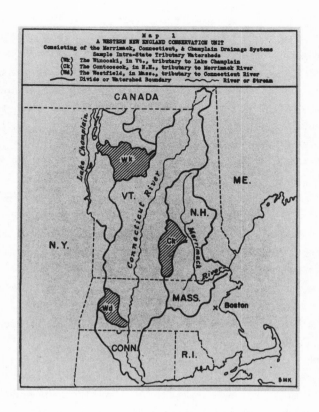

Map 1
A WESTERN NEW ENGLAND CONSERVATION UNIT
Consisting of the Merrimack, Connecticut, & Champlain Drainage Systems
Sample Intra-State Tributary Watersheds
(Wk) The Winooski, in Vt., tributary to Lake Champlain
(Ck) The Contoocook, in N.H., tributary to Merrimack River
(Wd) The Westfield, in Mass., tributary to Connecticut River
〰 Divide or Watershed Boundary   〰 River or Stream

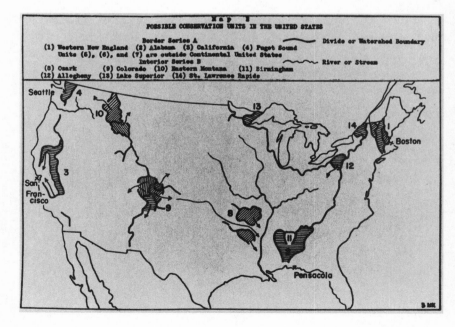

Map 2
POSSIBLE CONSERVATION UNITS IN THE UNITED STATES
Border Series A   〰 Divide or Watershed Boundary
(1) Western New England   (2) Alabama   (3) California   (4) Puget Sound
Units (5), (6), and (7) are outside Continental United States
Interior Series B   〰 River or Stream
(8) Osark   (9) Colorado   (10) Eastern Montana   (11) Birmingham
(12) Allegheny   (13) Lake Superior   (14) St. Lawrence Rapids

In a series of maps, MacKaye used the Contoocook River in New Hampshire and the Westfield River in Massachusetts, with the Winooski River, to illustrate how watersheds could be linked as organic "conservation units" into a larger network of "continental defense."[58] Munns thought enough of the report to have it reproduced for wider distribution. But MacKaye's ecologically inspired plan to reorganize the nation for war was probably almost incomprehensible to most policymakers and politicians; in fact, it received little official recognition beyond his own circle of acquaintances. As it happened, his singular and hopeful call to arms proved to be his last assignment for the Forest Service. Upon his return to Washington in June, he received word from an apologetic Munns that his position was about to be terminated. It was perhaps no coincidence that the news came shortly after his patron and friend, Bernard Frank, was reassigned to the Forest Service's Land Use office.[59]

As America followed its precarious path toward war, MacKaye could not prevent himself from looking for a ray of hope. He reminded his fellow Wilderness Society members, in an article for *Living Wilderness* titled "War and Wilderness," that their cause need not be entirely subordinated to military requirements. Wilderness embodied and symbolized the very freedoms Americans so cherished. "Our basic fight today, in 1941, is a fight for *the right to explore*," he wrote. "To explore and grow in the realm of understanding as a way of life all its own." More than ever, MacKaye predicted, in wartime the wilderness itself would be vital to the nation's spiritual and psychological strength. The nation's wilderness areas were a resource representing the "conservation of human energy. . . . Herein is contained our final reserve of human neural energy—a reserve to be heavily drawn upon—and soon—and increasingly—unless all present portents fail."[60] MacKaye learned that his notions were not merely high-minded sentimentality. An army private wrote him about his plans to hike the Appalachian Trail after his discharge: "WAR won't always be the trail this country will follow, so I'm looking to the AT for escape."[61]

FACING PAGE

In an April 1941 U.S. Forest Service report, "Conservation Units in National Defense," MacKaye envisioned how America's watersheds, such as those of New England, could be organized as "conservation units" in a larger network of "continental defense" to provide natural resources and materials. (Author's collection.)

In an effort to secure a position "in some other niche" of the government, MacKaye refashioned his "A.R.M." proposal as "a defense project pure and simple." His new plan for "Charting Defense Industry" depicted a "clearing house," operated by the federal government, that would oversee the nation's production network of raw commodities, factories, transportation systems, and manufactured goods. He sent his materials off to his old Hell-Raisers friend and roommate, Harry Slattery, who now headed the Rural Electrification Administration (REA).[62] Created by executive order in 1935, authorized by Congress a year later, then incorporated into the Department of Agriculture in 1939, the New Deal agency was charged with connecting the remote farmsteads of America's agricultural countryside to the nation's expanding electrical grid. (Only a tenth of the farms in the United States were electrified in 1933.) The REA made low-interest loans to local cooperatives to build power transmission lines and generating plants.[63] In February 1942, MacKaye was offered a post with the agency as an "industrial economist" in the agency's Technical Standards Division. His $3,800 salary ensured that he could maintain his sister's support.[64]

MacKaye joined the REA just weeks before its headquarters were relocated to St. Louis, Missouri, as part of the wartime effort to decentralize government operations. Taking up residence in the St. Louis YMCA, he reported for work at the REA's offices in the Boatman's Bank Building. M. M. Samuels, his chief at the Technical Standards Division, was an electrical engineer of agreeably visionary outlook; MacKaye described him to Stuart Chase as "a scientist, philosopher, poet, humorist, . . . —a cross between Will Rogers and Aristotle."[65]

MacKaye briefly promoted his "clearing house" as what he called a Resource Mobilization Service; but the idea—basically an elaborate series of flow charts depicting the centralized government control of raw materials and manufactured goods—was soon shelved.[66] The federal government instead instituted extensive wartime economic controls and production planning measures through such agencies as the War Production Board and the Office of Price Administration. There was one imminent military project which MacKaye was well prepared to promote, however. In March, Congress approved and work began on a joint U.S.-Canadian crash project to build a military highway connecting the lower forty-eight states and the Territory of Alaska—the Alaska Highway.

In the months after the bombing of Pearl Harbor, as fears grew about Japanese threats to other American ports and shipping routes, MacKaye was among the many sideline strategists looking northward. In the weeks just before joining the REA, he had been reformulating his 1926 proposal for a "new Northwest Passage," a transarctic air route between North America and Asia. Now, in a March 2, 1942, article in *The New Republic*, "An Alaska-Siberia 'Burma Road,'" MacKaye enthusiastically supported the potential military and strategic benefits of the Alaska Highway. But he also saw the road as just one link in a longer route connecting "the railway frontiers of Canada and Siberia." Traversing the Bering Strait by ferry or airplane, as the season demanded, the entire new road would extend "5,800 miles, 2,150 in North America and 3,650 in Siberia," connecting "the arsenals of Detroit to a front on Manchukuo." MacKaye saw the initiative as the basis for a postwar era of economic and political cooperation between the United States and the Soviet Union, and he envisioned the project as the backbone for the rational public colonization of Alaska that he had been contemplating since the years before World War I. "This Burma Road would open up new country—a stretch of wilderness forming half of the Arctic frontier, the last frontier of civilization left upon the planet," he concluded.[67]

MacKaye's proposal found its way into the hands of the Soviet ambassador in Washington, through his journalist friend, Tass correspondent Laurence Todd. But the diplomat, Todd reported back, dismissed the scheme as "impossibly remote in future time."[68] Later in 1942, after the Japanese attacked Dutch Harbor in the Aleutian Islands and the REA was consulted about the electrification of the Alaska Highway for "military purposes," Harry Slattery encouraged MacKaye to resurrect his Alaskan development ideas once again. But the breakneck pace of the road's construction limited REA involvement in the Alaska Highway.[69]

For the two veterans of Progressive-era conservation politics, the Alaska project harked back to the Ballinger-Pinchot controversy and other earlier battles over the territory's fate. As a patriotic government employee during wartime, MacKaye unabashedly promoted the Alaska road. As a leader of the still-modest movement to protect American wilderness, he offered no objection to the project's possible impact on "a stretch of wilderness forming half of the Arctic frontier."

❖

MacKaye's St. Louis routine—"city hermiting," he called it—did not match the excitement of the Knoxville years, but he did discover agreeable company among other REA staffers and some St. Louis journalists.[70] He struck up a friendship with *St. Louis Post-Dispatch* rural editor Paul Greer, with whom he shared a weekly dinner at the Hotel Jefferson. And an "unnamed lunch

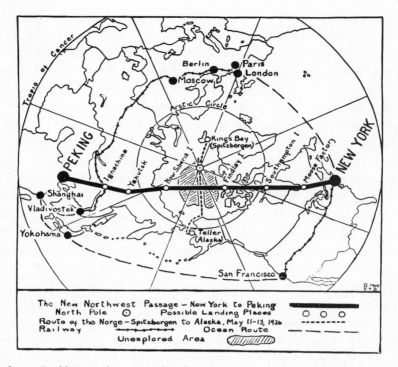

Explorers Roald Amundsen, Lincoln Ellsworth, and Umberto Nobile flew a dirigible across the North Pole from Spitsbergen, Norway, to Teller, Alaska in May 1926. In *The Nation,* June 2, 1926, MacKaye celebrated the air route as "the new Northwest Passage—strategic base line in the new exploration." In 1942, in response to wartime military needs, he revisited the idea of a transcontinental, northern-latitude transportation route in his proposal for an "Alaska-Siberia 'Burma Road.'" (See map on p. 305].)

bunch" that included Greer and such ideologically companionable REA "Elder Statesmen" as Udo Rall, M. M. Samuels, Franklin P. Wood, and John P. Duncan, gathered regularly to discuss the agency's troubles, the war, and other common interests.[71]

MacKaye's employment status was threatened in mid-1943, amidst intensifying political attacks on Harry Slattery. A bitter feud had developed between Slattery and Claude R. Wickard, the Secretary of Agriculture, who persistently sought the administrator's removal. Wickard also initiated an investigation into charges that Slattery had been hiring political cronies and old friends. MacKaye had joined the REA just as the controversy reached its climax. In July, he was among a few employees in the Technical Standards Division to receive layoff notices, supposedly for budgetary reasons. But his name may have turned up on lists of Slattery's alleged personal patronage favorites. Through the intervention of influential friends like Stuart Chase, however, and his own indignant demands for an investigation of the circumstances of his termination, MacKaye won reinstatement of his Civil Service status.[72] Transferred to the Generation and Transmission Section of the REA's Design and Construction Division, he was granted the new title of Industrial Engineer, despite his protest that he was "*not* an engineer as the term is generally understood." His new duties involved investigating locations for REA power sites, he explained to Stein, "the economic aspect thereof and not the engineering."[73]

After the transfer, MacKaye was allowed a certain degree of latitude, under his new supervisor, Franklin P. Wood, to promote missions for the agency

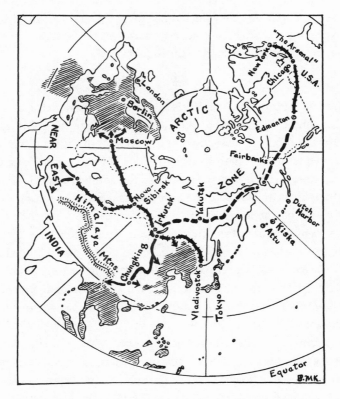

In a March 2, 1942, *New Republic* article, "An Alaska-Siberia 'Burma Road,'" MacKaye advocated extension of the quickly created Alaska Highway into Asia. In this draft version of the map that appeared with his article, he proposed an air-sea-rail link traversing 5,800 miles between Edmonton, Alberta, across the Bering Strait and Siberia, to Irkutsk in the Soviet Union.

more ambitious than just the production and distribution of low-cost public power. His studies of the REA's opportunities as a regional development agency in Vermont, New Hampshire, Arizona, New Mexico, the Arkansas River Valley, and "Dairyland" (Minnesota, Wisconsin, and Iowa) were all intended to infuse the REA with what he called "the 'Norris Doctrine'—namely to base power on water, to accept the watershed as the basic unit of rural development, [and] to associate the several elements of such development (especially electrification and flood control) to effect their mutual support."[74]

Among his REA colleagues, who constituted the only real audience for his reports, MacKaye was preaching to the converted. By the spring of 1944, supervisors like Samuels and Wood had apparently exhausted the tasks to which they could usefully assign their veteran staffer. MacKaye turned his thoughts to his own postwar future. He confessed to Mumford that he was homesick for the East and that "spiritually I'm ossified for the duration." His REA status and duties comprised "projects of my own making"; he did not even have an office to report to, simply working in the REA library.[75]

Among the satisfactions of his St. Louis sojourn, though, were the occasional visits of his long-time friend Horace Hildreth, during cross-country train rides to mining properties he oversaw in Colorado. In April 1944, the two inquisitive students of the American terrain added another symbolic landmark to their list of explorations when they visited the confluence of the Missouri and the Mississippi rivers. It was, MacKaye reported to Harvey Broome, "not a 'Cosmic' but a 'Continental Occasion.'"[76]

MacKaye's observations of the great American river system were not merely an amateur enthusiasm, though. During his St. Louis interlude, he became intensely involved in the effort to promote legislation to create a Missouri Valley Authority (MVA), modeled after the TVA. Congressional proposals to create an MVA seemed to have a reasonable chance of passage in the mid-1940s. Working with Paul Greer and some of his REA colleagues, MacKaye participated enthusiastically and publicly in a pro-MVA publicity campaign. Besides drafting versions of an MVA bill, he regularly contributed maps and letters to the editorial pages of St. Louis's newspapers. In the process, MacKaye attempted to address what he saw as two of the TVA's weaknesses: the lack of clearly focused, enforceable planning authority at the top of the agency; and the paucity of opportunities for genuine citizen involvement and grassroots control. He earnestly believed that these potentially conflicting administrative forces could be efficiently and democratically harmonized.

His proposed MVA bill called for the designation of a Planning Assistant "in complete charge under the Board of all surveys, setups, and plans," including the development, within nine months of the MVA's creation, of a "Provisional Master Plan" for the region. He also depicted the framework of a watershed government organized on a rigorously ecological matrix, namely, the divides of the Missouri River basin's major and minor rivers. In his scheme, the basin would first be divided into six regions, encompassing different tributaries and stretches of the main river along the Missouri's length. Each of these regions would be administered by a "Regional Operator" reporting to the MVA board; the regional operator could then subdivide his region, "on the basis of stream drainage," into as many as ten "Localities," each managed by a "Local Operator." Furthermore, MacKaye proposed the establishment of an advisory committee in each region and locality, to consult with the MVA board and other agencies in carrying out the broad powers of the act. He elaborated his watershed government scheme even further by proposing the creation of independent "local valley development associations," which could work with all the other elements of his intricate hierarchical structure to effect "a workable system of watershed development."[77]

MacKaye's scheme, though compelling and rational from a purely geographical and ecological point of view, characteristically ignored most existing local and state political boundaries and jurisdictions. Hence, it received no political sponsorship or support. Montana senator James Murray had offered legislation to create an MVA similar to the TVA. President Roosevelt, throughout the fall of 1944, expressed his support for a Missouri Valley Authority. But when the 1944 Flood Control Act was finally enacted in late December, language providing for an MVA had been dropped. Instead, the legislation provided the funding and authorization for the first stage of what came to be called the Pick-Sloan Plan.

In a political and bureaucratic maneuver to quash the MVA cause, the rival Corps of Engineers and the Bureau of Reclamation hastily reconciled their own previously competing plans for the Missouri River and developed a joint project acceptable to Congress. Named after the respective heads of the corps, Lewis A. Pick, and the bureau, W. G. Sloan, the program divided the generous bureaucratic and financial spoils granted by Congress to create "an ornate hydraulic regime" in the Missouri River basin. In the minds of critics like MacKaye's friend, Bernard Frank, this "fake MVA" represented a "patchwork of conflicting and some ill-conceived projects."[78] The Missouri basin, as historian Donald Worster describes it, "subsequently fell under complex, multiheaded federal regulation, and the Grand Missouri became a series of deadwater lakes."[79] There was no place in the region for MacKaye's vision of a working watershed democracy, linking the regionwide planning powers of a central agency with a network of locally controlled grassroots associations.

✣

M. M. Samuels had encouraged MacKaye to collaborate with Harry Slattery in writing a history of American conservation. In fact, MacKaye had already been contemplating a book about his "geotechnics." So Samuels's idea was quickly transformed into an outline for a book on MacKaye's pet subject. Stein summoned MacKaye to New York to discuss and promote the idea. With the support and encouragement of Samuels and Wood, MacKaye returned east that summer both to promote his book proposal and to be nearer his sister, whose health was rapidly failing.[80]

In August 1944, Benton received word of Hazel's death, at age 63, from what the doctors described as "chronic rheumatic heart disease" and "involutional depression." Hazel's "great release" freed her from psychological torment. And Benton, for virtually the first time in his adult life, was free from the burden of her financial support. During 1943, for example, he had spent $2,160 of his $3,812 income—almost 60 percent—for Hazel's care.[81] Perhaps just as important, her death ended her emotional dependence on him. He had not complained about his fate. Indeed, he had protected his sister with resolute filial love and devotion, especially during the troubled last twenty years of her life. After a memorial service for Hazel in Shirley Center in late August, Benton returned to St. Louis.

A few months later, eager to return east, he decided to retire from the REA. His decision may have been prompted partly by the resignation in December of his old friend and bureaucratic patron, the embattled Harry Slattery. "I guess it won't be comfortable for Ben, his friend 'Sammy,' or any of the others for very long," Stein predicted to Mumford. The REA's troubles were not the only events prompting MacKaye's somber spirits, however. The climactic year of the war brought the battlefield deaths of his nephew James, as well as the sons of Lewis Mumford and Allen Chamberlain. (The grieving Mumford found a measure of solace in Benton MacKaye's dependable fellowship, writing that theirs was "an association that has sweetened and enriched my life as few other friendships have.") "Under the 'festive circumstances,' I'll be damned if I'll wish anybody over 12 years old a 'Merry Christmas,'" Benton wrote his Shirley neighbor Harry Johnson. "It's just downright irreligious." A few days later, at the beginning of the new year, he submitted his resignation from the REA, effective on his sixty-sixth birthday, March 6, 1945.[82]

❖ 16 ❖

Wilderness in a
Changing World

1937–1950

In July 1945, MacKaye would be thrust into the presidency of the Wilderness Society. During his ensuing five-year tenure, he would preach an evolving gospel of worldwide conservation. This "terrestrial consciousness," as MacKaye came to call it, was based on principles of organic interdependence that he and other farseeing environmental thinkers, like the society's new vice-president, Aldo Leopold, had been promoting for decades. The potential threats to America's wildlands posed by economic growth, technological change, and mechanized recreation were those MacKaye had identified almost two decades earlier in *The New Exploration*. When, in that book, he had described "environment as a natural resource," he had also singled out the "primeval" environment as approaching "closest to the common mind of all humanity: it is the one thing which is agreeable to all our inner minds." For the ever-optimistic MacKaye, wilderness preservation offered a war-weary nation one modest opportunity to seek a unifying purpose.[1]

In the years immediately after creation of the Wilderness Society in 1935, the tight-knit fraternity had resisted the extension of skyline parkways in the eastern mountains, opposed CCC-built "truck trails" in Bob Marshall's beloved Adirondack wilderness, railed against a water tunnel through Rocky Mountain National Park, and supported the Park Service against the Forest Service in the battle for control of a new Olympic National Park.[2] The group won few clear victories, but its influence on government policy and public consciousness grew steadily.

Initially, the society had operated as virtually a private emergency committee, under the guidance of its prime benefactor and leader, Marshall, and its paid secretary-treasurer Robert Sterling Yard. But Marshall foresaw the need to perpetuate the group as something more than a vehicle of his own personal agenda and fortune. In April 1937, he summoned the Wilderness Society's council to Washington to adopt bylaws as a formal corporation. Marshall,

Yard, and Harold C. Anderson, collaborating on the preliminary organizational work, opposed including in the bylaws lengthy lists of prohibited and permitted wilderness uses, such as MacKaye and Harvey Broome had drafted two years earlier.

More important than any words on paper, Marshall insisted, was the unspoken common consciousness shared by the organization's leaders. "If there is any one thing which the whole history of the conservation movement in America has demonstrated," he declared to MacKaye, "it is that whoever runs a conservation organization interprets the officially established constitution or laws or objects however he desires. . . . The one major effective way to keep the Wilderness Society operating or functioning along the lines which those organizing it believed in is to set up a self-perpetuating council with members on it who believe thoroughly in these principles."[3] Given his own bitter experience with the Appalachian Trail Conference, MacKaye could well understand the internal politics of conservation organizations. Marshall's determination to maintain tight ideological control reflected the difference between the Wilderness Society's principal role as an advocacy group and the Appalachian Trail Conference's main mission as protector of a physical resource, the trail itself.

MacKaye did try to persuade Marshall to include in the society's bylaws a clear and specific definition of *wilderness*. He also made a tepid attempt to resign from the expanded thirteen-member council.[4] Marshall, Yard, and Anderson, writing to their fellow society organizers, resolutely rejected his offer: "Benton is such an integral part of the Society . . . that Benton must be a member of the Council whether he wants to or not."[5] Indeed, when the council met in April 1937, MacKaye was elected vice-president of the reorganized Wilderness Society, which by then had acquired 576 members. The other officers elected were Yard as president and permanent secretary, and Harold Anderson as treasurer. The new members of the expanded council included Marshall's brother, George, a New York attorney, and conservationists Dorothy Sachs Jackson, Olaus J. Murie, Irving M. Clark, and L. A. Barrett.[6]

As a step toward greater public financial support (and less reliance on Marshall), the new bylaws also provided for annual membership dues of one dollar. In response to the urgings of MacKaye and Broome, the group did amend its proposed redefinition of *wilderness*. For the Wilderness Society, the term would "apply to areas retaining their primeval environment or influence, or to areas remaining free from routes which can be used for mechanized transportation."[7]

❖

When Bob Marshall in 1937 became head of the Forest Service's Division of Recreation and Lands, his new position provided an avenue by which MacKaye could voice his continuing skepticism about the role of the federal government in promoting and protecting the Appalachian Trail. At the 1937 gathering of the Appalachian Trail Conference, in Gatlingburg, Tennessee, a Park Service field coordinator and trail enthusiast, Edward Ballard, offered a five-point proposal for "an Appalachian Trailway," a buffer zone of protected land on either side of the trail. He called for the "gradual extension of public holdings along the route of the Appalachian Trail" by federal and state governments. With ATC chairman Myron Avery's support, Ballard's resolution was adopted by the conference.[8]

A first step toward implementing Ballard's idea was the drafting of an interagency Appalachian Trailway Agreement between the Forest Service and the Park Service for the management of the more than 700 miles of the trail that traversed national forests and national parks. "As the father of this child which seems to have come somewhat under other influences," Marshall wrote MacKaye, "I would appreciate it very much if you could tell me confidentially your opinion about this proposed agreement."[9] Avery's obsession with promoting "a *continuous* trail (whether or not a wilderness trail)," MacKaye responded, had prevailed at the expense of his own conception of the AT as "a *wilderness* trail (whether or not a continuous trail). . . . I can see nothing to guarantee, on the National Parks and Forests, this basic idea of the Appalachian Trail."[10]

By this time, MacKaye had been cut out of the ATC's direct discussions with government officials about the fate of those sections of the trail on federal lands. His own persistent doubts about Myron Avery's dedication to preserving the AT's wilderness qualities could only have been reinforced by the agreement's lack of any expressly preservationist language describing the purpose and use of the trail. Nonetheless, the Appalachian Trailway Agreement, signed in 1938 by the Park Service and the Forest Service, did in fact include some measures which would protect MacKaye's basic idea. The federal agencies agreed not to build parallel roadways (excluding the Blue Ridge Parkway) in a zone that would extend a mile on either side of the trail. Logging was prohibited within 200 feet of the trail. The states through which the AT passed (except for Maine) agreed to similar road-building and logging restrictions in a corridor extending a quarter-mile on either side of the trail.[11] The Trailway Agreement laid the foundation for the trail's eventual protection and management under the 1968 National Trails System Act.

For reasons other than Ballard's trailway proposal, 1937 was a milestone year in the brief history of the Appalachian Trail and the Appalachian Trail Conference. At the Gatlinburg conference, Avery reported that the clearing of a two-mile span in western Maine would complete the uninterrupted Appalachian Trail from Katahdin in Maine to Mt. Oglethorpe in Georgia. That August, a CCC crew finished the link between Saddleback and Spaulding Mountains. The effect of this event would be short-lived, however. A New England hurricane in 1938, displacement of the trail by the Blue Ridge Parkway, disputes with landowners, and the demands of World War II soon began to interrupt the trail's continuity, which was not restored until the early 1950s.[12] If the trail's fleeting completion in 1937—just sixteen years after he first proposed the idea—even registered with MacKaye, his reaction may have been bittersweet.

Because Forest Service assignments in the years just after 1938 often required his presence in Washington, MacKaye was able to be in frequent personal contact with society leaders Marshall, Yard, Anderson, and Frank; he also maintained his regular correspondence with Harvey Broome. But the unexpected death of Robert Marshall, in November 1939 at the height of his influence and powers and at age 38, created a crisis for the fledgling Wilderness Society. The group had depended upon his ideas, his energy, his political contacts—and, not least, his money—for its influence and very existence.

The daily business of the Wilderness Society, before and after Marshall's death, was carried on by veteran conservationist Yard, who continued to run the organization from his Washington, D.C., home. Yard was passionately devoted to the wilderness cause, but his livelihood had depended directly upon Marshall's personal largess. In a letter to MacKaye, he worried that the council, with Marshall missing, would lack "men of eminence."[13] MacKaye was not likely to take offense at Yard's unintentionally patronizing comment. "I never knew anyone who was so completely the leader of a movement," he said of Marshall's pre-eminent role as a wilderness crusader.[14]

As it happened, Marshall had bequeathed a substantial share of his $1.5-million fortune to a Robert Marshall Wilderness Fund, the majority of whose five trustees were Wilderness Society Council members. (The rest of his estate was divided between trusts to support civil liberties and to promote an American economic system "based upon the theory of production for use and not for profit.") Marshall's generosity assured that the Wilderness Society, even after the death of its leading figure, would remain on the scene for years to come as a vigorous voice of the American conservation movement.[15]

✤

As a leader of the increasingly influential Wilderness Society, MacKaye assumed an inspirational, oracular role, like the one he had held in the Appalachian Trail Conference. He examined a fundamental dilemma of the wilderness cause in a 1939 article, "The Gregarious and the Solitary," which appeared in the magazine, *The Living Wilderness*, produced by the society. The appealing depictions of wilderness areas in the society's publications revealed an unsettling paradox. As Broome had once written to Yard, "The more publicity, the more tourists; and the more tourists the more roads; the more roads, the more publicity; and the more publicity the more tourists. It is an endless circle. Why don't we publicize the *idea* of wilderness and soft-pedal giving too much of an idea as to their location, etc.?"[16]

Part of the solution to the public demand for outdoor recreation, MacKaye believed, lay closer to the city. A thoughtful and democratic approach to outdoor recreation and wilderness protection, he proposed, required not that *all* places be accessible to *all* people. Rather, different kinds of facilities and resources should be made available for the varying interests and desires of Americans. Indeed, the protection of wilderness required the creation of parks and other intensive recreational facilities in more densely populated areas. "Every added gregarious acre in the urban foreground subtracts from the pressure on the solitary acres in the hinterland," he asserted.[17] For MacKaye, wilderness protection was only one element in a comprehensive vision of the entire physical, social, and economic landscape.

A year later, in a *Living Wilderness* article titled "The Spirit of Wilderness," he made a more philosophical point: Wilderness is both the physical outcome of evolutionary processes and the intellectual creation of the human mind. The "*spirit* of wilderness . . . is a human development and not terrestrial," he wrote. "Nevertheless it requires the touch of Mother Earth. We humans indeed need 'Mother' more than she needs us. Earth and wilderness will survive whether the human species does or not. They have survived a thousand species. Wilderness needs not man—that's sure. *But does man need wilderness?*"[18]

For MacKaye and his Wilderness Society colleagues, the answer to that high-minded question was obvious. Their efforts were beginning to have an effect in the down-to-earth realm of politics and public policy. In September 1939, under Marshall's prodding, the Forest Service had drafted and adopted so-called "U Regulations," which defined and classified national-forest wilderness areas, wild areas, and roadless areas according to size and permitted uses. The U Regulations replaced the decade-old "L-20 Regulation," which had provided for the administrative designation of "primitive areas" in national forests by the chief of the Forest Service. The U Regulations established three designations for wilderness reserves on some national forest lands: "wilderness areas," of more than 100,000 acres; "wild areas," of between 5,000 and 100,000 acres; and "roadless areas," which could be of any size and were to be administered for recreational use "substantially in their natural condition." The U Regulations provided that only the secretary of agriculture, not the Forest Service chief, could authorize, modify, or eliminate wilderness and roadless areas greater than 100,000 acres. But the new Forest Service regulations still did not guarantee permanent wilderness protection for national forest lands.[19]

Secretary of the Interior Harold Ickes in 1939 initiated bills in the House and Senate to authorize wilderness areas designated by the president within national parks and national monuments.[20] Though unsuccessful, the proposal pointed the way toward formal legal protection of federal wildlands. Furthermore, creation of Olympic National Park in 1938 and Kings Canyon National Park in 1940 emphasized the protection of wilderness qualities, limiting the sort of commercialization, road-building, and development taking place in other national parks. Such issues were debated amidst Ickes' persistent and controversial effort to create a new Department of Conservation, which would transfer the Forest Service from the Department of Agriculture to the Department of the Interior. His proposal was vigorously opposed and eventually overwhelmed by many interests, including Gifford Pinchot and his long-time loyalists in the Forest Service, the forestry profession, and the conservation movement itself.[21]

In fact, the clash between the Forest Service and the Park Service as guardians of American wilderness was a matter of intense concern to the Wilderness Society's founders. MacKaye, Marshall, Leopold, and Frank had all worked for the Forest Service. And some of the Wilderness Society's elder statesmen, notably Yard and MacKaye, were especially skeptical of the Park Service's commitment to wilderness preservation. MacKaye bluntly characterized the Park Service under Arno Cammerer as "a despoiler of the primeval."[22] Yard, a purist on the question of national park standards, was equally frustrated by Cammerer's reluctance to support a joint Wilderness Society–National Parks Association proposal for designation of National Primeval Parks, free of roads, vehicles, and commercial uses.[23] Marshall himself had criticized Park Service wilderness policies and taken credit for the designation of thirteen Forest Service "roadless areas" of 100,000 acres or more in six states.[24] But he also had urged the society to maintain an even-handed public stance in its dealings with the two crucial federal agencies, once scolding Yard, "Personally I feel your attitude is too pro–Forest Service. The Wilderness Society must continue neutral as between Park Service and Forest Service except where specific projects demand the support of one organization or the other."[25]

❖

The Wilderness Society was relatively quiescent following Marshall's death in 1939 and throughout the war years. Yard struggled with new organizational and financial realities, including the difficulty of securing a quorum of its far-flung members for council meetings. Indeed, the full council did not meet at all for three years beginning in 1942. Its five-member executive committee, according to Broome, "handled most of the problems" during that period.[26]

By 1943, the organization's membership had crept up to 821. "Bob Marshall would be proud of you," MacKaye encouraged Yard.[27] But the next year, Yard's health began to fail. When MacKaye briefly returned east from St. Louis during the summer of 1944, he met several times with George Marshall in New York City to discuss the Wilderness Society's problems and prospects. Marshall raised the possibility that MacKaye might be employed by the organization in some capacity. Benton may have imagined that he and Harvey Broome could somehow jointly lead the group. Immediately upon his retirement from the REA in early 1945, he visited Broome in Knoxville, where he suggested that his younger friend and fellow Wilderness Society co-founder assume Yard's role. However, when Broome traveled to Washington to meet with some of the society's other leaders, he did not receive an offer to take over Yard's position. At this moment in early 1945, Robert Sterling Yard lay dying, and MacKaye considered the society's "outlook very bad."[28]

But when Yard died in May, vice-president MacKaye quickly stepped into the breach, becoming acting president. As Broome later noted, MacKaye's "vigor and single-minded devotion to the interests of the Society" were crucial at this critical juncture in the group's history.[29] MacKaye, in a rapid-fire series of six letters to the society's council prior to its July 14 annual meeting, over which he would preside, outlined the broad goals of the society and of the wilderness crusade. He offered his colleagues both an overarching philosophical program and an immediate agenda for practical action. The particular role of their organization, he wrote, required a frank acknowledgment that public support for the wilderness-protection mission was limited, especially amidst other likely postwar priorities.[30] "Hence the need of education on this point. Our work in this deserves a place in post-war effort—not as a side-show but in the main show; our little Society should be part of a big movement. Otherwise I see our efforts doomed to dilettante uselessness."[31]

He reaffirmed that a commitment to the development of "outing areas" and parks near population centers was a necessary element of the society's program. "For thus only can we secure a sufficient range of untramelled [sic] 'wilderness areas' to provide the conditions of remoteness desired by the lesser number," he asserted. "This number though at present small will inevitably grow as more and more people, educated to higher appreciation of nature's forces, tend to demand a larger share of her repose."[32] He urged the society to look for such a constituency among particular varieties of potential wilderness users, such as fishermen, artists, and naturalists.[33]

Finally, he proposed action on two fronts. The "'wilderness' of post-war plans" taking shape in Washington demanded that the society designate an individual or committee to monitor and act on federal policies affecting the nation's wildlands. "In part this job is a negative one of preventing what we *do not want,*" he observed.[34] But he also proposed a positive program, founded on the principle that "ultimate power derives from the small and far communities." He envisioned a Wilderness Society comprising a network of local activists, inspired by the actions "in his particular bailiwick" of "each of our far-flung Councillors," scouting out potential outing areas and wilderness areas, and working with local activists and political officials to promote such "tangible projects."[35]

Though MacKaye held the unpaid position of president, his influence on the administrative and financial aspects of the society's operations was in fact quite modest. ("My judgment in matters financial is of no value," he humbly confessed to one council member, regarding his lack of expertise.)[36] The society's modest budget—$11,525 for 1945, $19,575 for 1946—was covered largely by the Robert Marshall Wilderness Fund, under the watchful eye of George Marshall and his fellow trustees, who essentially controlled the society's purse strings—and hence its fate.[37]

The future of the Wilderness Society also depended upon securing capable new leaders. To fill the newly created position of director, the council looked to one of its own members, the sensitive, scientifically scrupulous naturalist Olaus J. Murie. Though considered something of a bureaucratic maverick, Murie was renowned for his work in Alaska and the Rocky Mountains, where he studied wildlife for the Biological Survey and its successor, the Fish and Wildlife Service. Few Americans had as broad and sophisticated an understanding of the actual ecological dynamics of the remaining American wilderness. Murie was hired on the understanding that he could pursue his duties part-time from his Wyoming home in the shadow of the Grand Tetons. "Our lives just blossomed after that," recalled Murie's wife and partner in adventure, Mardy, many years later.[38]

Howard Zahniser, who had worked as an editor and publicist for such federal agencies as the Fish and Wildlife Service and the Department of Agriculture, was recruited to serve as executive secretary and as editor of *The Living Wilderness*. The thirty-nine-year-old Zahniser was an energetic, skilled wordsmith, who was willing to take a substantial cut in pay to pursue a career more consistent with his own conservationist principles. He quickly allayed the concerns of some council members, like Broome, who wondered whether the bookish, unassuming Zahniser possessed the skills and force of personality to fill Yard's shoes. In fact, Zahniser's literary talent, political judgment, persistence, and personal civility would prove instrumental in establishing the Wilderness Society as an important force in American conservation politics during the postwar years. MacKaye, as president, in early August 1945 helped negotiate and formally presented offers of employment to Murie and Zahniser; the pair, he predicted, would "make a real team—single in purpose and chemistry."[39]

❖

"ATOM BROKEN." On August 6, 1945, MacKaye noted in his diary the world-changing news that America had dropped an atomic bomb on Japan.[40] "The months after Hiroshima saw something approaching a national town meeting on the atomic bomb and its meaning," observes one historian of the era.[41] MacKaye wasted no time placing the epochal new issues of the dawning atomic era squarely before the other leaders of the Wilderness Society.

When the council held its annual meeting in June 1946 at Old Rag, Virginia, in Shenandoah National Park near the Appalachian Trail, President MacKaye presented a "Resolution on Atomic Energy." The wilderness environment, and indeed the entire "earth as a fitting abode for men," were threatened by atomic energy, he warned. The Wilderness Society should declare the control of atomic energy and weaponry "to be the 'number one conservation problem.'" Moreover, he proposed that the society take the lead in calling "an informal conference of conservationists" from around the world "to develop a practicable working program" to address the potentially horrific consequences of atomic energy.[42] MacKaye had intended to provoke discussion and thought, and his proposal won sympathy from some of his closest council colleagues, such as Broome and Anderson. But the council would not adopt such a provocative resolution without achieving consensus, and Robert Griggs, an eminent botanist, expressed opposition. When the council reported the outcome of its Old Rag meeting to members, no mention was made of atomic energy.[43]

Other society leaders also worried about the organization's priorities in the postwar era. Aldo Leopold wrote to MacKaye, "I am thoroughly convinced of one basic point: that wilderness is merely one manifestation of a change of philosophy of land use and that the Wilderness Society, while focusing on wilderness as such, cannot ignore the other implications and should declare itself on them, at least in general terms."[44] Bernard Frank, who still worked for the Forest Service as assistant chief of forest influences, fretted that the agency lacked resolve as a guardian of wilderness. "Wilderness affairs are in serious shape," he wrote MacKaye. "Perhaps some other agency should have charge of the Wild and Wilderness areas."[45] Frank's letter captured the Wilderness Society's shifting attitude toward the agency and its wilderness policies. In response to postwar booms in housing construction and outdoor recreation, the Forest Service pursued expansive new programs of commercial timber harvesting and recreational development.[46] The Forest Service's changing priorities would force the Wilderness Society to seek new allies, both inside and outside the federal government.

In the postwar era, MacKaye observed in "A Wilderness Philosophy," his March 1946 inaugural article in *The Living Wilderness*, wilderness protection was one element of the nation's unique leadership responsibility. The Wilderness Society's president insisted, as he had for years, that wilderness preservation was not an elitist cause. "We must widen the access to the sources of life," he wrote. "*Ipso facto,* we must hold said sources intact." The purpose of such a democratic mission, he concluded, was "not to grab off earldoms for *some* but to open up kingdoms for *all.*"[47]

To meet the practical challenges of wilderness preservation, MacKaye in early 1946 drafted a federal wilderness law. Such an idea was not new for him nor some of his Wilderness Society colleagues, of course. Leopold and Marshall had worked for years to secure administrative protection of federal wildlands. MacKaye's own 1936 draft "Scenic Resources Law" had called for the federal designation of wilderness areas and other outdoor recreational lands. Harvey Broome had occasionally speculated about legislation providing explicit designation and independent management of federal wilderness areas. "Do you think wildernesses would have more permanence if there were some new status, established by Congressional enactment?" he had asked Olaus Murie in 1940. Five years later, when the Wilderness Society's own future seemed in question, Broome, in his journal, speculated on whether "zoning on a nation-wide scale" might provide an approach to wilderness protection.[48]

As it happened, one of MacKaye's early AT allies, Daniel K. Hoch, a long-time leader of Pennsylvania's Blue Eagle Mountain Climbing Club, had been elected to Congress in 1942. With the ATC's support and encouragement, Hoch had filed a bill in 1945 to initiate a national system of foot trails, under the Forest Service's jurisdiction. Proposed as an amendment to the Federal-Aid Highway Act of 1944, H.R. 2142 called for the appropriation of only $50,000. It named the Appalachian Trail as the first path to be included in the proposed national trails system, and it provided that all the trails in the proposed 10,000-mile system be developed and maintained to "preserve as far as possible the wilderness values of the areas" they traversed.[49]

The bill received a perfunctory hearing before the Committee on Roads, at which Myron Avery and Harlean James, representing the ATC, presented testimony and the endorsements of many hiking clubs and outdoor organizations, including the Sierra Club and the American Forestry Association. The Federal Works Agency, the Bureau of the Budget, the Bureau of Public Roads, the Department of the Interior, and the Department of Agriculture all opposed the measure. However, Assistant Forest Service Chief L. F. Kneipp, who had long administered the agency's wilderness policies, gave sympathetic testimony, in which he cautiously acknowledged that his agency knew "definitely that there is a certain segment of the public thought and public desire which is strong in its support of a proposal of this kind."[50]

Without contacting Avery, the ATC, or any other interested organization, MacKaye drafted an alternative to Hoch's stalled initiative. His bill called for the establishment of "a national system of wilderness belts." Such interstate wilderness belts, he imagined, would be "located along the several mountain ranges of the country, including the Appalachian, Rocky Mountain, Sierra and Cascade Ranges, and along other natural resources including river courses."[51] He elaborated on his concept to Hoch:

> A wilderness belt is merely a linear wilderness area, but the term itself "wilderness area" (Aldo Leopold's) had not yet sunk into the public mind. People in the 1920's could visualize an Appalachian "Trail" but not an Appalachian "Wilderness Belt." . . . But now I believe they can. . . .

... The "belt" is a cross between line and area: it is a wide line and a narrow area; it allows the visualization of a system and it achieves our final object—area. The width would be anything to insure "insulation"—anywhere from a half mile up.[52]

Envisioning the legislation as a project the Wilderness Society might support, he presented his draft to both Hoch and Howard Zahniser in January 1946. He confessed that his bill was "a first shot only and doubtless hits wide of the bullseye." His proposal was a logical extension of his original conception of the Appalachian Trail, he explained, and showed "the progress made during this quarter century."

In MacKaye's view, Hoch's original trail-system bill, as supported by Myron Avery and the ATC, represented a vision of the Appalachian Trail and of wilderness protection that was literally too narrow, both geographically and conceptually. His alternative proposal outlined a decentralized, democratic program, in which local groups would nominate proposed interstate wilderness belts for consideration by a three-person "Federal Wilderness Project Committee," comprising representatives of the Forest Service, Park Service, and Fish and Wildlife Service. The committee would evaluate the proposals and their cost, then make recommendations to Congress. The federal government would hold title to the areas, but the local sponsors would be responsible for their administration.[53] MacKaye's vision of a national wilderness network built state by state, through the efforts of local organizations, was the product of long experience. The Appalachian Trail, of course, had been created by just such an approach.

MacKaye disseminated his draft wilderness belts bill to a select list of Wilderness Society colleagues and other friends. "Quite seriously," wrote the ever-supportive Mumford, "I regard your plan for *wilderness belts* as a masterstroke of regional statesmanship."[54] MacKaye's proposal included a list of permitted and prohibited uses in wilderness belts. Olaus Murie, who otherwise supported the proposal, criticized a provision allowing logging "for the sole purpose of maintaining and restoring a healthful primitive condition."[55] MacKaye's willingness to permit even limited logging on wilderness lands revealed his belief that protection of the wilderness environment might entail active management and "restoration."

By September, MacKaye had drafted another version of the bill. One modest change, reflecting a continuing preoccupation with wilderness terminology, substituted the term *wildland* for *wilderness*. Now, for example, he called for the creation of "wildland belts" and a "Federal Wildland Project Committee." It is not clear why MacKaye chose to use the term *wildland*, but his notion of federally designated wildland belts was significantly different in concept from the extensive "wilderness areas" his Wilderness Society colleagues like Aldo Leopold and the late Bob Marshall had long promoted. MacKaye's new draft also simplified the list of prohibited uses, taking up Murie's recommendation for an absolute ban on logging and other "industrial development."[56]

The revised wildland belts bill, unlike the earlier version, was reproduced by and identified as a Wilderness Society proposal. Zahniser circulated the document to his counterparts at other important American conservation groups. He received both supportive comments and blunt criticisms from, among others, Sierra Club secretary Richard Leonard, Izaak Walton League executive director Kenneth Reid, National Parks Association executive secretary Devereux Butcher, and National Audubon Society president John H. Baker. One testy response arrived from the ATC's Myron Avery, who complained that the wildland belt proposal competed with Hoch's original trail systems bill. "I shall be interest[ed] to learn the origin of this bill and whether the project of the Hoch bill inspired its thought," he pointedly commented to Zahniser.[57]

Representative Hoch did not take up MacKaye's proposal as a substitute for his trails bill. When Hoch lost his seat that November, MacKaye retreated from the effort to promote a federal wilderness bill, suggesting to Zahniser that the society instead concentrate on a state-by-state approach.[58] "It seems to me that we have not lost any time nor will we actually lose time by not going ahead right now with Congress," Zahniser judiciously replied when MacKaye announced the suspension of his effort to promote a federal wilderness law.[59] Zahniser himself had begun traveling the country extensively, to observe firsthand both the nation's wild terrain and the lay of the political landscape. "Olaus and I find more problems and threats everywhere we go," he reported to MacKaye after one sojourn in the West, "and also some encouragement."[60]

MacKaye's notion of a wildland belts law was quite different from the wilderness-protection legislation Zahniser would initiate in the 1950s. But the 1946 effort had provided Zahniser an opportunity early in his Wilderness Society tenure to work with other conservation leaders in assessing the practical political requirements for federal wilderness legislation.

During his years as Wilderness Society president, MacKaye took special pleasure in composing the "calls" for the annual meetings of the Wilderness Society Council, which were held amidst various of the nation's spectacular wilderness landscapes. For the four-day 1947 meeting, held at Ernest Oberholtzer's cabin on an island on Rainy Lake, Minnesota, MacKaye urged his fellow councillors to "relaunch our gentle campaign to put back the wilderness on the map of America. . . . We would reserve a sample plot of every wilderness community originally residing in the complete American primeval civilization." Meeting on the boundary between two North American countries, the council agreed to begin a campaign to include eleven varieties of natural communities in a continental wilderness system.[61] MacKaye credited the idea to Aldo Leopold. "All I did, in 1947, was to suggest a network of such areas, samples of the various types of wilderness—forest types, prairie types, desert and tundra," he later recalled. "There was nothing new in this suggestion; Leopold had been talking of it from the beginning."[62] Indeed, other conservation and scientific organizations, including the Ecological Society of America, the Ecologists Union (predecessor of the Nature Conservancy), and the Society of American Foresters, were simultaneously promoting the protection of representative "nature sanctuaries" and "natural areas."[63] By endorsing the ecological and educational significance of protecting representative land types throughout North America, the Wilderness Society had moved closer to some other American scientific and conservation organizations. But the "outstanding value" of wilderness areas, the society's council continued to maintain, was to "preserv[e] for all Americans the choice they now have of finding recreation in the wilderness if they so wish."[64]

Though MacKaye and Leopold had corresponded and been acquainted with each other for more than twenty years, the 1947 Rainy Lake meeting was one of the few occasions when they were able to talk at length and leisure. In the months after this inspiring northwoods encounter, MacKaye persisted in promoting a strategy for wilderness preservation which, as he explained to Olaus Murie, consciously integrated a "hometown approach," involving small-scale local efforts, with the "continental approach" represented by Leopold's proposal for a network of representative natural communities.[65] In a November 1947 memorandum describing the North American campaign, MacKaye explained that the "spots of wilderness" composing such a network "should be large and small," combining the efforts of the society's council and staff with those of "our individual members in their respective neighborhoods." Noting his own efforts along the Squannacook River, he urged the society's members to identify small "wildland patches" in their own hometowns and local regions. His proposal engendered a lively discourse in *The Living Wilderness*, where readers from across the country described their successes and setbacks in preserving local tracts of land.[66]

❖

The Wilderness Society's continental wilderness campaign was only one manifestation of an expansive new perspective that seized the thinking of American conservation leaders in the late 1940s. Such popular books as Fairfield Osborn's *Our Plundered Planet* and William Vogt's *Road to Survival,* both published in 1948, depicted a gloomy prospect for the future of humanity and the world's natural resources if current trends of population growth and resource exploitation persisted. MacKaye made the acquaintance of Vogt, a friend of Leopold's who then administered the conservation program of the Pan American Union and was promoting the creation of an Inter-American Conservation Congress. "After my forty years of trying to scare others you now scare me," MacKaye wrote Vogt upon reading his neo-Malthusian tract.[67]

MacKaye's own notion of the evolving science of ecology was not as rigorous or systematic as that espoused by academic ecologists. While embracing fundamental organic ecological principles, his idiosyncratic perspective also encompassed the realms of politics, social reform, and planning on a global scale. In a 1948 editorial in *The Living Wilderness,* he cited writings of Vogt, Frederick Law Olmsted, and Bernard Frank as depicting a shared vision of the "ultimate concept of one ecological world."[68] In another *Living Wilderness* article that year, "Primeval Security: The Impact of the Pan-American Wildlife Treaty," he optimistically portrayed the international effort to preserve natural habitats as a "common factor" among nations, distinct from their competing economic and political systems. The shared human interest "in their primeval home environment . . . in the course of time must belittle the petty conflicts leading to Gargantuan clashes," he wrote. "A world convention to further Nature Protection should sow the seed of a *global wilderness.*"[69]

For a retired man approaching his eighth decade, MacKaye was living an active and productive life during the late 1940s. "I don't know when I've seen Ben looking so well," Clarence Stein wrote Mumford after a visit from their friend in 1947.[70] But Benton also rued the deaths of such long-time compatriots as Horace Hildreth and Allen Chamberlain, who had died in 1945. Other close acquaintances, like Clarence Stein and Judson King, struggled with emotional and mental difficulties, for which MacKaye faithfully provided such moral support and companionship as he could. He was coping with his own chronic intestinal difficulties. In early 1947, staying at the home of Boston friends, he made a quick recovery from prostate surgery.[71] But such physical ailments barely affected his equanimity and hopeful outlook. Though his means remained modest, so were his needs. As in decades past, he spent weeks and months at a time in Shirley Center or at the homes of generous friends. Frequent and extended sojourns in Washington provided him the opportunity to meet with other conservation leaders and federal officials. In these years, he often stayed at the Cosmos Club, where Zahniser and other friends active in the capital's political, scientific, and cultural affairs were members. Their courtesies enabled him to reside economically at the club, where the accommodations and company suited his simple bachelor needs. He relished the opportunity to remain closely involved with the vital and varied conservation issues of the day. During the late 1940s, MacKaye could observe the gradual coalescence of efforts and ideas he had been promoting for decades.

The Wilderness Society council's 1948 meeting took place at Olaus Murie's home in Moose, Wyoming. "To put it bluntly," Murie later wrote MacKaye, "our family fell in love with you." Benton especially charmed the Muries' daughter Joanne, whom he called Little Sister. He amused his fellow councilors with his noisy early-morning ablutions in the dude cabin they shared and with the quality of his raconteurial talents.[72] At that meeting, MacKaye joined his colleagues in celebrating the life of Aldo Leopold, who had been felled that spring by a fatal heart attack while fighting a brushfire near his beloved Wisconsin retreat. "What is needed in the public mind is that thing imbedded [sic] in the mind of Aldo Leopold, namely *continental consciousness*," MacKaye wrote in his "call" to the council. "Every fight provides a chance to spread this consciousness."[73] MacKaye became an instant enthusiast for his friend's posthumous 1949 masterpiece, *A Sand County Almanac*. Leopold's elegant book, which introduced the notion of a "land ethic," became a sacred text of the incipient environmental movement. "The W[ilderness] S[ociety] should push Aldo's book to the limit," MacKaye exhorted Zahniser as soon as he had read it.[74]

Wilderness Society leaders were instrumental in producing a document that was also crucial in the movement to build popular and political support for the wilderness cause. Zahniser cultivated his Washington connections to win support for the study, prepared for the House Merchant Marine and Fisheries Committee; and council member and Library of Congress official Ernest Griffith oversaw production of the extensive publication, titled *The Preservation of Wilderness Areas: An Analysis of Opinion on the Problem* (1949), written and compiled by C. Frank Keyser. MacKaye enthusiastically described the report as "monumental," "a sort of 'war map' of strategy, and we must make the most of it."[75]

He soon had the opportunity to do just that. A long-time acquaintance, Morris Cooke, a politically astute engineer who had assisted Gifford Pinchot in promoting the "Giant Power" policies of the 1920s, had been named to chair the recently created President's Water Resources Policy Commission. Cooke asked MacKaye, in his capacity as Wilderness Society president, to offer his comments "with regard to wild life and other phases of natural life associated therewith."[76] Cooke may not have been prepared for the intensity of MacKaye's response. That winter, MacKaye had been working on a long, three-part manuscript titled "Why Wilderness?" "It is my same old scheme for a continental wilderness layout," he reported to Zahniser. "It is built on the 'Griffith-Keyser' report . . . , and on Aldo Leopold's book . . . , plus my own crackpot notions of half a century."[77]

Primed with such ready material, MacKaye urged Cooke to read Leopold's book and to distribute copies of it to the commission's members. "Until a would-be conservationist knows what Leopold has to say he has not yet graduated from the 'Grammar School' of conservation," he instructed the veteran conservationist, who in fact took the trouble to read the book.[78] MacKaye called one nine-page letter to Cooke his "attempt, crude and groping as it may be, toward formulating an 'esperanto' for the tribes of wilderness and water."[79] Cooke's cautious response suggested the difficulty he would face incorporating MacKaye's distinctive rhetoric and unconventional ideas into the official report of a federal commission. "I don't know when I [have] read anything," Cooke wrote, ". . . which made me feel so humble in the presence of not only a thoroughly competent writer, but one whose thoughts rise so far above the normal literature."[80] MacKaye quickly elaborated on his ideas in an article that appeared in the October 1950 issue of *Scientific Monthly*. His essay was a serious, sometimes awkward attempt to bridge the intellectual gap between the conservation philosophies of two eras: the conventional utilitarian ideas on which veteran progressives like himself and Cooke had cut their ideological and political teeth, and the biocentric environmental perspective informed by "the new science of ecology," as articulated by Leopold and represented by the growing wilderness movement.[81]

Leopold had urged his Wilderness Society colleagues and other conservationists to assert more boldly the scientific value of wilderness areas as ecological benchmarks. As interpreted by MacKaye, the "land organism," the "fountain of energy flowing through a circuit of soils, plants, and animals," provided the earth's evolutionary storehouse of information containing "the secret of land health."[82] MacKaye emphasized that Leopold's idea for a national system of wilderness areas had a practical, instrumental purpose beyond recreation or scenic preservation. "Wilderness is a reservoir of stored experience in the ways of life before man," he asserted.[83]

MacKaye's article and Cooke's commission were motivated in part by pending controversies and conflicts over proposals to build irrigation dams in Colorado's Dinosaur National Monument and Montana's Glacier National Park. The irrigation engineers were skilled and experienced in mapping dam sites, MacKaye observed. Ecologists needed to get busy generating maps "of a series of wilderness areas, or reservoirs of ecologic experience. Each area would be an exhibit of normal ecologic processes. I shall call it a norm area, or *norm site*." The battle of "dam site vs. norm site"—the title he gave his article—should be fought, he insisted, "on some common ground where benefits of widely different characteristics may be measured in fair and comparable terms. . . . Not till the ecologist has mapped his system of norm sites will he be on a par with the engineer who has already mapped his system of dam sites. Then, and not until then, can ecologist and engineer begin to talk the same technical language; not until then can they have anything like a common currency of measurement."[84]

MacKaye's proposal for a scientific, mediated process for comparing competing land uses at a particular location, as measured by the ultimate goal of "land health," was perhaps as politically naïve in the near term as it was rational and hopeful. "Dam Site vs. Norm Site" nonetheless illustrated the intellectual distance he had traveled in almost half a century. He never wholly abandoned the utilitarian "conservation" ideology of the Progressive heyday, as reflected in TVA-style multiple-purpose river development. But now, at mid-century, he insisted forcefully that, by whatever measure, the influence and value of undeveloped land and resources must be evaluated on an equal footing with other traditional land uses. MacKaye's 1950 article was itself an early indicator of the postwar transformation of the American conservation movement into what became known as the environmental movement.[85]

At its 1950 meeting in Colorado's Flat Top Wilderness, the Wilderness Society's council elected Olaus Murie, "the great populist of conservation," as president.[86] (He also retained the title of director.) MacKaye, who did not make the trip west, was elected to the newly created position of honorary president, which he held until his death twenty-five years later. Just before assuming MacKaye's title, Olaus Murie reported to the outgoing president his heartening experience testifying at a hearing about a proposed ski resort in the San Jacinto Primitive Area in California's San Bernardino National Forest. "Everybody spoke of *wilderness* with understanding, even some of our opponents," wrote Murie. "It has become a well-known concept."[87] And in March 1951, at the Sierra Club's Second Wilderness Conference, Howard Zahniser enthusiastically outlined the terms of a proposed bill "to establish a *national wilderness preservation system*," launching the legislative campaign that would culminate in 1964 with passage of the Wilderness Act.[88]

Murie and Zahniser could claim their own share of credit for the growing public understanding of the wilderness idea, but they well understood their debt to the Wilderness Society's pioneering founders, including the likes of Bob Marshall, Aldo Leopold, Robert Sterling Yard—and, not least, Benton MacKaye. Besides serving as the living link between two generations of wilderness leaders, MacKaye had persistently crusaded to reconcile wilderness preservation with a broad and inclusive social vision.

During his five-year tenure as president of the society, he had spent little time on the practical day-to-day aspects of the growing organization's operation. But in that lively, fast-moving era of postwar American conservation politics, his experience, ideas, vision, and energy galvanized veterans of the wilderness cause and inspired newcomers to join it. "Eternal vigilance is the price of wilderness," he had written his Wilderness Society colleagues.[89] MacKaye's timely and timeless exhortation embodied his own substantial legacy to a new generation of wilderness advocates.

✣ 17 ✣

"Geotechnics of North America"

1944–1972

"In 1945 I retired from officialdom," MacKaye reported to the Harvard class of 1900 in its fiftieth reunion year. "And I've never worked so hard as since I quit work."[1] The Wilderness Society would not be the only object of his attention during the years after he left the Rural Electrification Administration. In Shirley, he he would pamphleteer on behalf of the United World Federalist movement and the cause of world government. He would work with Clarence Stein and Lewis Mumford to revive the Regional Planning Association of America. And he would collaborate with others in conceiving and planning nature trails and natural areas in Shenandoah National Park, New England, and elsewhere.[2]

But these efforts during MacKaye's active later years were all subsumed within a grand literary project: composition of his final "opus," which he titled "Geotechnics of North America: Viewpoints of Its Habitability." Continually reconceived and revised, the project would occupy MacKaye, intermittently, from his last days at the REA until the early 1970s, by which time his eyesight had seriously deteriorated. By then he realized that his idiosyncratic, long-gestating manuscript on "the applied science of making the earth more habitable" would not be published—at least in his lifetime. In its final version, the one-of-a-kind work contained 800-plus pages of text and dozens of maps, charts, and exhibits. Variously taking the form of autobiography, historical narrative, legal treatise, textbook, even allegorical parable, "Geotechnics of North America" was the self-portrait of Benton MacKaye's singular mind, personality, and life experience.

The hopes engendered by the adoption of the United Nations charter in June of 1945 were confounded only a few weeks later by the detonation of the first atomic warhead and the bombings of Hiroshima and Nagasaki. Physicist Albert Einstein, essayist E. B. White, *Saturday Review of Literature* editor Norman Cousins, TVA chairman David Lilienthal, and educator Robert M. Hutchins were among the influential Americans advocating various designs for world government, international control of atomic energy, and the prevention of war.[3] The brief but intense postwar movement to promote world government and "world federalism" provided one of the new elements of MacKaye's evolving conception of "geotechnics."

Lewis Mumford was one of the most insistent public voices addressing the fearsome implications—cultural, political, and environmental—of atomic energy and weaponry. "I'm inspired by your bombshell at the bombshell," MacKaye wrote in 1946 after reading one Mumford screed, which called on the United States to halt production of the atomic bomb, dismantle existing bombs, and permit the United Nations to regulate atomic energy.[4] MacKaye was prodded into action by Mumford's appeal for grassroots efforts to press Congress for the international control of atomic weapons. Working with the Massachusetts Committee for World Federalism, he became an enthusiastic enlistee in the group's town-by-town campaign to place a world-government referendum on the November 1946 state ballot. MacKaye organized and spoke at small local informational meetings and tramped Shirley's roads gathering signatures for the referendum petition, which called on state legislators to appeal to Congress and the President to seek amendment of the United Nations charter, in order to "make it a World Federal Government able to prevent war."[5]

In Washington and New York that summer, MacKaye met with advocates of world government and atomic energy control, such as his old acquaintance Scott Nearing, future U.S. senator Alan Cranston, Norman Cousins, and California congressman Jerry Voorhis, who would be defeated that fall by the Red-baiting Richard Nixon. However, the fast-growing public and official concern about Soviet influence, real and imagined, foreign and domestic, soon quelled the momentum of the world-government movement. The onset of the Cold War was marked by such events as the hearings of the House Committee on Un-American Activities; passage of the National Security Act, which created the Central Intelligence Agency; establishment of loyalty oaths for federal employees; and the pursuit of "containment" and other the anti-Soviet foreign policies.[6]

An idealist like MacKaye was not easily discouraged by such intense manifestations of American autonomy and isolation, though. He cast his lot with the United World Federalists (UWF), founded in 1947.[7] Over the next few years, he frequented UWF's Washington and New York offices during his travels; and in Massachusetts he became a vocal activist for UWF, which also maintained an office in Boston. The World Federalists, at least in New England, saw an opportunity to link the region's local town-meeting tradition directly to the promotion of world government. Adopting "Federalist" as his pseudonym, MacKaye advanced the cause in the pages of the *Fitchburg Sentinel* with a series of letters addressed to the citizens of his local "Montachusett" region. (The term had been coined some twenty years earlier by his friend, *Sentinel* editor Samuel Hopley, to define the landscape "under the counseling of three mountains from which it takes its name, Monadnock, Watatic and Wachusett.")[8]

Invoking *The Federalist*, written by James Madison and Alexander Hamilton, MacKaye drew a parallel between the formative late-eighteenth-century era of American history and the circumstances of world history in the middle of the twentieth century. The infant United Nations was yet a "weak sister," comparable to the United States under the Articles of Confederation. "The job of our forefathers in the 1780's was to strengthen the said United States by changing its impotent setup (the Confederation) into a competent national government," he wrote. "The need of the 1940's is to strengthen the United Nations by changing its present impotent setup (the Charter) into a competent world government." His letters were gathered into a pamphlet titled "The Montachusett Federalist," published by the UWF's Massachusetts branch.[9]

But in an era of swelling anticommunist passions, world government quickly evaporated as a popular American cause. MacKaye recorded in his diary in September 1949 the recently disclosed news that the Soviet Union had tested an atomic bomb. The "dread destroyer of 1945 had become the shield of the Republic by 1950," historian Paul Boyer has observed of the American response to the Soviet atomic threat. MacKaye composed a second series of Montachusett Federalist letters a year later, in support of yet another UWF-inspired town meeting resolution. "If federation works on a continental basis," he asked his fellow citizens, "how about a planetary scale?"[10] MacKaye maintained his faith in the possibilities of some form of federated world government, but his own grassroots pamphleteering for world government ceased after his second series of Montachusett Federalist letters.

The atomic threat also provided a significant impetus for the efforts of Stein, MacKaye, Mumford, and others to bring the Regional Planning Association of America back to life. The issues of housing, transportation, land use, urban sprawl, and large-scale planning that had occupied the RPAA in the 1920s became even more intense in the booming years after World War II. Furthermore, the new prospects and fears of Soviet atomic attack created a flurry of official and public interest in "dispersal" of population and industry as a means of civil defense and national survival. Planners and policymakers debated schemes to decentralize population, government offices, military installations, industry, and public services across the American landscape and away from the nation's urban areas.[11]

MacKaye, Stein, and Mumford were all decentralists at heart. Stein, long frustrated in his hopes for developing "new towns" and "regional cities," took the lead in 1947 and 1948 to call back together some of the original members of the RPAA, as well as new and younger colleagues. Inspired by visits from Frederic J. Osborn, Britain's leading guardian of Ebenezer Howard's garden-city legacy, Stein initiated "a campaign for the building of new towns in America."[12] Mumford was initially skeptical that the spirit which had united them in the 1920s and 1930s could be recaptured, and he joined Stein in discouraging MacKaye's idea of recasting the RPAA as a "Geotechnic Association." "We just have to recognize that the old magic circle has busted up," he wrote MacKaye, "and what is left is only a ghost; and I fear that if we turned it into the Geotechnic group, that would be a ghost at the very moment it was formed."[13] Stein heard reservations from others, such as planning scholar Catherine Bauer Wurster, who urged him not to neglect such overarching issues as regional planning and the redistribution of population in his enthusiasm to make new towns the focus of a revived RPAA.[14]

Stein was not easily discouraged, though, and the RPAA was reconstituted in 1948. A year later, the group adopted a new name: the Regional Development Council of America (RDCA). The title was apparently chosen partly to avoid confusion with the well-established Regional Plan Association of New York. But these seasoned regionalists may also have hoped to skirt the negative political and ideological connotations that had come to be ascribed to "planning," especially since the days of the New Deal and the the ascendancy of centrally planned economies in the communist regimes of eastern Europe, the Soviet Union, and Asia.[15]

As in the RPAA's heyday, Stein's New York apartment became the usual meeting place for a series of wideranging talks and discussions, attendance at which usually ranged between ten and twenty. MacKaye and Mumford made presentations, but the group also heard from active new members, including architect Albert Mayer, New York planner Hugh Pomeroy, and an engineer friend of MacKaye's, Kenneth Ross. Those attending an October 1950 talk given by Mumford, titled "Regional Redistribution of Federal Government Functions," regarded the subject as worthy of further study. MacKaye's RDCA colleagues charged him with investigating the reorganization of the federal government's natural resource bureaucracy. In May 1951, after a winter of intensive work, he produced a lengthy report, which was distilled into a shorter memorandum for distribution to the RDCA's approximately 100 far-flung members.[16]

By MacKaye's account, most of the federal activities he examined had originally been administered along subject lines, in what he called "vertical administration," usually from a Washington headquarters. Examples of of this administrative structure included the postal service, agriculture, the census, and Indian affairs. By the 1900s, some activities were being administered by region. The Forest Service represented one such form of "horizontal administration." Individual national forests were organized into regions, whose supervisors then reported to the chief forester and his staff at Washington headquarters. The Soil Conservation Service and the Bureau of Reclamation were similarly organized. But this model "only salves the problems of regional coordination *within* each service or bureau," MacKaye observed. "It simply does not approach the wider problem of coordinating *all* activities in each part of the country." On the other hand, the "TVA bombshell," at least as originally conceived, represented "the first real 'regional' administration"; but the valley authority, he acknowledged, also represented a threat to existing agencies and interests.[17]

MacKaye offered a synthesis of the two regional administrative approaches. "Instead of each 'subject' being regionally administered," he suggested, "all 'subjects' in this field of habitability should be brought under one Department whose local programs would be administered through regional authorities." MacKaye was characteristically sanguine about the prospects for such a radical reorganization of the labyrinthine federal natural resource bureaucracy. He asserted that his proposal presented merely "a technical problem, not an insurmountable problem of conflicting lines of authority."[18]

In fact, as some RDCA members pointed out, MacKaye essentially ignored the prevailing political and economic forces that had created the existing, deeply entrenched federal administrative institutions. "The problems not covered appear to me to go to the heart of the question of feasibility," bluntly noted one planning scholar.[19] The Wilderness Society's Howard Zahniser worried that delegation of federal authority to regional authorities could make wilderness-area managers more susceptible to local pressures for "non-conforming purposes," which a Washington-based administrator might more easily resist.[20] MacKaye wrote several more memoranda for the RDCA on the subject of regional administration, promoting a vision of a federal "National

Resource Service, which would really be a habitability service."[21] But this effort proved to be the climax of the group's activities.

By 1952, Stein's attentions had turned instead to the design of a new town, at Kitimat, British Columbia, where the Aluminium Company of Canada was constructing a huge smelter and a townsite for its employees. Stein recruited MacKaye to offer a proposal for a planned Kitimat museum. In MacKaye's scheme, an indoor exposition at the community's planned civic center would illustrate the history, ecological processes, and economy of the local watershed. An outdoor exhibition, encompassing a footpath and educational "wayside stations leading from the tideland to the top of the divide," would depict "a cross section of the Kitimat watershed."[22] MacKaye was constitutionally wary of all commercial and financial interests; Kitimat, principally sponsored and financed by a huge corporation, was to be a company town. But unplanned development and settlement could lead to the despoliation of the wilderness "hinterland" surrounding Kitimat, he warned Stein, "whether the resource in question be considered as a public benefit, or as a private asset to be capitalized."[23]

The RDCA, for all the talent and vision of its individual members, faced obstacles to the acceptance of its ideas similar to those the RPAA confronted during the 1920s. Postwar economic forces, government policies, and consumer desires initiated an era of public housing, "urban renewal," and suburbanization often at odds with the decentralized vision of new towns and regional cities promoted by RDCA members. Such prevailing social trends included the Housing Act of 1949, easily available Veterans Administration and Federal Housing Administration mortgages, improvements in the nation's highway system, and the success of large-scale suburban developers (MacKaye himself in 1954 visited the most notorious of these new suburbs, Long Island's Levittown).[24] Nor did the RDCA's newcomers set off the intellectual and personal sparks that had illuminated the charmed circle of the original RPAA. For MacKaye, the demise of the short-lived Regional Development Council of America freed him to begin work in earnest on his long-considered book about geotechnics.

❖

"The book is on habitability, a subject too simple to understand," MacKaye declared in his 1954 winter solstice greeting card. Though he recognized with good humor the cryptic nature of his unique vocation, he was not discouraged from spending years trying to articulate "that weird subject 'geotechnics,' that sounds so like the latest brand of breakfast food."[25]

Since his talks with Patrick Geddes at the Hudson Guild Farm in 1923, MacKaye had adopted the word *geotechnics* to describe his own perspective on the relationship between humankind and the natural environment. The term suggested both a global compass (*geo*) and an active human role in shaping and using the environment (*technics*). (Eventually, *geotechnics* would customarily come to be used in a narrower technical sense, describing the application of geological science in civil engineering and mining.) For MacKaye, geotechnics also linked conceptually the various fields and disciplines in which he had pursued his own lengthy career.

For years MacKaye had remained cautious about using the little-known word. He had sometimes discussed with Mumford and Stein the possibility of writing a sequel to *The New Exploration,* based on his personal experiences as a practicing, self-styled "geotect." During his final year at the REA, there occurred an epiphanic moment on MacKaye's own geotechnic pilgrimage. One day in the spring of 1944, while working on an REA memorandum about the Arkansas River Valley, he

> . . . came to a paragraph wherein I just had to use the word "Geotechnics".
> ("Geography" would not do, nor any other word or phrase). . . . There was a big fat new Webster's dictionary by my side into which I frequently dipped. Says I to myself, says I,—"I know the damn word is not in that dictionary but it won't take long to *see* that it is not." I looked—and by God there it *was:* "Geotechnics, the applied science of making the earth more habitable."
>
> I had to bite my lips to keep from bellowing out loud and breaking up the composure of the REA Library.[26]

As MacKaye himself knew, Patrick Geddes had long used the term *geotechnics* to describe his own perspective on planning and the environment. But at last, as far as Benton was concerned, the American vocabulary had caught up with the idea he had been guarding for two decades. That summer of 1944, he drafted a grant proposal seeking assistance from the Guggenheim Foundation in writing his book on the subject. Despite the enthusiastic recommendations of Mumford, Stein, and Stuart Chase (whose letter of support described his friend as "one of America's most useful citizens as well as one of the most original"), he failed to win funding from the foundation.[27] But geotechnics remained a central focus of MacKaye's thinking and activity for the rest of his many energetic years.

During the late 1940s, as Benton struggled to find a literary framework for revealing the subject of geotechnics, Mumford urged him to take an autobiographical approach:

You've had a very colorful and varied existence; and the way you've handled your life, in this age of profiteering (alias! free enterprise) and mechanization and gadgeteering, should be a heartening example to many younger men, who feel that they have to compromise themselves into nullity, merely to keep alive. Begin with your childhood, your early explorations, and expedition Nine and come right down to date. Don't check yourself and don't try to be kind to the people you've known: act as if you were the Angel Gabriel yourself and just be your own honest, salty, straightforward self, talking as candidly to the reader as you would to your best friend. The result will be high literature and noble geotechnics; and the time to begin that book is Now!"[28]

Already, though, MacKaye had embarked on a grand detour from personal history, by "re-reading American history and re-visualizing it upon the map."[29] His deterministic interpretation of early American history, based in part on the archaic ideas of his one-time Cambridge neighbor John Fiske, provided some intriguing insights about the relationships among economics, politics, technology, and the reshaping of the American natural environment. His historical interests were inspired in part by the questions of national and political sovereignty raised in debates over world government and world federalism. As Mumford had warned, however, those historical digressions diluted the force of MacKaye's compelling personal observations of the dynamic American scene during his eventful lifetime.

MacKaye met and corresponded regularly with *Survey* editor Paul Kellogg about the possibility of publishing a series of articles on geotechnics. In September 1948, he sent Kellogg a 15,500-word manuscript, organized as a series of articles. "The whole thing makes an introduction to a book and also an epitome of one," he explained. "It is the story (in spots) of my working life during the last half century. It is the story (in spots) of an evolution during the past three half centuries. It is a thumb-nail text book. It combines solemn discussion with trivial anecdote. It adopts various dialects. So be prepared."[30] MacKaye skirmished with Kellogg about both the length of the articles and the editor's initial reluctance to use the term *geotechnics* in the title and promotion of the series. But finally the editor relented, publishing the seven-part series, "Geography to Geotechnics," in his venerable journal of reform in 1950 and 1951.

The series, especially its first few installments, generally heeded Mumford's advice to rely on his personal history. Thus, MacKaye introduced the term *geotechnics* at the outset by recalling his experiences at the feet of such mentors as William Morris Davis and Geddes. Then, through the course of his own education and career, he traced the evolution and connections of such fields as geography, forestry, conservation, regional planning, watershed management, and ecology.

His account of America's geotechnic history included a somewhat factually garbled account of George Washington's role, during the 1780s, in resolving the technical and political challenges of creating a navigable route between the watersheds of the Potomac and Ohio Rivers. Washington, "an unconscious 'geotechnist,'" hosted a Mount Vernon convention at which delegates from Maryland and Virginia deliberated the political problem of legal jurisdiction over commerce between the two colonies and future states. Washington's self-interested commercial efforts, by MacKaye's account, were among the key events leading to the adoption of the Constitution in 1787 and the inclusion in that document of the "commerce clause," which reserved to Congress the power to regulate commerce "among the several states." And the commerce clause, as confirmed by subsequent Supreme Court decisions, provided the legal linchpin for federal sovereignty over the regulation of the nation's navigable waterways—and thus over many aspects of the management and administration of their entire watersheds. Through the commerce clause, Mac-Kaye would conclude, "man's law was changed to conform to river's law."[31] He also celebrated Thomas Jefferson's role in envisioning and creating an American "folkland" on the new public domain comprising the Northwest Territory and the Louisiana Purchase.[32] But he bemoaned "the fee simple despot," whereby the American idolization of private property undermined the prospects for communal management and control of natural resources.[33]

In the series' last two articles, "From Continent to Globe" and "Toward Global Law," MacKaye professed the need for the development of a "terrestrial consciousness" to meet the challenges of natural resource management on an international scale.[34] He called for both a system of global law to regulate commerce and the administration of Antarctica as "a common global treasure trove" under UN jurisdiction.[35] "We must match nature's ecology with geotechnics, or perish," he warned. "Is there some rule of thumb for this vast consummation? Verily, the first and simplest rule on earth: *Give back to the earth that which we take from her.* Return the goods we have borrowed; in short, pay our ecological bills. Pay them in dirt, not dollars. It's the only currency the good earth accepts. Too long have we lived on dollar ecology."[36]

The *Survey* articles were provocative and colorfully written, but MacKaye did not consider them the full story of his geotechnics. The magazine reprinted the series in pamphlet form; and MacKaye attempted to promote them as the basis of a book, which would also include some of his prior writings on such subjects as the Appalachian Trail, the Townless Highway, and the

TVA. But the book plans fell through. "I have put five years on this job. I have no idea of sitting down now to do it all over again," he informed the director of the University of North Carolina Press, who rejected the proposal. "I may live to write another book, but it will be a sequel and not a substitute."[37]

A few years later, though, he did return to the project. In early 1954, he was staying in Port Washington, New York, at the home of his friend, Kenneth Ross. Emulating his Harvard geography professor Davis, MacKaye held a small globe in his hands and expounded on the geotechnic perspective, visualizing the earth, especially North America, as if from the distance of the moon. And "thus the 'opus' began," Ross later recalled of his friend's return to the literary undertaking.[38]

MacKaye prepared a one-page proposal outlining "a possible (or impossible) book, with the possible deadline, 1979" (his 100th year), titled "North America: Views of its Habitability."[39] He then set forth on an intensive campaign of reading, research, contemplation, mapping, and revision that stretched out for more than fifteen years. "I'm at the opus, day and night, and it's my breath of life," he once confessed to Mumford.[40] With the support of friends and family, he established an annual round that combined companionship, low expenses, and intellectual stimulation. Shirley Center remained the base station of his circuit. During the warmer months of the year, he shuttled among the Cottage and its "Sky Parlor," the Grove House, and the home of his next-door neighbors, Harry and Lucy Johnson, where he often took his meals. And he sojourned at length with nieces, nephews, and friends, who often transported him between their homes in Maine, New Hampshire, New York, suburban Washington, and elsewhere.

His seasonal preparations for "mobilizing" from place to place constituted an almost comic enterprise. "Rubber bands and paper clips just exude from your pockets and are always found in a chair when you get up from it," Anne Broome once wrote, describing MacKaye's habit of carrying his work not only in his head but on his person.[41] Another friend, Jack Durham, often welcomed MacKaye for long stays at his Virginia country home near the Great Falls of the Potomac. He recalled how the single-minded MacKaye, serenely unfazed, continued to work by oil lamp through a howling winter storm and a power outage that lasted for days. "At work he was oblivious to everything, moving only to consult this volume or that in the pile of books beside him," Durham recalled. "His geotechnics moved inexorably on."[42]

Washington's exclusive Cosmos Club also provided a rich resource and setting for his literary efforts. It was a gathering place for much of Washington's scientific and intellectual elite, although then still an all-male bastion. In 1950, Howard Zahniser and others had sponsored MacKaye for membership in the club. Stein made an initial financial contribution to the Wilderness Society, which used the funds to meet his friend's club expenses.[43] Throughout the 1950s and into the early 1960s, MacKaye was a fixture at the Cosmos Club during long periods of the fall, winter, and spring months. He held court at the dining room's Table 12, "the new capital of the U.S. bureaucracy in re Geotechnics," and greeted friends from a favorite chair on a landing at the top of the club's grand stairway.[44] He also contributed many brief, colorful articles and remembrances to the club's *Bulletin*. But while he enjoyed his extended periods of residency at the Cosmos Club, the cost still strained his modest resources. "At the Club, in order to keep the bill down to income, I just don't get enough to eat," he once confessed to Robert Bruère, his long-time friend and fellow club member. "Hence low 'blood count' and low feelings."[45]

By 1960, MacKaye had completed a sprawling thirteen-chapter draft of what he now called "Geotechnics of North America."[46] But he continued to revise his text and gather material, exploring new perspectives on his own decades-old notion of "environment as a natural resource." Air and water pollution, the issue of pesticides, which was dramatized by the publication of Rachel Carson's *Silent Spring* in 1962, global scientific endeavors like the International Geophysical Year, and the dawning era of space exploration were among the subjects he incorporated into his work. "Meanwhile I continue to get from other people so many new ideas that I feel I'm just only beginning to know something about 'my subject,'" he wrote Mumford in 1963. "So I seem to be just beginning to write another book (after 10 Opus years)."[47] During his final Cosmos Club sojourn, in the late spring and early summer of 1964, MacKaye gathered information about a remaining "101 questions." At the time, the new generation of leaders at the Wilderness Society—executive director Stewart Brandborg and *Living Wilderness* editor Michael Nadel, encouraged by Harvey Broome, who had become president—offered the services of the organization's secretaries to retype the manuscript.[48] A year later, MacKaye could report to Stein that "The Opus is done save for the maps and yet unanswered questions."[49]

❖

Several elements combined to attract serious interest in MacKaye's literary efforts during the 1960s: the conspicuous success of some of MacKaye's ideas, like the Appalachian Trail; the reflected illumination of Lewis Mumford's reputation; the growing environmental movement; and a burst of academic work on the legacy of the RPAA and the regional planning movement of the 1920s and 1930s. The University of Illinois for a time became a hotbed of scholarly activity involving MacKaye's intellectual legacy. Sherman Paul, a well-regarded scholar of American literature, urged the university's press to republish *The New Exploration*. MacKaye had at first been reluctant to return the work to print. "The book contains some wise principles and objectives, and these still hold," he wrote Mumford. "It contains also a layout of means and measures, and these are substantially obsolete. The main Regional American Demon was and is the Metropolitan Invasion. In 1928 that was a threat; in 1960 it's a fact. 'Prevention' was the word then; 'salvage' (or nothing) is the only word I see today."[50]

"Don't be a damn fool," Mumford responded. "The New Exploration by now is a classic, known only to a few people, but highly valued, and difficult to get hold of. No one imagines that it contains your latest thoughts. . . . The fact that you foresaw in 1927 all the things that have been happening now increases your present status: it doesn't make you obsolete."[51] The University of Illinois Press reprinted the book in 1962, with a perceptive and affectionate introductory essay by Mumford that will stand as the definitive personal portrait of MacKaye by an intimate contemporary.

Sherman Paul also encouraged one of his graduate students, Paul T. Bryant, to take up MacKaye's writing and career as the subject of his dissertation. Over a period of several years in the early 1960s, Bryant worked closely with MacKaye, visiting him and studying his papers in Shirley Center, interviewing many of his friends, even helping (with his wife's assistance) to edit and retype a draft of "Geotechnics." Bryant's taut, elegantly written dissertation took the form of a biography, bringing together for the first time much information about his subject's life and career. But MacKaye discouraged Bryant's consideration of certain personal and family matters. Indeed, it was Mumford, not MacKaye, who revealed to the young scholar the fact of his friend's brief, ill-fated marriage, which went unmentioned in Bryant's dissertation (which the author tried, unsuccessfully, to get published).[52]

The University of Illinois Press also took an interest in MacKaye's "Geotechnics of North America." At the end of 1965, MacKaye sent his manuscript along to the press's director, Donald Jackson, who enthusiastically expressed his intention to publish the work.[53] Jackson engaged a freelance copy editor, but the arrangement quickly descended into misunderstanding and disagreement. MacKaye's unique, even eccentric, manuscript could not be comfortably accommodated by conventional editorial procedures and standards. The proud author bridled at the scope and nature of the proposed revisions.

Mumford came to his friend's defense. "If a mistake was made, it was . . . in the belief that such a manuscript could be processed in the same fashion as a textbook is turned into the neuter language of a current education theory," Mumford scolded Jackson. "Part of the value of this book comes not just from the thought itself, but from the impression of this remarkable man, conveyed in his own picturesque, sometimes archaic, sometimes crabbed style. The man, the style, the thought are inseparable. The kind of editing your reader undertook to do would have made Carlyle sound like Macaulay."[54]

If only to buy time and preserve his friend's dignity, Mumford offered to edit the manuscript himself (with the assistance of his wife Sophia). But MacKaye's advancing age was taking its toll on his own ability to revise the work. When his vision began to deteriorate in the late 1960s, work on "Geotechnics" at last ground to a halt. In 1968, as he approached the age of 90, he told Stein that he was "winding up" the project after fifteen years.[55]

The faithful Mumford was left to explain that the manuscript would not be published, at least during MacKaye's lifetime. Mumford's nine-page memorandum was a businesslike and candid analysis of the work's strengths and weaknesses. But it was also a warm expression of genuine friendship and respect. "The combined work—four books in one—makes too heavy a demand upon even the most earnest and well qualified reader," Mumford concluded. "This is no reflection on you: it would have been a miracle if you could have succeeded equally well in handling all four themes." The first theme was the "intellectual autobiography" he had long urged on his friend. The second, "the geological and physiographic history of the American continent," had been capably addressed by many other writers and simply represented "a distraction" from MacKaye's own work. The third theme, MacKaye's "history of the land in relation to the political and economic history of North America, north that is of the Rio Grande," Mumford lauded as offering "a new twist," presenting "old historic facts in a new light." The fourth represented "a detailed expansion" of *The New Exploration*.[56]

Because it had been substantially conceived and written before 1960, as Mumford noted, MacKaye's tract, especially its practical proposals, did not acknowledge the work of many planners and scholars who had, over the ensuing decade, adopted and refined the ecological perspective that was at the heart of MacKaye's geotechnics. Mumford cited in particular the work of the University of Pennsylvania's Ian McHarg, whose just-published *Design with Nature*

(1969) formulated principles and techniques of "ecological planning" that would set the norm for a new generation of land-use planners and landscape architects.

Mumford made clear the insuperable editorial and practical obstacles to the work's publication. As an alternative, though, he offered to pay for producing photocopies of the manuscript, for deposit in several good scholarly libraries. "In the coming years," he predicted, "your opus would be sought by graduate students, Ph.D.'s, and professors, seeking direct knowledge from this fountain head of ecological thought and the geotechnic programs that are now sweeping the country."[57] MacKaye accepted Mumford's conclusions philosophically and appreciatively. Photocopies of the manuscript of "Geotechnics of North America" were sent to the Library of Congress and the Dartmouth College Library, the eventual repository for most of his other papers. And he published privately a sixteen-page synopsis of the larger work, in a pamphlet titled *A Two-Year Course in Geotechnics*.[58]

Benton MacKaye had completed the task he had set for himself, at his own deliberate pace and on his own terms. The product of a habitual, experienced, and well-educated lifelong writer, the nine-chapter manuscript was organized according to a logical, if singular, progressive historical and evolutionary scheme. But the work also reflected some of the flaws and eccentricities characteristic of a confirmed autodidact—and of a writer whose crusty, quirky prose style did not always capture his raconteurial fervor nor the force of his personality. "If he could only write the way he talks," Mumford had once lamented in a letter to Harvey Broome, "he would surpass Mark Twain."[59]

The pieces of the manuscript fit together awkwardly.[60] "Geotechnics covers a multitude of matters, just so they can be seen upon a map," MacKaye wrote in his introductory chapter, "The Viewpoints." The mission of geotechnics hinged on the control of "three kingpin despoilers of habitability: the land-killing bulldozer, the slum-spreading trailer, and the earth-threatening atomic comet." MacKaye was adopting these three symbols of environmental destruction from the discourse of the 1950s and 1960s, but they represented variations of alarms he had been sounding for decades about the threats of militarism, the "metropolitan invasion," and unbridled technology. Among other voices being heard was that of journalist William H. Whyte, whose *The Last Landscape* (1968) was a popular and compelling plea for rural and suburban open-space preservation. But the planning techniques Whyte offered, such as greenbelts and cluster zoning, closely resembled practical proposals

MacKaye had offered in *The New Exploration* forty years earlier.[61] Ever optimistic, though, MacKaye in "Geotechnics" identified a trio of corresponding counterforces to "the three kingpin despoilers," in each of which he had played a personal role: conservation (by way of forestry), regionalism, and world government.[62]

The first half of "Geotechnics" virtually forswears the autobiographical approach Mumford had recommended. Instead, the chapters titled "Prologue," "Founders and Frontiers," "The Path Builders," and "Creeping Science" recount MacKaye's distinctively expressed versions of the earth's natural history (beginning with the origins of the solar system 4.5-billion years earlier) and North America's cultural, geographical, and political history.

He chronicled as a romantic pageant the technological and political development of the American transportation system, including Conestoga wagon, stage road, canal boat, clipper ship, steamboat, railroad, automobile, and airplane. MacKaye's saga of the ongoing battle for the American "folkland"—the term he used to describe the land in the public domain—pitted "disposers" against "conservers."[63] By his account, the tide began to turn toward conservation in the latter half of the nineteenth century, the era that saw the emergence of the government-sponsored transcontinental railroad surveys, the Coast Survey, the Corps of Army Engineers, the Smithsonian Institution, and the Geological Survey. Such federal agencies and institutions provided opportunities for officials trained in science to assume important responsibilities for the democratic management and common use of the nation's resources.

In MacKaye's eyes, explorer and scientist John Wesley Powell, "a practical anthropic evolutionist," was the exemplary American geotect. He pronounced Powell's *Report on the Lands of the Arid Region* (1878) "the first great American classic on a geotechnic treatment of a particular species of latent human habitat."[64] MacKaye finally ushered himself into the story in the latter half of his manuscript, in chapters titled "The Conservers," "Roosevelt to Roosevelt," and "The Regionalists." He depicted his own experience as a bridge from the era of one Roosevelt, under whom he had served his apprenticeship in a school of conservation characterized by the "wise use" and efficient development of natural resources, to the New Deal of another Roosevelt, when conservation began to embrace more complex ecological and social perspectives.

It was in "The Regionalists" that MacKaye described, in spirited style, his own most significant ideas; he also attempted to integrate them into something resembling a coherent geotechnic method and program of action. "I can date myself in conservation from Gifford Pinchot and in regionalism from

Patrick Geddes," he wrote of his own role as link between two North American "geotechnic movements." Regionalism "deals with the human side of geotechnics," he asserted. "Conservation stresses physical habitability; regionalism, psychologic habitability."[65]

Drawing upon his experience as a regional planner, he described the "weapons of salvage" that together might yet preserve a balanced environment: townless highway, highwayless town, greenbelt, wilderness area, and regional city. Of the Appalachian Trail and his role in its creation, however, MacKaye wrote just a few paragraphs. Likewise, he did not mention his own essential role in the creation of the Wilderness Society; nor, in his passing reference to the 1964 Wilderness Act, did MacKaye discuss his own earlier conceptions of a federal wilderness law.[66] His understated account of his close involvement in these two important initiatives reveals his genuine modesty, but it also represents a missed literary opportunity. As Lewis Mumford understood, MacKaye was uniquely equipped by firsthand experience to write a rich memoir describing many important chapters of the American conservation and planning movements during a long sweep of the twentieth century. In the end, though, he lacked the necessary self-regard, detachment, and literary dexterity to succeed at an endeavor so vast, complex, and necessarily autobiographical.

MacKaye wished instead to engage new ideas and experiment with them. He set out in one chapter to redraw the map of the continental United States "into thirteen new might-be 'States,'" according to the logic of watershed management, regional planning, and decentralization. "Such a setup has all the odor and odium of a blueprint," he acknowledged of his depiction of the "New Thirteen" American regions. "Whatever you call it, it depicts *no* proposal, *no* prophecy, *no* likely possibility, and above all—*no plan*. It is a presentation aimed at sharpening our visualization of political regionalism."[67]

In his final chapter, "Global Frontiers," MacKaye visualized the prospects for worldwide habitability. His perspective was long-term, evolutionary, organic. The battle for the future, he foresaw, would not be fought between capitalism and communism, the United States and the Soviet Union, or West and East. Rather, the earth's destiny hinged on the outcome of a clash between two "wild tribes," the "*Practos*" and the "*Primos*." The first comprised the "machine-worship priesthood," the second "the freak fringe of conservers and regionalists (among them your author). Two sides, theirs and ours."[68]

"Nuclear chemistry" and "Malthusian biology" represented two imminent threats. "Will we pop from Earth in a single atomic flash or will we populate it in numbers of rampant suffocation?" he asked. He offered a rudimentary

"O.P. Formula," a way to calculate an "Optimum Population," expressing "the highest yield, in terms of human values, that would be sustainable by a region's resources. In other words, habitability is quantitative." He predicted that some form of "paramount law" would be inevitable to deal with global jurisdiction over the management of natural forces and environments unconfined by conventional human-defined boundaries. Well before popular recognition of such issues as global warming and destruction of the earth's ozone layer, MacKaye foresaw that human involvement in "climate making" would make unprecedented demands on human consciousness and institutions. "All we poor civilized can do," he concluded, "is trust to God that Homo may yet gather enough experience (and store it soon enough) to obtain sufficient geo-law to forestall geo-cide."[69]

In the concluding pages of his "Opus," MacKaye once again invoked the figure and the ideas of John Wesley Powell, whose example had inspired him since boyhood. The "Powell brand of evolution," MacKaye optimistically concluded his final "opus," represented "*the path toward hope.*" His own long personal and intellectual path toward the shaping of "the applied science of making the earth more habitable" detoured along many hard-to-follow side trails. But he did arrive finally at a definition of the ultimate human niche in the earth's evolutionary story and ecological plan: "*Habitant, steward of the planet.*"[70]

"Geotechnics of North America" was completed at the moment when *ecology* and *environmentalism* became popular, if somewhat nebulous, American causes and catchwords. By emphasizing human "habitability" and the necessity of active management of the natural environment on a vast scale, MacKaye, a prophet of wilderness preservation, was harking back to the utilitarian roots of his own career as an American conservationist. Thoreau and Muir figured in his geotechnic saga, but so did Powell and Pinchot. MacKaye's "opus" —indeed, his life and career—encompassed and attempted to reconcile various strands of the American conservation and environmental movements.

In the end, though, MacKaye's conception of geotechnics was essentially the intellectual province of one man alone. For all his efforts, he failed to create a crowning historical work comparable to Mumford's *The City in History* (1961). Nor, as a proponent and practitioner of ecologically informed planning and landscape-making, would his effort have an impact like McHarg's *Design with Nature.* As an ecological thinker, he did not possess the conceptual originality of an Aldo Leopold; neither did MacKaye's writing match the lyrical literary style of his friend's *A Sand County Almanac.* But few others, with the exception of kindred spirits like Geddes and Powell, possessed the instinctive powers of visualization to imagine recreated landscapes spanning watersheds, nations, and continents. Echoing Powell, MacKaye hoped that "Geotechnics of North America," his own ambitious, sometimes ungainly synthesis of a lifetime of work and thought involving American landscapes and communities, might guide others in the quest "to evolve a more perfect union of all organic form."[71]

Linking Action with Prophecy

1953–1975

Unlike most American utopia-builders, Benton MacKaye largely succeeded in fulfilling the lofty standard of his own philosophy. "Work and art and recreation and living will all be one," he had avowed in *The New Exploration*. The Appalachian Trail, among his many ideas, demonstrated not only the validity of that precept but the powerful effect of his personal example.[1] "Instead of looking for money and legislation, MacKaye treated this project as Thoreau might have organized a huckleberrying party," as Lewis Mumford described his friend's inimitable essence in an uncompleted essay written in 1969, the year MacKaye turned ninety:

> In old fashioned Yankee style he looked for help from his friends and neighbors, and he found all he needed. . . . The fact that the idea took on, that MacKaye, by correspondence and occasional personal appearance, evoked the spirit needed to build the trail, in little groups from Hanover, New Hampshire to Knoxville, Tennessee, . . . shows the sheer force of personality, as against the current forces of depersonalized power.[2]

As MacKaye exercised his "sheer force of personality" throughout his later years, his activities extended concentrically, from his Shirley Center "Empire" out across ever broader landscapes. In October 1953, he threw his energies into the two-day bicentennial celebration of Shirley's founding. An exhibit of his maps, depicting the community's history and development, was displayed at the town hall. Drawing from family tradition, he composed and directed an elaborate five-scene pageant, "The Story of Shirley." He narrated as a cast of fifty Shirley children, men, and women performed his dramatization of the town's history before a thousand people gathered around the town common. "Benton stood high on a platform," recalled one Shirley friend, "intoning in a strong stentorian voice: 'And here they come, a Holden, a Longley and a Farnsworth,' as present representatives of those three founding families came walking up the hill to the meetinghouse, dressed in 18th century attire and carrying muskets."[3] Some in the audience may have been puzzled, however, by the singularly "geotechnic" tone of MacKaye's text, which portrayed Shirley as "a homeland worthy of homely homogeneous human habitation."[4] His brother Percy read two poems for the festivities, one composed for a similar anniversary celebration a half century earlier. Benton's lecture about the landscape of Shirley closed the bicentennial gathering. "So for one day," as he related to friends, "the MacKaye boys ran the town."[5]

Shortly thereafter, in January 1954, the Trustees of Public Reservations honored MacKaye with the group's annual conservation award. In the words of the citation, written by Mumford, MacKaye was "a sort of nature-god, a Yankee Pan," who deserved a prominent place on the roll of distinguished Massachusetts conservationists, which included Thoreau, Frederick Law Olmsted, Charles Eliot, and so many others.[6] And he still had ideas and challenges to offer. In his acceptance remarks before the venerable land-preservation organization, MacKaye took the opportunity to suggest a plan for back road tours of Massachusetts, highlighting the state's colonial and natural history.[7]

When MacKaye traveled to Knoxville in the spring of 1954 to visit Harvey and Anne Broome, the couple surprised him, on his seventy-fifth birthday, with a basketful of letters and cards from 188 friends and acquaintances. In response, he composed a lengthy letter framed as a nine-"chapter" autobiography. This colorful chronicle exhibited the wit, old-fashioned sentiment, and anecdotal novelty that endeared him to so many. His "diamond milestone" letter also attempted to break down barriers of privacy and personal reserve he had built over a lifetime.[8] After the death in 1921 of his wife, Betty, MacKaye had "retired into his mind," as Mumford later put it.[9] "I've spent nine tenths of my adult life alone," MacKaye once confessed, "and it's the only kind of life that I can live."[10] For years, he had often kept separate the many circles in which he moved and lived. Now he wrote to his friends, "Together you've forged a chain of comradeship, one that binds, as never before, my separate little worlds. Long have I desired to share each of these life compartments with the inhabitants of each of the others." Hence, he included in his eighteen-page epistle references to each person who had written to him; and he signed the letter with ten different names—such as "Bent," "Mac," "Pop," "Oom Benton," "Nestor"—by which he was familiarly known.[11]

At the time, though, there was good reason for MacKaye to remain circumspect about certain aspects of his past. At least since the days of the Harvard Socialist Club, more than forty years earlier, he had maintained contacts with a wide variety of American activists, journalists, and organizations, including many on the left wing of the American political spectrum. During the Red Scare of 1919 and 1920, his housemate Stuart Chase had lost his job with the Federal Trade Commission, for his allegedly socialist leanings. In the late 1930s, Wilderness Society co-founder Bob Marshall was among those federal employees prominently named by the Special House Committee on Un-American Activities, chaired by Texas congressman Martin Dies, for their affiliations with organizations allegedly linked to communists.[12]

MacKaye himself did not go unnoticed by federal investigators. In 1941, the Dies Committee turned over to Attorney General Francis Biddle a list of 1,124 federal employees connected with allegedly "subversive organizations." MacKaye, then with the Forest Service, was listed as a "member" of the Washington Book Shop, regarded by some agencies as a Communist Party "front organization."[13] His name also turned up in a 1943 Naval Intelligence Service report on the war-time activities of Technocracy, Inc., with whose controversial leader, Howard Scott, MacKaye had been briefly associated in 1921 while working for the Technical Alliance. In 1950, MacKaye was questioned by FBI agents on several occasions, in Boston and in Washington. The government was then investigating the alleged Communist Party associations of some of MacKaye's younger acquaintances in the TVA and the Smoky Mountains Hiking Club during the mid-1930s. According to the reports of the FBI agents, he denied firsthand knowledge of any such Communist Party activity.[14] Like many Americans during the ferociously anticommunist early 1950s, MacKaye learned to guard carefully personal information about some of his own past associations.

MacKaye stepped down from the Wilderness Society Council in 1954. But as an elder statesman of the cause, he continued to influence conservation and environmental affairs at the national level. Holding forth at the Cosmos Club, the honorary president of the Wilderness Society offered both perspective and inspiration to a new generation of conservation leaders, writers, and key officials in federal natural-resource agencies. The roster of figures with whom MacKaye sometimes conferred included David Brower, the firebrand executive director of the Sierra Club, who collaborated closely with Zahniser; National Park Service directors Newton Drury and Conrad L. Wirth; the respected Department of Agriculture ecologist Walter Lowdermilk; and writers William Vogt, Freeman Tilden, David Cushman Coyle, Russell Lord, and Sigurd Olson.[15]

MacKaye was closely involved in the conception and planning of the well-publicized 1954 walk led by Supreme Court Justice William O. Douglas along the neglected towpath of the Chesapeake and Ohio Canal. When the editors of the *Washington Post* supported a proposed highway along the route, Douglas challenged them to accompany him on a 189-mile, eight-day hike, to reveal firsthand the canal's potential as a recreational resource.[16] MacKaye, working with his friend Ken Ross, promoted the idea of a rehabilitated C&O towpath as one of several potential "feeders" to the "main backbone hinterline" of the Appalachian Trail.[17] The seventy-five-year-old MacKaye joined the festive send-off for the hikers, who included Wilderness Society leaders Zahniser, Murie, and Broome, as they set out from Cumberland, Maryland, with Justice Douglas in the lead; and he joined them for some of the last miles as the "Douglas Expedition" returned to Washington. In the end, the newspaper and the Park Service relented in their support for the highway as originally proposed. Eventually, in 1971, the C&O Canal was designated a national historic park.[18]

MacKaye, of course, had since 1916 been promoting the idea of a national network of linear recreational parks, linking cities with their surrounding regions by following such natural features as rivers and mountain ridges. The widely publicized 1954 C&O Canal hike was a vivid demonstration that some of his ideas were moving closer to the cultural mainstream. Moreover, the C&O Canal event was just one sign of growing public support for protection of the nation's undeveloped landscapes. Another watershed event in the politics of the American conservation movement was simultaneously reaching a climax—the battle over a Bureau of Reclamation proposal to build a dam as part of the Colorado River Storage Project in Dinosaur National Monument, at an isolated canyon site called Echo Park on the Utah-Colorado border. Zahniser and the Sierra Club's Brower masterminded a political and public-relations campaign that united seventeen sometimes contentious conservation organizations. The principle at stake at Echo Park was the same one conservationists had failed in defending at Yosemite's Hetch Hetchy Valley some four decades earlier: if a dam and reservoir were built in one national monument or national park, what would prevent similar violations of other such treasured public landscapes? While Brower brilliantly mobilized public opposition to the project, Zahniser cultivated his connections in Congress. The legislation he drafted, as enacted in 1956, eliminated the Echo Park dam from the plans and ensured that the Colorado project's dams and reservoirs would not be built in national parks and monuments. In the Echo Park victory, observes historian Roderick Nash, "the American wilderness movement had its finest hour to that date."[19]

For conservationists, the successful Echo Park crusade represented a heartening victory over the development-oriented policies of the Eisenhower administration. (When MacKaye visited some of his friends at the Interior Department on the day after the 1952 election, he discovered a "house of gloom.")[20] The Echo Park victory also galvanized the forces of wilderness preservation, convincing Zahniser and others that the time was at last ripe for the introduction of federal wilderness legislation. In 1956, his draft of a federal wilderness bill was filed in the Senate by Hubert Humphrey and in the House by John P. Saylor, both Democrats.[21]

MacKaye remained in close contact with Zahniser during the latter's heroic, exhausting eight-year effort to win passage of the Wilderness Act. He also contributed his own rhetorical talents to the battle. Setting aside only two percent of the nation's land, the preservation of this small remnant of "original America" would be an act of true patriotism, he wrote in a letter to the editor of the *Washington Post*.[22] In a 1961 *Living Wilderness* article, "If This Be Snobbery," he defended American wilderness advocates against the charge of elitism. Wilderness opponents "say we are a self-elected sect that would exclude folks from our self-claimed wilderness," MacKaye wrote. "They accuse us of wishing to hog the wilderness, and of holding ourselves aloof from the rest of human kind. In short, we are 'snobs.'" He rejected the charge. "Our resistance to the abolition of wilderness is the real meaning of their word 'snobbery,'" he asserted, invoking the example of Virginia patriot Patrick Henry, "just as resistance to another evil was the real meaning of the word 'treason.'"[23]

On May 4, 1964, during MacKaye's final stay at the Cosmos Club, Zahniser, Jack Durham, and Paul Oehser joined him for a festive dinner in celebration of his annual Washington visit.[24] On the occasion, eighty-five-year-old MacKaye congratulated the fifty-eight-year-old Zahniser on the Senate's recent passage of the wilderness bill and the House's expected approval of the same measure. He was soon gratified he had done so, for at lunch the next day, Stewart Brandborg, Zahniser's deputy, arrived at MacKaye's table with somber news. "Benton, we have just lost Zahnie," Brandborg reported.[25] Their friend and colleague had died in his sleep that night. But Zahniser's legacy was secure. On September 3, 1964, the Wilderness Act was signed by Lyndon B. Johnson, thereby creating a National Wilderness Preservation System.[26]

MacKaye, who was then in Shirley, politely declined a last-minute telephone invitation to the White House ceremony for the legal enactment of an idea he had been among the first to propose: the permanent designation and protection of wilderness areas on America's federally owned lands. In 1936 and again in 1946 MacKaye had drafted and circulated his own versions of federal wilderness legislation. George Marshall once described him as "the first person to state in a developed way the concept of a continental wilderness system with all types of wilderness represented."[27] But MacKaye always downplayed his 1947 proposal to the Wilderness Society Council. He cited instead the inspiration of Aldo Leopold, and he never disputed the central and well-known contributions of Marshall's brother Bob, of Zahniser, and of many others to the eventual passage of the Wilderness Act.[28] MacKaye experienced the ambiguous fate of outliving many colleagues whose renown would outshine his own. His modest assessment of his own stature as a builder of the wilderness movement revealed his respect for departed friends. In fact, he had long been reconciled to his own role as primarily an inventor of ideas, not their implementer or administrator.

During the early 1950s, MacKaye also renewed friendly contacts with the leaders of the Appalachian Trail Conference. The door opened for such a reconciliation soon after Myron Avery's death in 1952. The leadership and operation of the ATC and the Potomac Appalachian Trail Club were inseparable during Avery's lifetime (and for many years thereafter). But during MacKaye's Washington sojourns in the 1950s, he was often invited to ATC/PATC gatherings, where newer members were eager to learn about the early history of the Appalachian Trail directly from the project's "Nestor." In several articles for the PATC's magazine, he related in colorful detail some of the events and personalities from those earlier years, making sure to credit especially Clarence Stein and Charles Harris Whitaker, whose contributions had pre-dated Avery's involvement in the project. He also preached the ecological principles that had not been emphasized in his original conception of the trail. "Not until there is a broader cultural interest in nature and its process shall we stir true economic interest in the conservation of nature's resources," he now stressed.[29]

In his view, though, the fundamental economic, social, and political aspects of the trail environment were inseparable from the recreational and ecological values it embodied. Indeed, in a *Living Wilderness* article titled "Wilderness as a Folk School," he explicitly linked the development of a sensitive local ecological consciousness to the civic and political health of communities. In the seemingly innocent pastime of "nature study," he saw a potent opportunity for citizens to create "a self-taught, self-governing body" prepared to challenge the influence of the nation's vast natural resource management bureaucracy. Such local citizens, he wrote, would constitute "a school of ad-

ministration, a school of planning and management, planning of natural resources based on knowledge of natural processes. Its student body—the local
folk community; its laboratory—the local earth community, or wilderness; its
full if fearsome name—*Wilderness Folk School of Natural Resource Management.*" A long-time federal bureaucrat himself, MacKaye saw no irony in his
plea for community empowerment through nature study. By such means, after all, he had inspired one of the nation's great grassroots environmental initiatives.[30]

❖

If MacKaye himself was an "educator without a classroom," as Frederick Gutheim once remarked, his many friends looked upon his survival and upkeep
as something of a project in preservation—much like salvaging a historic
building, a valuable manuscript, or the habitat of an endangered species.[31]
"I'm reminded of a little meeting in the Cosmos Club where a member introduced me to a gang of old foresters as a 'mythical character,'" MacKaye commented to one PATC leader. "It's as hard to live up to one's own myth as to live
up to his epitaph."[32] But even his closest friends sometimes expressed good-
natured exasperation with MacKaye's passionate speculations, propositions,
and expectations. "The dinner conversation was high-minded, as you can imagine," Anne Broome once recalled of a gathering that included her husband
Harvey, Bernard Frank, and MacKaye. "At one point, Bernie burst forth, 'God
damn it, Benton, the trouble with you is you're twenty years ahead of your
time.'"[33]

In 1959, still working steadily on his "Geotechnics" at age 80, he laid out his
financial situation to Stein. It cost him $175 a month (exactly matching his
pension check) to live at the Cosmos Club, $55 a month in Shirley. Twelve of
his Cosmos Club friends fêted him on his birthday and made a substantial deposit to his club account. For years, Clarence Stein somehow managed to discover what he claimed were surplus funds from the defunct Regional Development Council of America, and he would send them along a few hundred
dollars at a time. Mumford likewise forwarded money on occasion.[34] The
grateful MacKaye never expressed any embarrassment at the seemingly one-
sided nature of these transactions; nor did his good friends treat him in a patronizing manner. If anything, the friendship among MacKaye, Stein, and
Mumford was cemented ever more firmly and expressed even more openly as
the men grew older. MacKaye was particularly faithful to Stein, who for years
struggled with periods of mental depression and collapse that sometimes resulted in hospitalization and electroshock therapy. Benton was on hand dur-

ing several of his friend's most difficult ordeals.[35] In certain respects, his friends came to envy MacKaye's persistent optimism and idealism, as well as his genuine accomplishments. "Go on dreaming, dear Ben," Mumford wrote in the mid-1960s. "There is plenty of time, and you have had far better luck in your dreams than the rest of us have had, for most of mine have been turning into nightmares, or have been lost like water in the sand."[36] MacKaye's unique variety of landscape art, the plaintive Mumford seemed to acknowledge, might outlast his own literary endeavors.

⁕

Benton also continued to nurture connections with his MacKaye relatives. His relationship with his brother Percy had not always been easy, but these last two MacKayes of their generation, both widowers, were gradually reconciled. During Benton's visits to New York City, usually to visit or stay with Clarence Stein, he also frequently saw Percy, who sometimes resided at the Players Club or the National Arts Club. Benton often made summertime visits at Hillsdale, New York, and "Hilltop," near Cornish, New Hampshire, with the families of Arvia Ege and Christy Barnes, Percy's daughters. Arvia recalled a reunion her father and uncle had shared at Hillsdale in the summer of 1956: "The two sat out together overlooking the valley for hours on end, day after day, reminiscing of the marvelous times of their youth, which they alone could know and share." In Cornish later that summer, Benton enjoyed a long evening's talk with Percy. That night, Benton was awakened by a commotion in the night. He "found Christy talking to [Percy] in a way that told me what was happening," he related to Mumford. "I am right here with you," Christy said to her father in his last moments. Then she turned to her Uncle Benton with the words, "he's gone."[37]

"He was irritating to many people, myself included," Benton confessed to Mumford after Percy's death. "He was demanding of others, and his going gives a deep release to his two daughters. Christy has been an angel to him."[38] Over the remaining decades of Benton's own life, Percy's daughters, especially Christy, assumed increasing responsibility for their aging uncle's welfare. Benton also maintained close and friendly contact with the family of his late brother Harold. The Maine homes of Benton's nephew, William Payson MacKaye, and of his grandnephew, William, Jr., were also occasional way stations on his annual round of travel.

One financial burden was lifted in 1960 when Benton sold the Grove House in Shirley, his sister, Hazel's, legacy to him. The sale provided a modest financial cushion, netting him $3,000. Thereafter, the Cottage served as his home base and its Sky Parlor as his summer "office."[39] Benton came to rely even more on his friends and next-door neighbors Harry and Lucy Johnson, who for years had looked after the properties of the MacKaye "Empire" during his absences. "[Harry] and Lucy are my adopted family, you might say," Benton once reported to another old Shirley friend. "Lucy feeds me gorgeous suppers, and I live in their house about as much as in my own."[40]

MacKaye, working steadily on his opus, did not make his annual Cosmos Club sojourn during the winter of 1962-1963. He was nursing a broken wrist, suffered in a fall during his routine lunchtime walk between the Cottage and the Johnsons.[41] But expenses and physical maladies were not the only reasons why, by the early-1960s, the scope of his peregrinations from Shirley Center began to contract. This long-time student and critic of the American transportation system also felt acutely the gradual decline of passenger railroad service, which he had enjoyed and depended upon throughout his lifetime. "I've got to take immediate chance of such railroad accommodations as still endure," the seventy-nine-year-old MacKaye had told Mumford in 1958. Five years later, in a letter to Harvey Broome, he wrote, "I don't move much these days unless folks fetch and carry me." And on January 17, 1965, he noted in his diary that the last passenger train had stopped in Shirley, ending service that had been continuous since 1849.[42]

By 1964, MacKaye's eyesight had begun to deteriorate. But not until a year later did his various physical ills—stiff legs, failing vision, intestinal distress—finally provoke mild complaint. "I'm doomed to mild pale toast in lieu of my long cherished blackened Carboniferous brand," he told a Harvard classmate.[43] When Harry Johnson died in early 1966, Benton's friendship and companionship with Lucy Johnson deepened. Although a semi-invalid himself, MacKaye could accurately say to Aline Stein of Lucy Johnson that his widowed caretaker "needs me."[44] The Johnson home essentially became his own home. Lucy's depiction of her eighty-eight-year-old boarder and companion portrays their mutual affection and concern. "Benton keeps very well," she reported to the Broomes. "His eyes have changed very little. He keeps his typewriter going every day. He writes letters and 'articles' all the time. He is asleep now in the Morris chair."[45]

There were genuine compensations for the burdens of age, however. By the mid-1960s, events and the nation's consciousness were finally beginning to follow MacKaye down the trail of ideas he had been blazing since the beginning of the century. Like many of the era's conservationists, MacKaye found a measure of hope and satisfaction in the policies of the Kennedy and Johnson administrations of the 1960s. In fact, several of his acquaintances held high positions under Secretary of the Interior Stewart Udall; such long-time younger disciples or Wilderness Society colleagues as Stanley Cain, Joseph Kaylor, and Theodor Swem were key officials in the National Park Service and the Bureau of Outdoor Recreation. MacKaye's place in the history of the American conservation movement thus came directly to Udall's attention. Udall inscribed a copy of his popular 1963 book, *The Quiet Crisis,* for Benton, who in turn described his eminent admirer as "a big man, the best head of Interior since those two great Secretaries Garfield and Schurz."[46] MacKaye did not waste the opportunity to exploit his high-level government connections.

On an early June day in 1966, three National Park Service officials arrived in Shirley Center to present Benton with the Department of the Interior's Conservation Service Award. With a few local reporters in attendance, the animated and alert MacKaye gratefully acknowledged Udall's apt recognition of his "ability to link action with prophecy."[47] "Then came my turn," MacKaye wrote to Harvey Broome the next day. "I handed the gents copies of my spiel and scheme for combining nationwide wilderness Trails with nationwide wilderness Areas, the same illustrated with a map depicting said combination along the Continental Divide and labelled 'Cordilleran Trail.'"[48] MacKaye's four-page memo, "Of Wilderness Trails and Areas: Steps to Preserve the Original America," called for the creation of a wilderness trail along the crest of the Continental Divide, connecting more than a dozen national parks, national forests, national monuments, and federally designated wilderness areas between the borders of Canada and Mexico.[49]

Geographically and conceptually, the Cordilleran Trail paralleled the Appalachian Trail and its proven legacy. In fact, as Udall himself responded, the Department of Interior was already studying a proposed Continental Divide Trail, along essentially the same route.[50] But MacKaye also sought to dramatize two important conservation issues of the 1960s: the designation of federal wilderness areas, as called for by the 1964 Wilderness Act, and the passage of proposed legislation to protect the Appalachian Trail and create a national trails system. He hoped to link the efforts of two federal agencies whose origins he had witnessed, the Forest Service (1905) and the Park Service (1916), with those of the two durable conservation organizations he had been instrumental in founding, the Appalachian Trail Conference (1925) and the Wilderness Society (1935). "The Trails can help the areas, and the Areas can help the Trails," he wrote to Paul Bryant. "I'm working on a scheme to put these two jobs together."[51] MacKaye was keenly aware that his two most important legacies were

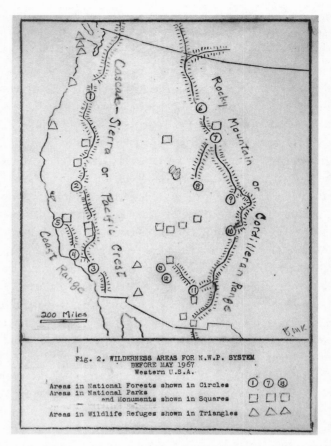

Fig. 2. WILDERNESS AREAS FOR N.W.P. SYSTEM
BEFORE MAY 1967
Western U.S.A.

Areas in National Forests shown in Circles ① ⑦ ⑧
Areas in National Parks
and Monuments shown in Squares ▢ ▢▢
Areas in Wildlife Refuges shown in Triangles △ △△

When he received the Interior Department's Conservation Service Award in 1966, Mac-Kaye presented the Park Service officials with a proposal for a "Cordilleran Trail" along the Continental Divide. One of his maps accompanying the proposal illustrated how national parks, national forests, national monuments, wilderness areas, and federal wildlife refuges could be linked to form "wilderness ways" that would follow the Rocky Mountains and the Cascade-Sierra ranges.

exemplified by the trails and wilderness movements, but he may have understood better than many younger conservation activists that those two causes were complementary elements of a larger environmental and social vision.

A controversy then unfolding in Great Smoky Mountains National Park provided another telling example of MacKaye's efforts to encourage connections among the various conservation groups and initiatives he had helped to create. The Appalachian Trail traversed the park for its full length. Much of the park was also under consideration for inclusion in the new National Wilderness Preservation System. But in 1965, plans were announced to study a new transmountain parkway, which would slice across the western half of the park from North Carolina to Tennessee. When 600 outdoor activists gathered at Clingmans Dome on October 23, 1966 to participate in the Save-Our-Smokies Wilderness Hike, ATC chairman Stanley Murray kicked off the event by reading a message from MacKaye, who addressed his thoughts across time to "Unborn Friends." He counseled the gathered hikers that their vigilance as wilderness defenders was a responsibility owed to future generations. "There are, of course, on any particular date, many more souls unborn than there are in the flesh," he observed. "And so, expressed in sheer numbers, their weal transcends ours."

"We are told these mountains need people more than they need trees," he continued his argument for the utility of wilderness. "They surely do. Especially the unborn people. We care no more for trees than they care for us. No affection lost on either side. All we want of trees is what they can give us—their service. And the more people to be served the more trees needed." After several more years of uncertainty and debate, carrying over into the Nixon administration, the proposed transmountain highway would be abandoned.[52]

Throughout the 1960s, the Appalachian Trail Conference turned its attentions to the promotion of federal legislation to protect the Appalachian Trail. During a 1963 Maine hiking trip with three other trail activists, ATC chairman Stanley Murray suggested the idea of a federal trail-protection law. The initiative was a response to the steady onslaught of residential development, new highways, ski resorts, and land fragmentation then threatening the trail. The next year, Wisconsin senator Gaylord Nelson introduced a bill aimed specifically at the acquisition of land to protect the trail corridor. Eight other House members also offered versions of bills to protect the Appalachian Trail. The legislation stalled, but in 1965 President Johnson directed Secretary Udall to conduct a study of the potential of a national trail system, "to copy the great Appalachian Trail in all parts of America."[53]

In 1966, the Bureau of Outdoor Recreation published *Trails for America*, a report with which Murray and other ATC leaders were closely involved. The report and accompanying legislation, filed in the Senate by Henry Jackson of Washington State and in the House by Representative Roy Taylor of North Carolina, called for creation of a national trails system, which would include the Appalachian Trail.[54] The congressman from Benton's home district in Massachusetts, Democrat Philip Philbin, also filed a trails bill. (Benton had been in frequent and friendly contact for many years with Cliff Gaucher, a trails enthusiast on Philbin's staff.)[55] From Shirley, MacKaye closely followed the progress of the trail bills, maintaining a constant flow of correspondence with Murray, Gaucher, and others. He was not bashful about criticizing both the overall concepts and the specific language in various legislative proposals. One PATC activist, inquiring skeptically about the need for a 200-foot trail right-of-way, received a sternly worded response. "Many people," MacKaye explained, "are blind to the difference between the actual *path* for the actual foot and the needed *belt* of wilderness on both sides. The path width should be three feet or *less;* the belt should be 200 feet or *more.*"[56]

"Never for a moment (or for a sentence) forget what the A.T. *is,*" he likewise scolded Gaucher, about the significance of seemingly innocuous legislative terminology, "and what it has been for nigh half a century (since 1921). It is a *foot* trail and nothing else. I note the word *primarily* in your Act. The A.T. is not 'primarily' a foot trail, it is wholly and totally and only a foot trail."[57] The word *primarily* remained in the law as enacted; but he was pleased that the term *trailway*—which always suggested to him the prospect of a route for buses and automobiles—did not appear.[58]

MacKaye was gratified by the passage in 1968 of the National Trails System Act. Besides authorizing the creation of a national system of trails, the law also designated the Appalachian Trail and the Pacific Crest Trail as the first two National Scenic Trails. It delegated to the Interior Department, in cooperation with the Forest Service, the principal authority for designating and acquiring the Appalachian Trail's route; but it also authorized agreements between the department and private organizations, such as the ATC, for the operation and maintenance of the trail. MacKaye warned ATC leaders that the trail's being placed under federal authority might undermine its grassroots vitality and heritage. "The A.T. has thus far been a soul without an adequate protective body; now it is about to obtain such a body," he observed to Murray. "My worry has been that a soul without a body might become a body without a soul."[59]

Though invited, the elderly MacKaye could not attend the October 2, 1968, White House ceremony at which Lyndon Johnson signed the trails law, as well as the Wild and Scenic Rivers Act. Nonetheless, passage of the national trails and rivers laws provided at least the conceptual and legal framework for an idea he had first suggested in 1916, when he had envisioned a connected "national recreation ground which would reach from ocean to ocean."[60]

There was one close friend with whom MacKaye would have liked to but could not share this satisfying accomplishment. Harvey Broome had died in

March 1968. His death was an acute shock and loss to MacKaye. Broome had anticipated such a response. Indeed, when he experienced a heart attack several months earlier, Broome had instructed friends not to report the news to MacKaye. Since their first correspondence in 1931, the personal lives and the deepest passions of MacKaye and his "adopted son" Broome had been fatefully intertwined.[61] When Broome was elected Wilderness Society president in 1957, he described to his older friend the mission they had shared since their own pivotal effort to create the organization. "We are discovering for people things that, deep-down, they want for themselves," Broome attested with no false modesty.[62] The long personal friendship between MacKaye and Broome had been enriched by their high-minded sense of public mission—and by their success—as wilderness advocates.

Like the Park Service officials who came to present the Interior Department's award in 1966, many others—scholars, journalists, biographers, Appalachian Trail hikers—made the pilgrimage to visit the oracle of Shirley Center, the living memory of the American conservation movement. His once black hair had gone white; his hawkish countenance was etched with age; his nearly sightless eyes were focused on his memories. Nonetheless, MacKaye still retained his intense curiosity, lively wit, and clarity of mind. "I'm making more speeches now, and getting more response, than I ever did before," he observed to author and conservationist Freeman Tilden, "sitting right here on my Empire, without the fuss of facing a crowd."[63] During these years, moreover, MacKaye's trove of widely scattered writings introduced his ideas to a new generation attuned to his broad-scale environmental perspective. Besides reprinting *The New Exploration,* the University of Illinois Press also published *From Geography to Geotechnics,* a 1968 collection of articles, selected by Paul Bryant, spanning MacKaye's career.

MacKaye responded to the tumultuous national events of the 1960s and early 1970s with a steady stream of letters to newspaper editors and public officials. He protested to Udall the proposed SST, the high-speed commercial jet which had the potential, he said, to transform wilderness into "a super sonic dumping ground." And he opposed Richard Nixon's 1969 appointment of Walter Hickel, development-minded governor of Alaska, as Udall's successor at the Department of the Interior.[64] In fact, the new Republican administration arrived at a climax of growing public and political concern for ecology and the environment. Stewart Brandborg, while reporting to MacKaye the Nixon administration's initial resistance to such initiatives, took heart in "the increasing involvement of people everywhere in environmental issues."[65] This new civic impulse led, in the later 1960s and early 1970s, to a burst of federal environmental legislation. Such political achievements reflected more than the public's strong support for protecting recreational and scenic resources; the new laws also demonstrated an increasing concern with pollution and a more expansive public comprehension of ecological principles. But the fundamental motivation of the post–World War II environmental movement was more personal and parochial, according to historian Samuel P. Hays. "Environmental organization," he observed, "began with a desire to control one's immediate surroundings."[66]

MacKaye had identified and grappled with many of these issues decades earlier. On the occasion of the first Earth Day, in April 1970, he recalled for one inquiring journalist that the "environment as a natural resource" had been the subject of a chapter of his *New Exploration*. In the prescient, sometimes enigmatic 1928 book, MacKaye had adopted the term *environment* to encompass the complex relationship of humans with their own natural habitat. "The nation as a whole needs towns and roads and industries and a great deal of other material plant," he had written, "but it takes more than these to make a pleasant land to live in. Mere 'shelter,' therefore, will not suffice. We need a further category. This is environment: it is a particular kind of environment—*indigenous, innate, symphonious environment.*"[67]

Perhaps as satisfying to MacKaye as any national initiatives, though, were some of the efforts taking place right in his own Shirley neighborhood to protect the "indigenous" regional environment that had been his original inspiration. During the 1960s, the state of Massachusetts began acquiring land along the undeveloped banks of the pristine Squannacook River, which skirted Shirley. MacKaye and other Massachusetts conservationists had proposed preserving the river as a "wilderness way" in the 1920s. "Here is the fulfillment of my own boyhood dream," he now wrote to Stein, "and of the scheme that I and [William] Wharton and Josh Reynolds worked on forty years ago."[68] He was also asked to be a corporator of the Nashua River Watershed Association, a local grassroots group that later won national acclaim for its successful campaign to clean up one of the nation's most grossly polluted waterways.[69]

In an effort to build support for what he called "a hometown wilderness movement," MacKaye wrote a series of articles for a local newspaper. Recollecting his boyhood expeditions, he instructed readers in the fundamentals of ecology, as represented in the region's swamps, streams, geological features, and wildlife. Mumford, during a visit to Shirley, prodded him to gather his recent articles into a book, suggesting the obvious title. MacKaye had been urging the Wilderness Society to publish a "Homeland Wildland Series."[70] And in 1969, on the occasion of MacKaye's ninetieth birthday, the slim, handsome *Expedition Nine* was published by the society. Subtitled "Return to a Region," the book was the culmination of MacKaye's long-time desire to write a primer of nature study, a guide to what he once called the "open book of primeval nature."[71] Describing a proposed local nature trail, he envisioned in his hometown a "path of endless expeditions, never complete so long as kept open to what nature has to tell."[72] His own cosmic pilgrimage in quest of environmental understanding had come full circle, back to his own Shirley backyard.

❖

When a 1968 *Christian Science Monitor* article about the Appalachian Trail had mentioned "the late Benton MacKaye," he, like his literary hero Mark Twain, had enjoyed refuting the exaggerated reports of his own demise. However, then approaching the age of ninety, the physical toll of his years was becoming increasingly apparent to friends. Lucy Johnson worried about his gaunt physique, pallid aspect, and slowing gait. He abandoned his daily walk to the Shirley Center post office.[73]

He suffered a more serious break of health in 1969. That summer his usual prolonged visit to Hilltop with the family of his niece Christy turned into an ordeal. His severe coughing and other symptoms suggested a brush with pneumonia, but he and Christy apparently collaborated in resisting hospitalization or other intensive medical intervention. Although he weathered the immediate crisis, the episode left him too weak to return to Lucy Johnson's at summer's end. Christy made arrangements for his care at a nursing home not far from Shirley Center.[74]

The move seemed barely to disrupt his routine or equanimity, though. Lucy Johnson visited him almost daily, reading his mail to him and attending to other chores. "I'm able to attend to business," he reported to Mumford. "I have a big suitcase filled with important files. These I am studying and thus placing myself up to date. I am also reading for the same purposes."[75] In fact, during his nursing home stay, MacKaye turned out several short manuscripts of reminiscence and autobiography, and he soon began to regain his health and a measure of strength. After six weeks, he returned to Lucy Johnson's good care in Shirley Center.

The scope of his movements and activities continued to shrink, however, as his eyesight dimmed and his legs stiffened. But he retained his curiosity about history and current affairs, listening to books on tape and abandoning his lifelong aversion to the radio in order to monitor events like the Watergate scandal and the resignation of Richard Nixon. Harley Holden, a Shirley neighbor, described the playful spirit with which the old railroad enthusiast approached his exercise regimen. As he circled Lucy Johnson's dining room table, "each perambulation . . . represented to him a stop on the Boston [railroad] line to the west," Holden wrote. "Around he would go, past Cambridge, Baker Station, Kendall Green, Concord, Acton, Ayer, Shirley and Fitchburg. If he felt energetic he would go on to Greenfield, through the Hoosac Tunnel and on to North Adams. A really good day of exercise would take him in imagination all the way to Mechanicsville, New York."[76]

The systematic habits and disciplines of a lifetime equipped Benton MacKaye with the resources to cope with his physical limitations. His years of diary keeping, note taking, letter writing, storytelling, and visualizing now provided him the ability to "train his mind and depend upon it completely for all knowledge," Holden observed.[77] He could draw on a rich store of facts, lore, and recollection, which he related with impeccable accuracy to the eclectic variety of visitors who arrived at his Shirley doorstep. "My callers this week included a professor from Zurich, Switzerland and a janitor from Providence, R.I.," he reported in 1971 to Michael Nadel, the Wilderness Society's assistant executive director.[78]

The accounting of so full a life continued to occupy MacKaye for the rest of his days. Several local women came regularly to read and take dictation for his still considerable correspondence. He remained the honorary president of the Wilderness Society until his death. The Cosmos Club, the American Institute of Architects, the Society of American Foresters, the Appalachian Trail Conference, the Appalachian Mountain Club, and the Potomac Appalachian Trail Club were among the organizations that awarded him honorary or lifetime memberships.

"Between 1965 and 1975 an unprecedented increase in hiking filled the once quiet mountains with nature seekers," observe Laura and Guy Waterman, historians of northeastern hills and trails.[79] The era's "backpacking boom" was fueled by the environmental movement, back-to-nature enthusiasm, and the large cohort of young adventurers representing an affluent baby boom generation. For the Appalachian Trail, the impact was swift and dramatic, in terms of both trail use and public consciousness. Lester Holmes, the first paid executive director of the Appalachian Trail Conference, reported to MacKaye in early

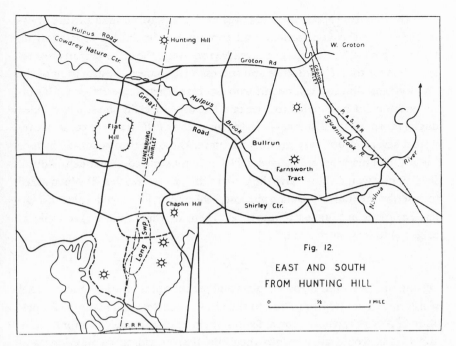

MacKaye's 1969 book, *Expedition Nine: A Return to a Region,* included a dozen maps and charts describing the ecology of the region surrounding Shirley Center. "East and South from Hunting Hill" encompasses the landscape he had described on his original "expedition nine" in June 1893. (Courtesy The Wilderness Society.)

1973 that the organization was growing by 250 members a month. In 1965, the ATC had fewer than 1,000 individual members; in late 1973, 8,740 had joined. An estimated four million hikers were afoot on the trail each year.[80]

For more and more of its users, a hike on the Appalachian Trail came to represent a rite of passage and a spiritual quest. In 1948, a hiker named Earl V. Shaffer had been the first to walk the trail end-to-end in one uninterrupted hike. When Shaffer appeared in Shirley Center five years later, the two men talked late into the night about his trail adventure. MacKaye "told me he had believed a trail, like a chain, was no better than its weakest link," Shaffer later related, "but had changed his mind. The best way was to strengthen the weak links as much as possible."[81] Shaffer's 1948 feat, the Watermans write, "had an impact on the thinking of a lot of hikers, almost comparable in psychological effect to the first four-minute mile six years later."[82] Even so, by 1963 the number of end-to-enders numbered just fifty-three. Then the explosion hit. In 1973 alone, the peak year until then, 166 hikers hiked the trail end-to-end.[83]

As the literature about the trail burgeoned, MacKaye became a figure of living legend to the Appalachian Trail community. (It did his reputation no harm, of course, that he long outlived Myron Avery, the other person most responsible for the AT's existence and success.) He enjoyed the attention of the many visitors and correspondents who expressed their appreciation for his inspired idea; but he worried that certain trends—especially the single-minded quests for speed and distance—violated his vision of the trail's purpose. He repeated several cautionary maxims whenever the opportunity arose. "I hope the A.T. will never become a race track," he wrote one correspondent. "But if so, I for one would vote to give the prize to the *slowest* traveler."[84] When asked to describe the trail's "ultimate purpose," the almost blind wilderness sage offered a cryptic but suggestive response: "There are three things: 1) to walk; 2) to see; 3) to *see* what you see."[85]

During the 1960s and 1970s, scholars and practitioners of urban and regional planning also found inspiration in the ideas of the Regional Planning Association of America, whose central figures, including MacKaye, Mumford, Stein, and Chase, would all live into their nineties, providing recollections and materials for a variety of capable scholars then writing about the RPAA's history and influence, such as Roy Lubove, Carl Sussman, and Walter Creese. Lewis Mumford's 1961 *The City in History* was a crowning statement of the garden city and regional planning ideals promoted by the RPAA. But other urban scholars during the 1960s, notably Jane Jacobs in her own popular 1961 book, *The Death and Life of Great American Cities,* fiercely criticized as antiurban these leading American proponents of garden cities and regional planning. The RPAA "Decentrists," charged Jacobs (borrowing Catherine Bauer's term), "hammered away at the bad old city. They were incurious about successes in great cities. They were interested only in failures. All was failure."[86]

According to Jacobs, America's "deeply reactionary" garden city advocates—she singled out Mumford and Stein—promoted a romanticized "idea of simple environments that were works of art by harmonious consensus," concealing "a more dominant theme of harmony and order imposed and frozen by authoritarian planning."[87] MacKaye, himself the most unauthoritarian of personalities, shared with his friends Mumford and Stein certain planning notions that were schematic and idealized. Of course, he had usually addressed the regionwide planning of wildland and rural landscapes, as well as their relationships to cities—not the detailed planning of cityscapes. Jacobs found hope in the disorderly vitality and diversity of cities, not in what she depicted as the RPAA's inflexible, darkly utopian prescription for urban revival. The troubled physical and social urban environment of the United States in the 1960s nonetheless provided stark evidence of many pessimistic trends and forces that the RPAA had accurately identified during the 1920s.

Mumford's own writing career in fact continued to flourish, as he produced both angry jeremiads and more reflective memoirs. But he remained faithful to his elder friend, with regular phone calls, written testimonials and greetings, and occasional visits. For instance, on MacKaye's eighty-ninth birthday, in 1968, Mumford wrote to him, saying, "How happy the thought of our long friendship and our active partnership in ideas makes me. No matter how widely we have been separated in space we have stayed close together in thought. My only regret—far too late to mend!—is that we never spent even a couple of days camping out on the A.T. together! More than anyone else in our time you have given us back the wilderness we had almost lost, by showing us the kind of man it could produce."[88]

Their close but guarded half-century friendship never waned during the final gathering-in of Benton MacKaye's years. In early February 1975, Aline Stein called him early one morning to report the death of her ninety-two-year-old husband, Clarence. When Mumford phoned from Cambridge that night, the two "talked about the RPAA days," as MacKaye recorded in his diary, "of which he and I are now the only survivors." Later that month, MacKaye himself was hospitalized for several weeks after a fainting spell. But he then returned to his Shirley routine.[89]

Mumford paid a call in late April, bringing Carl Sussman, a scholar then at work on a book about the RPAA. Among other things, the two long-time friends talked about MacKaye's planned message for the 1975 Appalachian Trail Conference.[90] In early June, two ATC staffers came to Shirley to film MacKaye in his native surroundings. Later that month, when the brief movie was shown at the conference in Boone, North Carolina, the thousand trail enthusiasts in attendance responded to the image of the "Father" of the Appalachian Trail as they might to "the shroud of Christ," according to one bemused observer. MacKaye was spending his last summer at Cornish with his niece Christy, but Lucy Johnson traveled to the ATC meeting, where she delivered his message to an enthusiastic ovation.[91]

In May, Benton had spent a "lovely day" talking with a writer for *Backpacker* magazine; the journalist was awed by MacKaye's detailed knowledge of the nation's geography and his intense curiosity about current environmental and trail issues.[92] During the final months of his life, he received visits from both Stewart Brandborg, Howard Zahniser's successor at the Wilderness Society, and Paul Pritchard, the new executive director of the Appalachian Trail Conference. Both conservation leaders were still eager to receive the opinions and blessing of the venerable progenitor of their respective organizations. Brandborg described some of the society's organizational and staff problems, a harbinger of difficulties the group would experience for years to come, as it struggled to find a role after its great victory, the Wilderness Act. Pritchard talked with MacKaye about the proposal for an Appalachian Trail "Greenway," a belt of protected land surrounding the trail.[93]

Benton carried on his routine serenely until one Monday evening in early December. After dinner and the reading of the day's mail, Benton retired to bed earlier than usual. When Lucy asked whether he was sick, he replied, "Oh no, I'm alright, it's just a little gas. That's all, I'm alright." But over the next two days, he continued to experience discomfort, requiring a visit by a doctor.

"Thursday the picture changed," Lucy later related to Stuart Chase. "He was miserable. He could not get comfortable. He turned on one side and then the other, sat up, laid down, etc. In the afternoon he had difficulty with breathing, and as the day progressed it became much worse. From ten o'clock to eleven o'clock it was very heavy. At eleven he went to sleep and in a minute he stopped breathing. He passed at 11:10 on December 11th, 1975." He was ninety-six years old.[94]

The immediate tributes to MacKaye's life were modest but fitting. His Shirley neighbors honored him at a memorial service held in the First Parish Church on the common. The next March, more than a hundred friends and acquaintances gathered in the John Wesley Powell Auditorium at the Cosmos Club to celebrate his life. The Wilderness Society's *Living Wilderness* published a 1976 memorial issue rich in biographical and personal detail, with eloquent contributions from Mumford, Chase, Robert Howes, Frederick Gutheim, Paul Oehser, Paul Bryant, and other friends and colleagues.[95] Although Benton MacKaye's domestic life had often been marked by tragedy and solitude, in his final decade with Lucy Johnson, his genuine gift for companionship had been amply rewarded. "He was a dear man, and it was a pleasure to have him here," his carekeeper and friend reflected. "I miss him very much."[96]

Epilogue

A "Planetary Feeling"

In 1964, the Appalachian Trail Conference met near Stratton Mountain in southern Vermont. Benton Mackaye's message to the assembly, read in his behalf by his friend Jack Durham's twenty-one-year-old daughter Nancy, recalled his experience on that very mountain in the summer of 1900, when the notion of the Appalachian Trail was first taking shape in his mind's eye. He and his friend Horace Hildreth had climbed trees to take in the view. "It was a clear day with a brisk breeze blowing," as he vividly recalled more than six decades later. "North and south sharp peaks etched the horizon. I felt as if atop of the world, with a sort of 'planetary feeling.' I seemed to perceive peaks far southward, hidden by old Earth's curvature. Would a footpath someday reach them from where I was then perched?"[1]

When I hiked to the summit of Stratton Mountain exactly one hundred years after MacKaye's treetop revelation, I followed a stretch of the Appalachian Trail. At the 3,936-foot summit, I climbed a forest-fire lookout tower, constructed by the Civilian Conservation Corps in 1934 but now used primarily as an observation tower by hikers. For a few minutes, before being joined by other hikers, I had the summit and the tower to myself. It was a cool, overcast day "with a brisk breeze blowing," as I scanned the same panorama MacKaye had found so inspiring on that July day in 1900.

His influence was discernible, in substantial and subtle ways, in every direction. Several lean, trail-weathered AT thru-hikers, en route from Springer Mountain in Georgia and other "peaks far southward," stopped only momentarily at the summit before moving briskly on towards Maine's Katahdin. There is a bronze plaque at Stratton's summit which reads, "Cradle of the Appalachian and Long Trails," commemorating the mountain's "unique role in hiking path history." (MacKaye's was not the only claim to the mountaintop as an inspirational site. It was here, according to Green Mountain Club lore, that James Taylor in 1909 first conceived of the Long Trail, which traverses Vermont north-south.) Today, the summit of Stratton lies within the Green Mountain National Forest. MacKaye's own watershed surveys for the Forest

Service in the White Mountains in 1912, immediately after passage of the Weeks Act, played a small but significant role in affirming the legal basis for the creation of other eastern national forests, including Vermont's. Just northwest of the mountain, I gazed across the 15,680-acre Lye Brook Wilderness, designated in the Eastern Wilderness Act of 1975 and expanded in the 1984 Vermont Wilderness Act. To the south of Stratton, Somerset and Harriman reservoirs, created by the construction of hydrolectric dams in the 1920s, contained the headwater runoff of MacKaye's beloved Deerfield River. Just months before my visit, the private power company that owns large tracts of land surrounding the reservoirs had conveyed a 16,000-acre conservation easement to the Vermont Land Trust, permanently protecting the reservoir shorelines from development while allowing for sustained forestry and public recreation. Piecemeal, over the course of a century, MacKaye's vision of a regional public landscape, managed and preserved to provide for economic uses, such as electrical power and timber products, as well as recreational uses, including hiking in rugged mountain and wilderness terrain, had been at least partially realized in this Vermont domain.

Not everything in sight was as MacKaye had envisioned, of course. Stratton Mountain itself perfectly illustrates the often uneasy coexistence on the American landscape of public and private uses. The mountain's North Peak serves as the summit station of a chair lift for a popular ski resort. Some of my fellow hikers included nattily dressed retirees in sneakers, who had ridden up the mountain, then sauntered along the ridge over a short side trail. And the planned community at the base of the ski trails was not the communally owned logging village MacKaye had proposed in the 1920s but a sprawling private recreational development of hotels and trailside condominiums.

MacKaye's influence and legacy—intellectual, social, political, and material—is evident in other ways, and far beyond the landscape encompassed by the view from Stratton Mountain. His innovative ideas about the relationship between humanity and the natural environment still reverberate—in a vibrant grassroots trails and greenways movement; a boundary-challenging, decentralist bioregionalism; the New Urbanism and Neotraditionalism championed by a vocal cadre of architects and community planners; a resurgent response to urban and suburban sprawl, embracing provocative alternative approaches to transportation, zoning, and land use; the disciplines of conservation biology, landscape ecology, and ecological restoration, which emphasize the protection of connected land and water corridors on regional and continental scales; ongoing academic and political debates about the meaning of wilderness in an age of leisure, economic affluence, sophisticated technology, and burgeoning population; an environmental justice movement that emphasizes issues of class, gender, and race; a "civic environmentalism" that strives to connect urban, suburban, and rural communities; and the renewed attention to the significance of place and community as the sources of a rich and durable civic life.

MacKaye's contribution to this braided stream of American thought and activism illustrates the precept of visualization offered in his own *New Exploration,* where he declared that "ultimate human aspiration consists in one form or another of broadening our mental and spiritual horizon."[2] Benton MacKaye surmounted many personal and professional impediments during his resolutely unconventional American life, which was zestfully pursued on a plane of extraordinary idealism, hope, and vision. Even when his eyesight finally failed him, he retained a "planetary feeling" enjoyed by few—the ability, at least in spirit and imagination, to see beyond the farthest horizon.

✛

APPENDIX

Reprinted here is the article in which Benton MacKaye originally proposed the Appalachian Trail. It was published in *The Journal of the American Institute of Architects*, in October 1921. The article was illustrated with the map reprinted on page 149.

An Appalachian Trail: A Project in Regional Planning

BY BENTON MACKAYE

Something has been going on in this country during the past few strenuous years which, in the din of war and general upheaval, has been somewhat lost from the public mind. It is the slow quiet development of a special type of community—the recreation camp. It is something neither urban nor rural. It escapes the hecticness of the one, the loneliness of the other. And it escapes also the common curse of both—the high powered tension of the economic scramble. All communities face an "economic" problem, but in different ways. The camp faces it through cooperation and mutual helpfulness, the others through competition and mutual fleecing.

We civilized ones also, whether urban or rural, are potentially as helpless as canaries in a cage. The ability to cope with nature directly—unshielded by the weakening wall of civilization—is one of the admitted needs of modern times. It is the goal of the "scouting" movement. Not that we want to return to the plights of our Paleolithic ancestors. We want the strength of progress without its puniness. We want its conveniences without its fopperies. The ability to sleep and cook in the open is a good step forward. But "scouting" should not stop there. This is but a faint step from our canary bird existence. It should strike far deeper than this. We should seek the ability not only to cook food but to raise food with less aid—and less hindrance—from the complexities of commerce. And this is becoming daily of increasing practical importance. Scouting, then, has its vital connection with the problem of living.

A New Approach to the Problem of Living

The problem of living is at bottom an economic one. And this alone is bad enough, even in a period of so-called "normalcy." But living has been considerably complicated of late in various ways—by war, by questions of personal liberty, and by "menaces" of one kind or another. There have been created bitter antagonisms. We are undergoing also

the bad combination of high prices and unemployment. This situation is world wide—
the result of a world-wide war.

It is no purpose of this little article to indulge in coping with any of these big questions. The nearest we come to such effrontery is to suggest more comfortable seats and more fresh air for those who have to consider them. A great professor once said that "optimism is oxygen." Are we getting all the "oxygen" we might for the big tasks before us?

"Let us wait," we are told, "till we solve this cussed labor problem. Then we'll have the leisure to do great things."

But suppose that while we wait the chance for doing them is passed?

It goes without saying we should work upon the labor problem. Not just the matter of "capital and labor" but the *real* labor problem—how to reduce the day's drudgery. The toil and chore of life should, as labor saving devices increase, form a diminishing proportion of the average day and year. Leisure and higher pursuits will thereby come to form an increasing proportion of our lives.

But will leisure mean something "higher"? Here is a question indeed. The coming of leisure in itself wll create its own problem. As the problem of labor "solves," that of leisure arises. There seems to be no escape from problems. We have neglected to improve the leisure which should be ours as a result of replacing stone and bronze with iron and steam. Very likely we have been cheated out of the bulk of this leisure. The efficiency of modern industry has been placed at 25 per cent of its reasonable possibilities. This may be too low or too high. But the leisure that we do succeed in getting—is this developed to an efficiency much higher?

The customary approach to the problem of living relates to work rather than play. Can we increase the efficiency of our *working* time? Can we solve the problem of labor? If so we can widen the opportunities for leisure. The new approach reverses this mental process. Can we increase the efficiency of our *spare* time? Can we develop opportunities for leisure as an aid in solving the problem of labor?

An Undeveloped Power—Our Spare Time

How much spare time have we, and how much power does it represent?

The great body of working people—the industrial workers, the farmers, and the housewives—have no allotted spare time or "vacations." The business clerk usually gets two weeks' leave, with pay, each year. The U. S. Government clerk gets thirty days. The business man is likely to give himself two weeks or a month. Farmers can get off for a week or more at a time by doubling up on one another's chores. Housewives might do likewise.

As to the industrial worker—in mine or factory—his average "vacation" is all too long. For it is "leave of absence *without* pay." According to recent official figures the average industrial worker in the United States, during normal times, is employed in industry about four fifths of the time—say 42 weeks in the year. The other ten weeks he is employed in seeking employment.

The proportionate time for true leisure of the average adult American appears, then, to be meagre indeed. But a goodly portion have (or take) about two weeks in the year. The industrial worker during the estimated ten weeks between jobs must of course go on eating and living. His savings may enable him to do this without undue worry. He could, if he felt he could spare the time from job hunting, and if suitable facilities were provided, take two weeks of his ten on a real vacation. In one way or another, therefore, the average adult in this country could devote each year a period of about two weeks in doing the things of his own choice.

Here is enormous undeveloped power—the spare time of our population. Suppose just one percent of it were focused upon one particular job, such as increasing the facilities for the outdoor community life. This would be more than a million people, representing over two million weeks a year. It would be equivalent to 40,000 persons steadily on the job.

A Strategic Camping Base—The Appalachian Skyline

Where might this imposing force lay out its camping ground?

Camping grounds, of course, require wild lands. These in America are fortunately still available. They are in every main region of the country. They are the undeveloped or under-developed areas. Except in the Central States the wild lands now remaining are for the most part among the mountain ranges—the Sierras, the Cascades, and Rocky Mountains of the West and the Appalachian Mountains of the East.

Extensive national playgrounds have been reserved in various parts of the country for use by the people for camping and kindred purposes. Most of these are in the West where Uncle Sam's public lands were located. They are in the Yosemite, the Yellowstone, and many other National Parks—covering about six million acres in all. Splendid work has been accomplished in fitting these Parks for use. The National Forests, covering about 130 million acres—chiefly in the West—are also equipped for public recreation purposes.

A great public service has been started in these Parks and Forests in the field of outdoor life. They have been called "playgrounds of the people." This they are for the Western people—and for those in the East who can afford time and funds for an extended trip in a Pullman car. But camping grounds to be of the most use to the people should be as near as possible to the center of population. And this is in the East.

It fortunately happens that we have throughout the most densely populated portion of the United States a fairly continuous belt of under-developed lands. These are contained in the several ranges which form the Appalachian chain of mountains. Several National Forests have been purchased in this belt. These mountains, in several ways rivalling the western scenery, are within a day's ride from centers containing more than half the population of the United States. The region spans the climates of New England and the cotton belt: it contains the crops and the people of the North and of the South.

The skyline along the top of the main divides and ridges of the Appalachians would overlook a mighty part of the nation's activities. The rugged lands of this skyline would form a camping base strategic in the country's work and play.

Seen from the Skyline

Let us assume the existence of a giant standing high on the skyline along these mountain ridges, his head just scraping the floating clouds. What would he see from this skyline as he strode along its length from north to south?

Starting out from Mt. Washington, the highest point in the northeast, his horizon takes in one of the original happy hunting grounds of America—the "Northwoods," a country of pointed firs extending from the lakes and rivers of northern Maine to those of the Adirondacks. Stepping across the Green Mountains and the Berkshires to the Catskills he gets his first view of the crowded east—a chain of smoky bee-hive cities extending from Boston to Washington and containing a third of the population of the Appalachian drained area. Bridging the Delaware Water Gap and the Susquehanna on the picturesque Allegheny folds across Pennsylvania he notes more smoky columns— the big plants between Scranton and Pittsburgh that get out the basic stuff of modern industry—iron and coal. In relieving contrast he steps across the Potomac near Harpers Ferry and pushes through into the wooded wilderness of the Southern Appalachians where he finds preserved much of the primal aspects of the days of Daniel Boone. Here he finds, over on the Monongahela side, the black coal of bituminous and the white coal of water power. He proceeds along the great divide of the upper Ohio and sees flowing to waste, sometimes in terrifying floods, waters capable of generating untold hydro-electric energy and of bringing navigation to many a lower stream. He looks over the Natural Bridge and out across the battle fields around Appomatox. He finds himself finally in the midst of the great Carolina hardwood belt. Resting now on the top of Mt. Mitchell, highest point east of the Rockies, he counts up on his big long fingers the opportunities which yet await development along the skyline he has passed.

First he notes the opportunities for recreation. Throughout the Southern Appalachians, throughout the Northwoods, and even through the Alleghenies that wind their way among the smoky industrial towns of Pennsylvania, he recollects vast areas of secluded forests, pastoral lands, and water courses, which, with proper facilities and protection, could be made to serve as the breath of a real life for the toilers in the bee-hive cities along the Atlantic seaboard and elsewhere.

Second, he notes the possibilities for health and recuperation. The oxygen in the mountain air along the Appalachian skyline is a natural resource (and a national resource) that radiates to the heavens its enormous health-giving powers with only a fraction of a percent utilized for human rehabilitation. Here is a resource that could save thousands of lives. The sufferers from tuberculosis, anemia, and insanity go through the whole strata of human society. Most of them are helpless, even those economically well off. They occur in the cities and right in the skyline belt. For the farmers, and especially the wives of farmers, are by no means escaping the grinding-down process of our modern life.

Most sanitariums now established are perfectly useless to those afflicted with mental disease—the most terrible, usually, of any disease. Many of these sufferers could be cured. But not merely by "treatment." They need comprehensive provision made for them. They need acres not medicine. Thousands of acres of this mountain land should be devoted to them with whole communities planned and equipped for their cure.

Next after the opportunities for recreation and recuperation our giant counts off, as a third big resource, the opportunities in the Appalachian belt for employment on the land. This brings up a need that is becoming urgent—the redistribution of our population, which grows more and more top heavy.

The rural population of the United States, and of the Eastern States adjacent to the Appalachians, has now dipped below the urban. For the whole country it has fallen from 60 per cent of the total in 1900 to 49 per cent in 1920; for the Eastern States it has fallen, during this period, from 55 per cent to 45 per cent. Meantime the per capita area of improved farm land has dropped, in the Eastern States, from 3.35 acres to 2.43 acres. This is a shrinkage of nearly 18 percent in 20 years; in the States from Maine to Pennsylvania the shrinkage has been 40 per cent.

There are in the Appalachian belt probably 25 million acres of grazing and agricultural land awaiting development. Here is room for a whole new rural population. Here is an opportunity—if only the way can be found—for that counter migration from city to country that has so long been prayed for. But our giant in pondering on this resource is discerning enough to know that its utilization is going to depend upon some new deal in our agricultural system. This he knows if he has ever stooped down and gazed in the sunken eyes either of the Carolina "cracker" or of the Green Mountain "hayseed."

Forest land as well as agricultural might prove an opportunity for steady employment in the open. But this again depends upon a new deal. Forestry must replace timber devastation and its consequent hap-hazard employment. And this the giant knows if he has looked into the rugged face of the homeless "don't care a damn" lumberjack of the Northwoods.

Such are the outlooks—such the opportunities—seen by a discerning spirit from the Appalachian skyline.

Possibilities in the New Approach

Let's put up now to the wise and trained observer the particular question before us. What are the possibilities in the new approach to the problem of living? Would the development of the outdoor community life—as an offset and relief from the various shackles of commercial civilization—be practicable and worth while? From the experience of observations and thoughts along the sky-line here is a possible answer:

There are several possible gains from such an approach.

First there would be the "oxygen" that makes for a sensible optimism. Two weeks spent in the real open—right now, this year and next—would be a little real living for thousands of people which they would be sure of getting before they died. They would get a little fun as they went along regardless of problems being "solved." This would not damage the problems and it would help the folks.

Next there would be perspective. Life for two weeks on the mountain top would show up many things about life during the other fifty weeks down below. The latter could be viewed as a whole—away from its heat, and sweat, and irritations. There would be a chance to catch a breath, to study the dynamic forces of nature and the possibilities of shifting to them the burdens now carried on the backs of men. The reposeful study of these forces should provide a broad gauged enlightened approach to the problems of industry. Industry would come to be seen in its true perspective—as a means in life and not as an end in itself. The actual partaking of the recreative and non-industrial life— systematically by the people and not spasmodically by a few—should emphasize the distinction between it and the industrial life. It should stimulate the quest for enlarging the one and reducing the other. It should put new zest in the labor movement. Life and study of this kind should emphasize the need of going to the roots of industrial questions and of avoiding superficial thinking and rash action. The problems of the farmer, the coal miner, and the lumberjack could be studied intimately and with minimum partiality. Such an approach should bring the poise that goes with understanding.

Finally there would be new clews to constructive solutions. The organization of the cooperative camping life would tend to draw people out of the cities. Coming as visitors they would be loath to return. They would become desirous of settling down in the country—to *work* in the open as well as *play*. The various camps would require food. Why not raise food, as well as consume it, on the cooperative plan? Food and farm camps should come about as a natural sequence. Timber also is required. Permanent small scale operations could be encouraged in the various Appalachian National Forests. The government now claims this as a part of its forest policy. The camping life would stimulate forestry as well as a better agriculture. Employment in both would tend to become enlarged.

How far these tendencies would go the wisest observer of course can not tell. They would have to be worked out step by step. But the tendencies at least would be established. They would be cutting channels leading to constructive achievement in the problem of living: they would be cutting across those now leading to destructive blindness.

A Project for Development

It looks, then, as if it might be worth while to devote some energy at least to working out a better utilization of our spare time. The spare time for one per cent of our population would be equivalent, as above reckoned, to the continuous activity of some 40,000 persons. If these people were on the skyline, and kept their eyes open, they would see the things that the giant could see. Indeed this force of 40,000 would be a giant in itself. It could walk the skyline and develop its varied opportunities. And this is the job that we propose: a project to develop the opportunities—for recreation, recuperation, and employment—in the region of the Appalachian skyline.

The project is one for a series of recreational communities throughout the Appalachian chain of mountains from New England to Georgia, these to be connected by a walking trail. Its purpose is to establish a base for a more extensive and systematic development of outdoor community life. It is a project in housing and community architecture.

No scheme is proposed in this particular article for organizing or financing this project. Organizing is a matter of detail to be carefully worked out. Financing depends upon local public interest in the various localities affected.

Features of Project

There are four chief features of the Appalachian project:

1. *The Trail—*

The beginnings of an Appalachian trail already exist. They have been established for several years—in various localities along the line. Specially good work in trail building has been accomplished by the Appalachian Mountain Club in the White Mountains of New Hampshire and by the Green Mountain Club in Vermont. The latter association has built the "Long Trail" for 210 miles through the Green Mountains—four fifths of the distance from the Massachusetts line to the Canadian. Here is a project that will logically be extended. What the Green Mountains are to Vermont the Appalachians are to eastern United States. What is suggested, therefore, is a "long trail" over the full length of the Appalachian skyline, from the highest peak in the north to the highest peak in the south—from Mt. Washington to Mt. Mitchell.

The trail should be divided into sections, each consisting preferably of the portion lying in a given State, or subdivision thereof. Each section should be in the immediate charge of a local group of people. Difficulties might arise over the use of private property—especially that amid agricultural lands on the crossovers between ranges. It might sometimes be necessary to obtain a State franchise for the use of rights of way. These matters could readily be adjusted, provided there is sufficient local public interest in the project as a whole. The various sections should be under some form of general federated control, but no suggestions regarding this form are made in this article.

Not all of the trail within a section could, of course, be built at once. It would be a matter of several years. As far as possible the work undertaken for any one season should complete some definite usable link—as up or across one peak. Once completed it should be immediately opened for local use and not wait on the completion of other portions. Each portion built should, of course, be rigorously maintained and not allowed to revert to disuse. A trail is as serviceable as its poorest link.

The trail could be made, at each stage of its construction, of immediate strategic value in preventing and fighting forest fires. Lookout stations could be located at intervals along the way. A forest fire service could be organized in each section which should tie in with the services of the Federal and State governments. The trail would become immediately a battle line against fire.

A suggestion for the location of the trail and its main branches is shown on the accompanying map.

2. *Shelter Camps—*

These are the usual accompaniments of the trails which have been built in the White and Green Mountains. They are the trail's equipment for use. They should be located at convenient distances so as to allow a comfortable day's walk between each. They should be equipped always for sleeping and certain of them for serving meals—after the fashion of the Swiss chalets. Strict regulation is essential to provide that equipment is used and not abused. As far as possible the blazing and constructing of the trail and building of camps should be done by volunteer workers. For volunteer "work" is really "play." The spirit of cooperation, as usual in such enterprises, should be stimulated throughout. The enterprise should, of course, be conducted without profit. The trail must be well guarded—against the yegg-man, and against the profiteer.

3. *Community Camps*—

These would grow naturally out of the shelter camps and inns. Each would consist of a little community on or near the trail (perhaps on a neighboring lake) where people could live in private domiciles. Such a community might occupy a substantial area—perhaps a hundred acres or more. This should be bought and owned as a part of the project. No separate lots should be sold therefrom. Each camp should be a self-owning community and not a real estate venture. The use of the separate domiciles, like all other features of the project, should be available without profit.

These community camps should be carefully planned in advance. They should not be allowed to become too populous and thereby defeat the very purpose for which they are created. Greater numbers should be accommodated by *more* communities, not *larger* ones. There is room, without crowding, in the Appalachian region for a very large camping population. The location of these community camps would form a main part of the regional planning and architecture.

These communities would be used for various kinds of non-industrial activity. They might eventually be organized for special purposes—for recreation, for recuperation, and for study. Summer schools or seasonal field courses could be established and scientific travel courses organized and accommodated in the different communities along the trail. The community camp should become something more than a mere "playground"; it should stimulate every possible line of outdoor non-industrial endeavor.

4. *Food and Farm Camps*—

These might not be organized at first. They would come as a later development. The farm camp is the natural supplement of the community camp. Here in the same spirit of cooperation and well ordered action the food and crops consumed in the outdoor living would as far as practicable be sown and harvested.

Food and farm camps could be established as special communities in adjoining valleys. Or they might be combined with the community camps by the inclusion of surrounding farm lands. Their development would provide tangible opportunity for working out by actual experiment a fundamental matter in the problem of living. It would provide one definite avenue of experiment in getting "back to the land." It would provide an opportunity for those anxious to settle down in the country; it would open up a possible source for new, and needed, employment. Communities of this type are illustrated by the Hudson Guild Farm in New Jersey.

Fuelwood, logs, and lumber are other basic needs of the camps and communities along the trail. These also might be grown and forested as part of the camp activity, rather than bought in the lumber market. The nucleus of such an enterprise has already been started at Camp Tamiment, Pennsylvania, on a lake not far from the proposed route of the Appalachian trail. This camp has been established by a labor group in New York City. They have erected a sawmill on their tract of 2000 acres and have built the bungalows of their community from their own timber.

Farm camps might ultimately be supplemented by permanent forest camps through the acquisition (or lease) of wood and timber tracts. These of course should be handled under a system of forestry so as to have a continuously growing crop of material. The object sought might be accomplished through long term timber sale contracts with the Federal Government on some of the Appalachian National Forests. Here would be another opportunity for permanent, steady, healthy employment in the open.

Elements of Dramatic Appeal

The results achievable in the camp and scouting life are common knowledge to all who have passed beyond the tenderfoot stage therein. The camp community is a sanctuary and a refuge from the scramble of every-day worldly commercial life. It is in essence a retreat from profit. Cooperation replaces antagonism, trust replaces suspicion, emulation replaces competition. An Appalachian trail, with its camps, communities, and spheres of influence along the skyline, should, with reasonably good management, accomplish these achievements. And they possess within them the elements of a deep dramatic appeal.

Indeed the lure of the scouting life can be made the most formidable enemy of the lure of militarism (a thing with which this country is menaced along with all others). It comes the nearest perhaps, of things thus far projected, to supplying what Professor James once called a "moral equivalent of war." It appeals to the primal instincts of a fighting heroism, of volunteer service and of work in a common cause.

These instincts are pent up forces in every human and they demand their outlet. This is the avowed object of the boy scout and girl scout movement, but it should not be limited to juveniles.

The building and protection of an Appalachian trail, with its various communities, interests, and possibilities, would form at least one outlet. Here is a job for 40,000 souls. This trail could be made to be, in a very literal sense, a battle line against fire and flood—and even against disease. Such battles—against the common enemies of man— still lack, it is true, the "punch" of man vs. man. There is but one reason—publicity. Militarism has been made colorful in a world of drab. But the care of the country side, which the scouting life instills, is vital in any real protection of "home and country." Already basic it can be made spectacular. Here is something to be dramatized.

⁜

ABBREVIATIONS

The following abbreviations of names and sources
are used throughout the notes.

MacKaye Family Members

BMK   Benton MacKaye
HMK   Hazel MacKaye (sister)
HSMK   Harold Steele MacKaye (brother)
JHSMK   Jessie Hardy Stubbs MacKaye (wife)
JMK   James Medbery MacKaye (brother)
MMK   Mary Medbery MacKaye (mother)
PMK   Percy MacKaye (brother)
SMK   (James) Steele MacKaye (father)

Other Individuals, Collections, Organizations; Publications Not by Benton MacKaye

AB   Anne (Pursel) Broome
AC   Allen Chamberlain
AL   Aldo Leopold
ATC   Appalachian Trail Conference
ATCA   Appalachian Trail Conference Archives, Harpers Ferry, West Virginia
ATN   *Appalachian Trailway News*
BF   Bernard Frank
CCB   *Cosmos Club Bulletin*
CHW   Charles Harris Whitaker
CS   Clarence S. Stein
CSP   Clarence S. Stein Papers, Rare Book and Manuscript Collections, Carl A. Kroch Library, Cornell University, Ithaca, New York
ED   Earle S. Draper
Epoch   Percy MacKaye. *Epoch: The Life of Steele MacKaye, Genius of the Theatre, in Relation to his Times and Contemporaries.* 2 vols. New York: Boni & Liveright, 1927.
FDR   Franklin Delano Roosevelt
FG   Frederick Gutheim

GP   Gifford Pinchot

GPP   Gifford Pinchot Papers, Library of Congress, Manuscripts Division, Washington, D.C.

HA   Harold C. Anderson

HB   Harvey Broome

HBP   Harvey Broome Papers, McClung Historical Collection, East Tennessee Historical Center, Knoxville

HH   Horace Hildreth

HPH   Harley P. Holden

HS   Harry A. Slattery

HSP   Harry A. Slattery Papers, Rare Book, Manuscript, and Special Collections Library, Duke University, Durham, North Carolina

HTC   Harvard Theatre Collection, Marion and Percy MacKaye Collection, Houghton Library, Cambridge, Massachusetts

HUA   Harvard University Archives, Cambridge, Massachusetts

HZ   Howard Zahniser

JAIA   *Journal of the American Institute of Architects*

JK   Judson King

LM   Lewis Mumford

LMP   Lewis Mumford Papers, Rare Book and Manuscript Library, University of Pennsylvania, Philadelphia

LP   Louis F. Post

LPP   Louis Post Papers, Library of Congress, Manuscripts Division, Washington, D.C.

LW   *The Living Wilderness*

MA   Myron Avery

MAM   Mable Abercrombie (Mansfield)

MKCP   MacKaye Cottage Papers, Robert Adam personal collection

MKFP   MacKaye Family Papers, Dartmouth College Library, Hanover, New Hampshire

ML   *The Milwaukee Leader*

OM   Olaus Murie

PB   Paul T. Bryant

RDCA   Regional Development Council of America

REA   Rural Electrification Administration

RF   Richard T. Fisher

RH   Robert M. Howes

RHP   Robert M. Howes personal collection

RM   Robert Marshall

RMP   Robert Marshall Papers, Bancroft Library, University of California, Berkeley (BANC MSS 79/94 c)

RPAA   Regional Planning Association of America

RSY   Robert Sterling Yard

RT   Raymond Torrey

RZ   Raphael Zon

SC   Stuart Chase

SCP   Stuart Chase Papers, Library of Congress, Manuscripts Division, Washington, D.C.

SHS   Shirley Historical Society, Shirley, Massachusetts

SP   James Sturgis Pray

TA   Tracy B. Augur

TVA   Tennessee Valley Authority

TVACL   Tennessee Valley Authority Corporate Library, Knoxville, Tennessee

USFS   United States Forest Service

WS   Wilderness Society

WSP/DC   Wilderness Society Papers, Washington, D.C.

WSP/DEN   Wilderness Society Papers, Western History and Genealogy Collection, Denver Public Library, Denver, Colorado

---

Selected Writings by Benton MacKaye

ATrail   "An Appalachian Trail: A Project in Regional Planning." *Journal of the American Institute of Architects* 9 (October 1921): 325–330.

ENR   U.S. Department of Labor. *Employment and Natural Resources: Possibilities of Making New Opportunities for Employment through the Settlement and Development of Agricultural and Forest Lands and Other Resources.* Washington, D.C.: GPO, 1919.

Ex9   *Expedition Nine: A Return to a Region.* Washington, D.C.: The Wilderness Society, 1969.

FGG   *From Geography to Geotechnics.* Edited by Paul T. Bryant. Urbana: University of Illinois Press, 1968.

GG   "Geography to Geotechnics." Parts I–VII. *The Survey* 86 (1950): I. "Growth of a New Science," 439–442; II. "From Homesteads to Valley Authorities," 496–498; III. "Genesis and Jefferson," 556–559; *The Survey* 87 (1951): IV. "Folkland as Nation Maker," 14–16; V. "Washington—and the Watershed," 172–175; VI. "From Continent to Globe," 215–218; VII. "Toward Global Law," 266–268, 285.

GNA   "Geotechnics of North America: Viewpoints of Its Habitability." Unpublished manuscript (misc. drafts, c. 1954–c. 1971). (The draft cited is identified and catalogued as "Wilderness Society Edition, Copy #3," MacKaye Family Papers, Dartmouth College Library, box 196, folders 24–39.)

NE   *The New Exploration: A Philosophy of Regional Planning.* New York: Harcourt, Brace and Company, 1928. Reprint, with an introduction by Lewis Mumford, Urbana: University of Illinois Press, 1962. Reprint, with a foreword by David N. Startzell, Urbana-Champaign: University of Illinois Press and Appalachian Trail Conference, 1991. (The 1991 edition is the one cited in notes).

❖

NOTES

Correspondence, diaries, and manuscripts for which no collection is indicated are from the MacKaye Family Papers, Dartmouth College Library.

Introduction: "Expedition 9"

1. BMK, "Geographical Hand Book," 12 June 1893.

2. *Ex9*, 1–2; *GG* (I), 440.

3. *Ex9*, 2.

4. BMK, "Geographical Hand Book."

5. Ibid.

6. Ibid.

7. BMK to LM, 16 May 1962.

8. *NE*, 228.

9. BMK diary, 7 Sept. 1957.

10. BMK, foreword to *The Appalachian Trail*, by Ronald M. Fisher (Washington, D.C.: National Geographic Society, 1972), 5.

11. "Sizing Up the Hiker Population," *Inside ATC* 3 (autumn 1996): 9–10; "Appalachian Trail Conference, Annual Report 1995," *ATN* 57 (May–June 1996): 11; "Appalachian Trail Conference, 1997 Annual Report," *ATN* 59 (July–Aug. 1998): 17–24; Brion O'Connor, "When the Going Gets Tough," *AMC Outdoors* 65 (June 1999): 18–19.

12. LM quoted in *Epoch*, 2:475.

13. *ATrail*, 329; BMK, "The Appalachian Trail," *MCM* (Bulletin of the Mountain Club of Maryland), Apr.–June 1943, 8–9.

14. See Earl V. Shaffer, *Walking with Spring: The First Thru-Hike of the Appalachian Trail* (Harpers Ferry, W.Va.: ATC, 1983); Bill Bryson, *A Walk in the Woods: Rediscovering America on the Appalachian Trail* (New York: Broadway Books, 1998). For overviews of AT literature and history, see David Emblidge, ed., *The Appalachian Trail Reader* (New York: Oxford University Press, 1996); Ian Marshall, *Story Line: Exploring the Literature of the Appalachian Trail* (Charlottesville: University Press of Virginia, 1998); and Gerald B. Lowrey Jr., "Benton MacKaye's Appalachian Trail as a Cultural Symbol" (Ph.D. diss., Emory University, 1981).

15. President's Commission on Americans Outdoors, *Americans Outdoors* (Washington, D.C.: Island Press, 1987), 142. See also Charles E. Little, *Greenways for America* (Baltimore: Johns Hopkins University Press, 1990). For one idealistic example of the AT's

386 · Notes to Pages 7–12

persistent, evolving influence, see Jamie Sayen, "The Appalachian Mountains: Vision and Wilderness," *Earth First!* 1 May 1987, 26–30. Sayen envisions the AT as the "wilderness backbone" for "a contiguous Appalachian Wilderness reuniting the Florida Keys with the Maritimes of Canada and beyond."

16. David R. Brower with Steve Chapple, *Let the Mountains Talk, Let the Rivers Run* (New York: HarperCollins West, 1995), 179.

17. BMK, "Progress Toward the Appalachian Trail," *Appalachia* 15 (Dec. 1922): 245.

18. FG, "Saying It and Doing It," *LW* 39 (Jan.–Mar. 1976): 28–30.

19. BMK to HB, 5 Sept. 1932.

20. C. J. S. Durham, "A View from the Earldom," *LW* 39 (Jan.–Mar. 1976): 27.

21. Tony Hiss, *The Experience of Place* (New York: Vintage Books, 1991), 192. See also Robert D. Yaro and Tony Hiss, *A Region at Risk: The Third Regional Plan for the New York–New Jersey–Connecticut Metropolitan Area* (Washington, D.C.: Island Press, 1996).

22. Robert McCullough, *The Landscape of Community: A History of Communal Forests in New England* (Hanover, N.H.: University Press of New England, 1995), 297.

23. Keller Easterling, *Organization Space: Landscapes, Highways, and Houses in America* (Cambridge: MIT Press, 1999), 29, 32.

24. Robert L. Dorman, *Revolt of the Provinces: The Regionalist Movement in America, 1920–1945* (Chapel Hill: University of North Carolina Press, 1993), 318.

25. Paul S. Sutter, "Driven Wild: The Intellectual and Cultural Origins of Wilderness Advocacy during the Interwar Years" (Ph.D. diss., University of Kansas, 1997), 13, 235.

26. Robert Gottlieb, *Forcing the Spring: The Transformation of the American Environmental Movement* (Washington, D.C.: Island Press, 1993), 8.

27. PB, "MacKaye as Writer," *LW* 39 (Jan.–Mar. 1976): 33.

28. See John L. Thomas, *Alternative America: Henry George, Edward Bellamy, Henry Demarest Lloyd and the Adversary Tradition* (Cambridge: Harvard University Press, 1983), 354–366.

29. LM, "Benton MacKaye," unpublished review of *Ex9*, 24 Mar. 1969, attached to LM to BMK, 6 Mar. 1974.

30. BMK, "Vision and Reality," speech to ATC, 30 May 1930.

Chapter 1 / The MacKaye Inheritance

1. SMK to MMK, 11 Mar. 1879, in *Epoch*, 1:298.

2. *Epoch*, 1:302.

3. Ibid., 307–308.

4. Ibid., 116–118.

5. Ibid., 94.

6. Ibid., 18–22; James Morrison McKaye, "Autobiography," Library of Congress, Manuscripts Division, Washington, D.C.

7. *Epoch*, 1:21–22, 53, 131, 257.

8. Ibid., 22; Lucius Beebe and Charles Clegg, *U.S. West: The Saga of Wells Fargo* (New York: E. P. Dutton, 1949), 27.

9. *Epoch*, 1:53–78.

10. Ibid., 100–111; George R. Bentley, *A History of the Freedmen's Bureau* (New York:

Octagon Books, 1970), 25–26; James McKaye, *The Mastership and Its Fruit: The Emancipated Slave Face to Face with His Old Master* (New York: W. C. Bryant, 1864), 4, 34–38.

11. William James to PMK, 1908, in *Epoch,* 1:95–96.

12. *Epoch,* 1:132–227.

13. Howard M. Feinstein, *Becoming William James* (Ithaca, N.Y.: Cornell University Press, 1984), 27.

14. *Epoch,* 1:288–289.

15. James Morrison McKaye to SMK, 6 June 1879, in *Epoch,* 1:308–309.

16. *Epoch,* 2:74–93.

17. Ibid., 141.

18. Ibid., 169.

19. MMK to Emily von Hesse, 13 Feb. 1889; *Epoch,* 1:128–129, 2:177.

20. HPH, "The Shirley Influence," *LW* 39 (Jan.-Mar. 1976): 18–23.

21. *NE,* 58–61. About Shirley and its history, see Ethel S. Bolton, *Shirley Uplands and Intervales* (Boston: George Emery Littlefield, 1914), and Forrest Bond Wing, *The Shirley Story* (Shirley, Mass.: n.p., 1981).

22. BMK, "Some Shirley Ctr. Post Office history," memorandum to Lucy Longley, 25 May 1963; BMK to HB and AB, 29 June 1962.

23. BMK to Emily von Hesse, 27 Dec. 1888.

24. *Epoch,* 2:194–196.

25. Ibid., 181.

26. HPH, "The Shirley Influence."

27. PMK to BMK, 13 Feb. 1889.

28. MMK to JMK, [?] Mar. 1890, SHS.

29. *Epoch,* 2:276–279; PMK diary, 24 Dec. 1890.

30. BMK diary, Dec. 1890, Jan.-Apr. 1891; BMK, "Wilderness as a Folk School," *LW* 17 (winter 1952–53): 3.

31. BMK, letter to editor, "Turns Back Clock," *Washington Evening Star,* 1 May 1959.

32. PMK diary, 1890, 1891; BMK diary, 1890, 1891; BMK, "Turns Back Clock."

33. BMK diary, 1891; BMK to James G. Deane, 4 Mar. 1968.

34. BMK diary, 13 Feb. 1891.

35. Ibid., 17, 19 Feb. 1891.

36. Ibid., 27 Feb. 1891.

37. Ibid., 12–13 Mar. 1891.

38. Ibid., Apr. 1891.

39. BMK, letter to editor, *CCB,* Dec. 1967, 10; BMK diary, 11 Apr. 1891.

40. BMK to Gilbert H. Grosvenor, 11 Dec. 1961.

41. *Epoch,* 2:280; PMK, "Half Hour Happenings in the Teeles Family," Dec. 1890–Nov. 1893, ATCA.

42. BMK diary, Feb.-Apr. 1891.

43. Walter Lippmann, "All the MacKayes," *The International* 3 (Jan. 1911): 29.

44. BMK, "Constitution of 'The Rambling Boys' Club,'" c. 1891.

45. BMK, "A Mountain and a Man," *Fitchburg (Mass.) Sentinel,* 2 Feb. 1949.

46. About the Spectatorium, see Larry Anderson, "Yesterday's City: Steele MacKaye's

Grandiose Folly," *Chicago History* 16 (fall–winter 1987–88): 104–114, and *Epoch,* 2:311–453.

47. BMK to SMK, 27 July 1892.

48. BMK to Harry M. Treat, 24 Sept. 1952.

49. *Ex9,* 7–8; BMK diary, 1 Oct. 1892.

50. *GNA,* ch. 8, 2.

51. BMK to SMK, 9 Nov. 1891.

52. *Epoch,* 2:474; BMK to SMK, 15 Jan. 1893.

53. BMK to SMK, 1 Feb. 1893.

54. BMK, "Cosmopolitan Survey," c. 1892–1893.

55. BMK diary (looseleaf pages), Sept. 1893; PMK diary, 21–26 Sept. 1893; *Epoch,* 2:421–426.

56. BMK to SMK, 20 Oct. 1893.

57. Gertrude Nutting Spaulding to BMK, 12 Jan. 1949.

58. BMK, "Report for Harvard College Class of 1900" (draft), 26 Oct. 1964.

59. SMK to BMK, 14 Dec. 1893.

60. BMK to Jennie M. Stein, 19 Jan. 1965.

61. HSMK to MMK, 23 Mar. 1894.

62. BMK diary, 10 Mar. 1894; BMK to MMK, 14 Mar. 1894.

63. BMK diary, Sept.–Nov. 1894.

64. HPH, "The Shirley Influence."

65. BMK diary, 29 June 1895.

66. BMK, "Edward Clayton Sherman, 1877–1961," *CCB,* Sept. 1961, 2–4; BMK to Elsie Whittemore, 29 Nov. 1966.

67. BMK diary, Sept.–Dec. 1895.

68. Edwin Osgood Grover, ed., *Annals of an Era: Percy MacKaye and the MacKaye Family, 1826–1932* (Washington, D.C.: Dartmouth College / Pioneer Press, 1932), 505; *The Harvard University Catalogue, 1895–1896* (Cambridge, 1895), 201–218; HSMK to MMK, 25 June 1896.

69. "Report of the Class of 1900: I–Z," HUA; HSMK to JMK, 3 Oct. 1896.

Chapter 2 / From Harvard Yard to the "Primaevial Forest"

1. *GG* (I), 439.

2. Samuel Eliot Morison, *Three Centuries of Harvard: 1636–1936* (Cambridge: Harvard University Press, 1936), 368.

3. Donald Fleming, "Harvard's Golden Age?" in *Glimpses of the Harvard Past,* by Bernard Bailyn et al. (Cambridge: Harvard University Press, 1986), 77–95; *Harvard University Catalogue, 1896–97* (Cambridge, 1896), 226.

4. *Harvard University Catalogue, 1896–97,* 265–271; Morison, *Three Centuries of Harvard,* 419–422.

5. *Harvard Crimson,* 9 and 17 Oct. 1896; BMK to MMK, 6 Dec. 1898; BMK to HH, 25 July 1898; BMK to Hoffman Miller, 18 Aug. 1900.

6. George Santayana, "The Harvard Yard (1882–1912)," in *The Harvard Book,* rev. ed., ed. William Bentinck-Smith (Cambridge: Harvard University Press, 1982), 67.

7. Donald Fleming, "Some Notable Harvard Students," in *Glimpses of the Harvard Past*, 141; *Secretary's First Report. Harvard College Class of Nineteen Hundred* (Cambridge, 1902), 12, 16–17.

8. Fleming, "Harvard's Golden Age?" 90.

9. "Record of the Class of 1900: I–Z," HUA; L. B. R. Briggs to MMK, 27 Feb. 1897.

10. Donald Fleming, "Eliot's New Broom," in *Glimpses of the Harvard Past*, 72; David N. Livingstone, *Nathaniel Southgate Shaler and the Culture of American Science* (Tuscaloosa: University of Alabama Press, 1987), 7.

11. Nathaniel S. Shaler, "The Betterment of Our Highways," *Atlantic Monthly*, Oct. 1892, 506.

12. Rollo W. Brown, *Harvard Yard in the Golden Age* (New York: Current Books / A. A. Wyn, 1948), 118; N. S. Shaler, "The Landscape as a Means of Culture," *Atlantic Monthly*, Dec. 1898, 777–778.

13. William Morris Davis, "The Geographical Cycle," in *Geographical Essays* (Boston: Ginn, 1909), 249.

14. Davis, "Geographical Cycle," 253–254; BMK, "Letter to the Editor," *CCB*, Nov. 1966, 6–7.

15. BMK, papers for English C, 13 Mar. 1899, 10 Apr. 1899; BMK, "Effects of the Drainage of the English Midland Lowland," paper for Geology 7, 1900; Stephen J. Pyne, *Grove Karl Gilbert: A Great Engine of Research* (Austin: University of Texas Press, 1980), 258.

16. BMK, "Letter to the Editor," 6–7; BMK to Albert F. Durrin, 7 Jan. 1954.

17. BMK to HB, 5 Sept. 1932.

18. BMK diary, 12 Aug. 1897.

19. Ibid., 13 Aug. 1897.

20. On White Mountain history see: Charles D. Smith, "The Mountain Lover Mourns: Origins of the Movement for a White Mountain National Forest 1880–1903," *New England Quarterly* 33 (1960): 37–56; Charles D. Smith, "The Movement for Eastern National Forests—1899–1911" (Ph.D. diss., Harvard University, 1956); Frederick W. Kilbourne, *Chronicles of the White Mountains* (Boston: Houghton Mifflin, 1916); Alfred Chittenden quoted in C. Francis Belcher, *Logging Railroads of the White Mountains*, rev. ed. (Boston: Appalachian Mountain Club, 1980), 8.

21. Albert Bushnell Hart, "The Protest of the Mountain Lover," *The Nation*, 4 June 1896, 430–431.

22. Kilbourne, *Chronicles*, 345–359; Laura Waterman and Guy Waterman, *Forest and Crag: A History of Hiking, Trail Blazing, and Adventure in the Northeast Mountains* (Boston: Appalachian Mountain Club, 1989), 189–192.

23. BMK diary, 14 Aug.–2 Sept. 1897.

24. BMK to HB, 5 Sept. 1932; BMK to Maud Hosford, 21 Aug. 1897.

25. BMK diary, 16 Aug. 1897.

26. Ibid., 16–17 Aug. 1897.

27. BMK to M. Hosford, 21 Aug. 1897.

28. BMK diary, 17 Aug. 1897.

29. "Record of the Class of 1900"; BMK, paper for Economics 6, 23 Jan. 1900.

30. BMK to PB, 26 July 1964.

31. BMK, paper for English 22, 25 Feb. 1898.

32. BMK to MMK, 2 Apr. 1900; Emily von Hesse to BMK, 21, 26 June 1900.

33. BMK to JMK, 15 Aug. 1898.

34. BMK, diary of Vermont trip, July 1900; BMK, "Some Early A.T. History," *Potomac Appalachian Trail Club Bulletin* 26 (Oct.–Dec. 1957): 91; Larry Anderson, "A Classic of the Green Mountains: The 1900 Hike of Benton MacKaye and Horace Hildreth," *Harvard (Mass.) Post,* 19 Dec. 1986, 22–25; Waterman and Waterman, *Forest and Crag,* 351–373.

35. HH to BMK, 14 Sept. 1941.

36. BMK to JMK, 15 Aug. 1898; BMK to Frank Buxton, 30 May 1965; BMK to MMK, 19 Sept. 1899.

37. Regarding Muir and Pinchot, see: *Harvard University Catalogue, 1896–97,* 594; John Muir, "The National Parks and Forest Reservations," *Harper's Weekly,* 5 June 1897, 563–567; "Forestry as a Profession," *Harvard Crimson,* 3 Mar. 1900; BMK, speech to Harvard class of 1900, June 1960.

38. BMK to JMK, 1 Aug. 1898.

39. MMK to JMK, Sept. 1900; MMK to JMK, 11 Mar. 1901.

40. JMK to BMK, 5 Mar. 1901.

41. BMK to JMK, 7 Apr. 1901.

42. MMK to JMK, 8 Oct. 1901.

43. BMK to HMK, 22 May 1902.

44. BMK to James B. Craig, 23 Oct. 1964.

45. BMK to JMK, 8 Apr. 1902; *Camp Penacook* (1903), brochure, ATCA; BMK to HH, 30 June 1902.

46. Florence Pray to BMK, 9 Nov. 1902.

47. BMK to MMK, 30 Oct. 1902.

48. Virgil Prettyman to BMK, 28 Mar. 1956; BMK to V. Prettyman, 15 Apr. 1956.

49. *GG* (II), 497.

50. Waterman and Waterman, *Forest and Crag,* 375–377; "James Sturgis Pray," *National Cyclopedia of American Biography* (New York: James T. White, 1939), 27:399–400.

51. James Sturgis Pray, "The New Swift River Trail and its Bearing on the Club's Policy," *Appalachia* 10 (May 1903): 173–179.

52. BMK to ATC, 25 Aug. 1948. Pray described his principles of trail design in a pamphlet, *Beauty in Trails,* Publication No. 9 (Boston: New England Trail Conference, 1923).

53. Dorothy M. Martin, "Interview with Benton MacKaye," *Potomac Appalachian Trail Club Bulletin* 22 (Jan.–Mar. 1953): 12–13.

54. BMK to Frederick Hooper, 16 Feb. 1935.

55. BMK to CS, 2 Sept. 1963.

56. BMK to JMK, 20 Oct. 1903.

57. Ibid.; BMK to Harris A. Reynolds, 28 May 1946.

58. BMK to JMK, 20 Oct. 1903.

59. BMK to Nancy Durham, 13 June 1964; BMK to HB, 5 Sept. 1932.

Chapter 3 / The Education of a Progressive Forester

1. Richard G. Lillard, *The Great Forest* (New York: Alfred A. Knopf, 1947), 267; BMK, speech to Harvard class of 1900, June 1960; Edmund Morris, *The Rise of Theodore Roosevelt* (New York: Ballantine Books, 1980), 739–741.

2. Harold K. Steen, *The U.S. Forest Service: A History* (Seattle: University of Washington Press, 1976), 61–64; Curt Meine, *Aldo Leopold: His Life and Work* (Madison: University of Wisconsin Press, 1988), 75–76.

3. Donald R. Theoe, letter to author, 3 Dec. 1986.

4. "Scientific School," *Harvard Graduates Magazine* 12 (Dec. 1903): 252–253; "President Eliot's Report," *Harvard Graduates Magazine* 12 (Mar. 1904): 419–420.

5. *Harvard University Catalogue, 1903–04* (Cambridge, 1903), 425; BMK undergraduate folder, HUA; Samuel Eliot Morison, *The Development of Harvard University* (Cambridge: Harvard University Press, 1930), 342, 514–15.

6. "Development of Forestry at Harvard," *Harvard Graduates Magazine* 14 (Dec. 1905): 257; RF, "The Year in the Forestry Department," *Harvard Graduates Magazine* 14 (June 1906): 607.

7. Morison, *Development of Harvard*, 444, 448–450.

8. BMK, "Report on the Operations of the Goodyear Lumber Company of Pennsylvania," c. 1905.

9. *Harvard University Catalogue, 1904–05* (Cambridge, 1904), 238, 535.

10. BMK to PB, 26 July 1964.

11. JMK, *The Economy of Happiness* (Boston: Little, Brown, 1906), vii.

12. Ibid., 117.

13. Ibid., 183–184.

14. Ibid., 432.

15. Ibid., 414.

16. Ibid., 405–406.

17. Ibid., 417.

18. Ibid., 532.

19. Ibid., 326–327.

20. BMK to MMK, 10 Mar. 1905.

21. BMK to PB, 25 July 1964.

22. Bill Ballard to JMK, 25 Mar. 1928.

23. Emily von Hesse to BMK, 29 June 1905.

24. BMK to MMK, 10 Mar. 1905.

25. BMK to HB, 9 Sept. 1932.

26. BMK, "Our White Mountain Trip: Its Organization and Methods," in *The Log of Camp Moosilauke 1904* (n.p., c. 1905), 4–11.

27. Ibid., 10–11.

28. Ibid.

29. *U.S. Statutes at Large* 33 (1905): 628; *GG* (I), 440; Steen, *U.S. Forest Service*, 74–75, 99.

30. Gifford Pinchot, *Breaking New Ground* (1947; Seattle: University of Washington

Press, 1972), 291; BMK, USFS questionnaire, 27 Nov. 1940; BMK diary, 31 Aug. 1905.

31. BMK diary, 1–6 Sept. 1905; BMK to Charles L. Tebbe, 7 Nov. 1955; Norman J. Schmaltz, "Raphael Zon: Forest Researcher," parts 1 and 2, *Journal of Forest History* 24 (Jan., Apr. 1980): 24–39, 86–97.

32. BMK diary, 10 Sept. 1905.

33. Gifford Pinchot, *Practical Assistance to Farmers, Lumbermen, and Other Owners of Forest Lands,* U.S. Department of Agriculture, Forest Service Circular No. 21 (Washington, D.C.: GPO, 1898); Pinchot, *Breaking New Ground,* 141–143; Steen, *U.S. Forest Service,* 53–55.

34. BMK diary, 14 Sept. 1905; BMK to "Friends, neighbors, colleagues, kin," 6 Mar. 1954.

35. BMK to Howard Davenport, 12 Sept. 1905.

36. RF to BMK, 26 May 1906; RF to BMK, 12 June 1906.

37. RF, "Progress in Forestry," *Harvard Graduates Magazine* 15 (Dec. 1906): 286–287; BMK, forest survey of J. E. Henry and Sons, Aug.–Sept. 1906.

38. GG (I), 441; BMK, reports on eight forest tracts, Aug.–Sept. 1907.

39. AC to BMK, 5 Mar. 1907; AC, "Reports of the Councillors for the Autumn of 1907. Exploration and Forestry," *Appalachia* 11 (June 1908): 397–402.

40. BMK, "Chamberlain of Monadnock," *Fitchburg Sentinel,* 27 June 1945; AC, "Reports of the Councillors for the Autumn of 1908. Exploration and Forestry," *Appalachia* 12 (July 1909): 81–85.

41. BMK, testimony by RF to the Committee on Agriculture, U.S. House of Representatives, 22 Jan. 1908.

42. BMK to MMK, 21 Mar. 1908; BMK to Samuel T. Dana, 14 July 1975; Samuel T. Hays, *Conservation and the Gospel of Efficiency: The Progressive Conservation Movement, 1890–1920* (Cambridge: Harvard University Press, 1959), 127–133.

43. GG (I), 441.

44. RF to BMK, 18 July 1908.

45. BMK to HMK, 3 Sept. 1908, SHS.

46. *Harvard University Catalogue, 1908–09* (Cambridge, 1909), 414–415; Granville Hicks, *John Reed: The Making of a Revolutionary* (New York: Macmillan, 1936), 32–33.

47. Ronald Steel, *Walter Lippmann and the American Century* (New York: Vintage Books, 1981), 15; Minutes of the Harvard Socialist Club, 1908–1915, HUA.

48. Minutes of the Harvard Socialist Club; Steel, *Walter Lippmann,* 23–32.

49. Carl Binger, "A Child of the Enlightenment," in *Walter Lippmann and His Times,* ed. Marquis Childs and James Reston (New York: Harcourt, Brace, 1959), 34; BMK to Carl Binger, 8 Dec. 1959.

50. BMK to C. Binger, 8 Dec. 1959; BMK to Helen Northup, 28 Oct. 1963; Lincoln Steffens, *The Autobiography of Lincoln Steffens* (New York: Harcourt, Brace, 1931), 2:644–647.

51. BMK to MMK, 8 Mar. 1909; BMK to MMK, 26 Oct. 1909; "Lecture by P. MacKaye '97," *Harvard Crimson,* 17 Feb. 1909; PMK, *The Playhouse and the Play* (New York: Macmillan, 1909), 123–154; PMK, *The Civic Theatre in Relation to the Redemption of Leisure* (New York: Mitchell Kennerley, 1912), 15.

52. Hicks, *John Reed,* 40–41.

53. BMK to HMK, 18 Mar. 1909.

54. Jane S. Knowles, Radcliffe College Archivist, letter to author, 10 Mar. 1986; BMK to HMK, 6 Mar. 1907.

55. HSMK to MMK, 16 Sept. 1909.

56. BMK to HMK, 23 Dec. 1909.

57. (Mabel) "Lucy" Abbott to HMK, 4 Jan. 1910.

58. Pinchot, *Breaking New Ground,* 325–326.

59. Hays, *Conservation,* 47; Pinchot, *Breaking New Ground,* 252; William H. Harbaugh, *The Life and Times of Theodore Roosevelt,* rev. ed. (London: Oxford University Press, 1975), 304–319.

60. For the differences between Pinchot and Taft, see Hays, *Conservation,* 147–174.

61. For background on the Ballinger-Pinchot controversy, see James Penick Jr., *Progressive Politics and Conservation: The Ballinger-Pinchot Affair* (Chicago: University of Chicago Press, 1968); Hays, *Conservation,* 165–174.

62. BMK to HMK, 11 Jan. 1910.

63. M. F. Abbott, "A Brief History of the Conservation Movement," parts 1–5, *Twentieth Century Magazine,* Mar. 1910, 537–543; May 1910, 143–148; June 1910, 228–234; Aug. 1910, 410–415; Sept. 1910, 511–515.

64. RF to BMK, 13 Mar. 1910; BMK to RF, 16 Mar. 1910; BMK to PMK, 2 Apr. 1910.

65. Jerome D. Green to BMK, 2 Apr. 1910; BMK to PMK, 2 Apr. 1910; MMK to HMK, 21 May 1910, SHS; BMK to HSMK, [?] Apr. 1910.

66. HMK to MMK, 24 July 1910; HMK to MMK, 25 Aug. 1910.

67. JMK to MMK, 11 Jan. 1911.

68. HMK to MMK, 27 Feb. 1911.

69. JMK to MMK, 11 Jan. 1911.

70. H. S. Graves to BMK, 9 Jan. 1911; MMK to PMK, 27 Dec. 1910.

71. HMK to MMK, 3 Mar. 1911; HSMK to MMK, 6 Mar. 1911.

72. BMK to MMK, 29 Apr. 1911.

73. BMK, "A Theory of Forest Management," c. 1910–1911, 3–4; "The Case Method in Forest Organizing," c. 1910–1911, 94–95; "Forest Management," c. 1910; "Working Plan Report for Forest Land of the MacDowell Memorial Association, Peterboro, N.H.," c. 1910–1911.

74. M. F. Abbott to MMK, 1 May 1911.

75. M. F. Abbott, "Maine—A Power-House or a Factory?" *Collier's,* 4 Mar. 1911, 32, 34.

76. BMK to MMK, 7 Mar. 1911; M. F. Abbott to MMK, 1 May 1911; GP diary, 28, 31 May, 1 June 1911, GPP.

77. M. F. Abbott, "The Latest in Alaska: Controller Bay and Its Control of the Alaskan Situation," *Collier's,* 6 May 1911, 19.

78. Alleged postscript to Richard S. Ryan's letter to Richard A. Ballinger, 13 July 1910, as published in *Philadelphia North American,* 7 July 1911, and reprinted in Alpheus T. Mason, *Brandeis: A Free Man's Life* (New York: Viking Press, 1956), 283. Mason's biography of Brandeis includes a chapter titled "The Controller Bay Fiasco, 1911," 282–289, in which he misidentifies Mabel Abbott as "Myrtle" Abbott, an error repeated in some

subsequent accounts. As she bylined her articles "M. F. Abbott," there was some confusion at the time of the controversy about Abbott's gender and personal background.

79. Mason, *Brandeis*, 284–287. For Brandeis's own views of the Controller Bay controversy, see his published correspondence in *Letters of Louis D. Brandeis: Volume II
(1907–1912): People's Attorney*, ed. Melvin I. Urofsky and David W. Levy (Albany: State
University Press of New York, 1972), including, among others, his letters to James M.
Graham, 14 July 1911, 460–463, and 21 July 1911, 466; to Robert M. LaFollette, 29 July
1911, 467–472; and to Amos R. E. Pinchot, 2 Aug. 1911, 476–479, and 18 Sept. 1911, 490–
494.

80. Senate, *Chugach National Forest Lands in Alaska*, 62d Cong., 1st sess., 1911, S. Doc.
77, 13–15; "Fight to Involve Taft in Scandal," *New York Times*, 12 July 1911; "Taft Refutes
Alaskan Charges," *New York Times*, 27 July 1911.

81. "Fight to Involve Taft," *New York Times*, 12 July 1911; Gilson Gardner to GP, 3 July
1911, GPP; GP to G. Gardner, 5 July 1911, GPP; GP diary, 28 July 1911, GPP.

82. MMK to PMK, 17 July 1911; BMK to MMK, 28 July 1911.

83. Almost fifty years later, when MacKaye was sorting family papers, he noted in his
diary (3 Aug. 1960) that he was "burning old letters of vintage 1907–1910"—that is, the
period of his relationship with Mabel Abbott.

Chapter 4 / Raising Hell

1. Frank J. Harmon to James Deane, 20 Jan. 1976, incorporating notes from inter-
views with BMK in May and June 1973, conducted by Robert Monahan, WSP/DC.

Hays's *Conservation and the Gospel of Efficiency* provides the standard interpretation
of the Progressive conservation movement. Paul S. Sutter, in his dissertation "Driven
Wild," 14–18, has effectively argued that the era's conservation ideology was not as
monolithic as the Hays thesis suggests. For a concise overview of MacKaye's activities
and ideas during this period of his career, see also Paul Sutter, "'A Retreat from Profit':
Colonization, the Appalachian Trail, and the Social Roots of Benton MacKaye's Wilderness Advocacy," *Environmental History* 4 (Oct. 1999): 553–577.

2. Steen, *U.S. Forest Service*, 137.

3. Harold T. Pinkett, *Gifford Pinchot: Private and Public Forester* (Urbana: University
of Illinois Press, 1970), 75–80, 130–132.

4. Senate, *Preliminary Report of the Inland Waterways Commission*, 60th Cong., 1st
sess., 1908, S. Doc. 325, iv.

5. BMK, "Powell as Unsung Lawgiver," *CCB*, Nov. 1969, 2–4.

6. Hays, *Conservation*, 91–109.

7. Ibid., 203–204.

8. Ibid., 192–195; Roderick Nash, *Wilderness and the American Mind*, 3d ed. (New
Haven: Yale University Press, 1982), 161–181.

9. *GG* (I), 441.

10. U.S. Constitution, art. 1, sec. 8; Steen, *U.S. Forest Service*, 122–129; Gordon B.
Dodds, ed., "H. M. Chittenden's 'Notes on Forestry Paper,'" *Pacific Northwest Quarterly*
57 (1966): 73–74.

11. M. O. Leighton to BMK, 20 Nov. 1912; M. O. Leighton to BMK, 21 Jan. 1913; BMK

to C. Edward Behre, 9 Nov. 1930; BMK to Warren T. Murphy, 26 June 1961; BMK to Robert Monahan, 15 Aug. 1974.

12. BMK to W. T. Murphy, 26 June 1961; BMK to HMK, 1 Oct. 1912, MKCP; BMK to HMK, 20 Oct. 1912, MKCP; BMK, "Message to ATC," 15 June 1972.

13. M. O. Leighton, et al., "The Relation of Forests to Stream Flow," administrative report, U.S. Department of the Interior, Geological Survey, 1913; BMK to W. T. Murphy, 26 June 1961.

14. BMK to MMK, 31 Oct. 1912, MKCP.

15. BMK to Marie Stoddard, "Solsticetide" (Dec.) 1965.

16. BMK to MMK, 4 July 1913; *GG* (I), 442.

17. Eugene M. Tobin, *Organize or Perish: America's Independent Progressives, 1913-1933* (New York: Greenwood Press, 1986), 3–11; Charles W. Ervin, *Homegrown Liberal: The Autobiography of Charles W. Ervin*, ed. Jean Gould (New York: Dodd, Mead, 1954), 113–115.

18. JMK, "An Act Proposed as a Solution of the Conservation Problem in Alaska, with Explanation of the Same," (Boston, c. 1911); "A Bill to authorize the President of the United States to provide transportation and coal-mine development in the Territory of Alaska," S.R. 2714, 63d Cong., 1st sess.; H.R. 7085, 63d Cong., 1st sess.; BMK to Michael Nadel, 2 Aug. 1972.

19. S.R. 2714.

20. BMK to Mardy Murie, 18 Feb. 1963; William L. Stoddard, "Conditional Compensation: Correlating Work and Wage," *Survey* 10 (Jan. 1914): 442.

21. *U.S. Statutes at Large* 38 (1914): 741-745, 305-307; *ENR*, 25-27.

22. Christine A. Lunardini, *From Equal Suffrage to Equal Rights: Alice Paul and the National Woman's Party, 1910-1928* (New York: New York University Press, 1986), 28; BMK to SB, 19 July 1965; BMK to Alice Paul, 13 Dec. 1974; David Glassberg, *American Historical Pageantry: The Uses of Tradition in the Early Twentieth Century* (Chapel Hill: University of North Carolina Press, 1990), 135-136, 323 n. 55; Karen J. Blair, "Pageantry for Women's Rights: The Career of Hazel MacKaye, 1913-1923," *Theatre Survey* 31 (May 1990): 34-38.

23. BMK to HMK, 19 Oct. 1913.

24. BMK to HMK, 10 Nov. 1913; BMK to "Friends," 6 Mar. 1954.

25. BMK to Fanny Villard, 8 May 1921.

26. BMK to JHSMK, 12 July 1914.

27. BMK, "Foreword," 8 Oct. 1922.

28. Frances England, "Woman's Is the Responsibility for Renovating the Race's Morals, and to Accomplish This Jessie Hardy Stubbs Works for Suffrage as a Means," *New York Tribune*, 5 Apr. 1915.

29. JHSMK, notes for proposed book, "The Sexual Revolution," in notebook, c. 1920-1921; England, "Woman's Is the Responsibility."

30. England, "Woman's Is the Responsibility."

31. Helena Hill Weed, letter to editor, *Norwalk (Conn.) Hour*, 22 Apr. 1921.

32. Lunardini, *From Equal Suffrage to Equal Rights*, 59-67; H. H. Weed letter, 22 Apr. 1921; *Washington Evening Star*, 9 May 1914.

33. BMK to PMK, 18 July 1914, HTC.

34. BMK to PMK, 26 July 1914, quoted in *Epoch*, 2:468.

35. BMK to MMK, 27 July 1914, SHS.

36. Hazel H. Reinhardt, "Social Adjustments to a Changing Environment," in *The Great Lakes: An Environmental and Social History,* ed. Susan L. Flader (Minneapolis: University of Minnesota Press, 1983), 205–219.

37. Ibid., 209.

38. BMK, "Powell as Unsung Lawgiver," 3.

39. BMK, notes of interview with Vincent Was, 26 July 1915.

40. BMK, "Summary Statement of the Problem of Clearing and Settling the Cut-over Timber Lands of the Northern Lake States," c. 1915, 2.

41. Vincent Was to BMK, 3 Sept., 3 Dec. 1915.

42. BMK to HMK, 7 Sept. 1914.

43. BMK to PMK, 2 Oct. 1914, HTC.

44. BMK to HMK, 2 Nov. 1914.

45. BMK to JHSMK, 16 Aug. 1914.

46. BMK to PMK, 30 Nov. 1914, 9 Dec. 1914, HTC; BMK, memorandum, "(Confidential)," 1914.

47. BMK to PMK, 30 Nov. 1914.

48. H.R. 20147, 63d Cong., 3d sess.; David M. Kennedy, *Over Here: The First World War and American Society* (Oxford: Oxford University Press, 1982), 97.

49. Page Smith, *America Enters the World,* (New York: McGraw-Hill, 1985), 442–448.

50. "'Red' Agitator Stirs Hearers to Yawns," *Washington Herald,* 3 May 1915.

51. "Benton MacKaye to Wed Suffragist," *Washington Times,* 29 May 1915.

52. BMK to PMK, 26 May 1915, HTC.

53. "Cupid Not to Hinder Mrs. Stubb's Work," *Washington Times,* 30 May 1915.

54. "Women Can Be Good Suffragists and Good Housewives, Too!" June 1915, clipping from unidentified newspaper; England, "Woman's Is the Responsibility."

Chapter 5 / Reclaiming America's Wild Lands for Work and Play

1. JHSMK to HMK, 14 Sept. [1915?].

2. BMK to HMK, 16 Aug. 1915.

3. Steen, *U.S. Forest Service,* 104–122, 137–140; Ashley L. Schiff, *Fire and Water: Scientific Heresy in the Forest Service* (Cambridge: Harvard University Press, 1962), 7–8; William G. Robbins, *American Forestry: A History of National, State, and Private Cooperation* (Lincoln: University of Nebraska Press, 1985), 17–18.

4. Steen, *U.S. Forest Service,* 107–113; Michael Williams, *Americans and Their Forests: A Historical Geography* (Cambridge: Cambridge University Press, 1989), 425–430, 440–443.

5. William B. Greeley, *Some Public and Economic Aspects of the Lumber Industry* (Washington, D.C.: GPO, 1917), 4–5; William B. Greeley, *Forests and Men* (Garden City: Doubleday, 1951), 235–242.

6. GP quoted in Greeley, *Forests and Men,* 118; BMK, "The National Forest Timber Policy," c. 1914–1915.

7. Department of Labor, *Reports of the Department of Labor, 1915, Report of the Secretary of Labor and Reports of Bureaus* (Washington, D.C.: GPO, 1916), 43–44.

8. Ibid.

9. Ibid., 45.

10. *GG* (II), 496; John Lombardi, *Labor's Voice in the Cabinet* (New York: Columbia University Press, 1942), 90.

11. BMK to LP, 27 Dec. 1915; BMK, "A Government Policy for Reclaiming and Colonizing Wild Lands," c. 1915–1916, 16.

12. *GG* (II), 496.

13. BMK, "Powell as Unsung Lawgiver," 3.

14. LP, unpublished autobiography, "Living a Long Life Over Again," 375–377, LPP; *National Colonization Bill,* 64th Cong., 1st sess., H.R. 11329.

15. *National Colonization Bill.*

16. Ibid.

17. House Committee on Labor, *National Colonization Bill: Hearings on H.R. 11329,* 64th Cong., 1st sess., 18, 22, 25 May, 5, 15 June 1916, 32.

18. Ibid., 105.

19. Ibid., 37.

20. Ibid., 125.

21. House Committee on Labor, *National Colonization Bill: Hearings on H.R. 11329: Part 2,* 64th Cong., 1st sess., 15, 20 Dec. 1916, 86, 113.

22. Ibid., 86.

23. Lombardi, *Labor's Voice in the Cabinet,* 157–159; LP to Carl Vrooman, 2 June 1916; BMK, "Settling Timber Lands (Suggestions for Clearing and Settling the Cut-over Timber Lands of the Northern Lake States)," June 1916, 11.

24. Hays, *Conservation,* 237–238; H.R. 15284, 64th Cong., 1st sess.

25. H.R. 14619, 64th Cong., 1st sess.; Judson King, *The Conservation Fight from Theodore Roosevelt to the Tennessee Valley Authority* (Washington, D.C.: Public Affairs Press, 1959), 46; Jerome G. Kerwin, *Federal Water-Power Legislation* (New York: Columbia University Press, 1926), 7–11.

26. *Federal Aid Road Act of 1916, U.S. Statutes at Large* 39 (1916): 355–359; Smith, *America Enters the World,* 866–869.

27. Robert Shankland, *Steve Mather of the National Parks* (New York: Alfred A. Knopf, 1952), 150–151; Henry S. Graves quoted in Donald N. Baldwin, *The Quiet Revolution: Grass Roots of Today's Wilderness Preservation Movement* (Boulder: Pruett Publishing, 1972), 80.

28. National Park Service policies regarding automobiles, roads, and tourism in these years are discussed by Sutter, "Driven Wild," 167–174, and Richard West Sellars, *Preserving Nature in the National Parks: A History* (New Haven: Yale University Press, 1997), 22, 26–27, 58–64.

29. J. Horace McFarland quoted in Shankland, *Steve Mather,* 6.

30. Alfred Runte, *National Parks: The American Experience,* 2d ed., rev. (Lincoln: University of Nebraska Press, 1987), 82–87; Nash, *Wilderness and the American Mind,* 169.

31. *National Park Service Act, U.S. Statutes at Large* 39 (1916): 535; Runte, *National Parks,* 103–104.

32. H. S. Graves quoted in Steen, *U.S. Forest Service,* 119.

33. BMK, "Recreational Possibilities of Public Forests," *Journal of the New York Forestry Association* 3 (Oct. 1916): 4–10, 29–31.

34. AC, "Recreational Use of Public Forests in New England," *Proceedings of the Society of American Foresters* 11 (July 1916): 271–280.

35. Ibid., 276.

36. Ibid., 273.

37. BMK, "Recreational Possibilities of Public Forests," 5.

38. Ibid., 10.

39. Ibid., 10, 29.

40. Ibid., 29–30.

41. Frank A. Waugh, *Recreation Uses on the National Forests* (Washington, D.C.: GPO, 1918), 28–29, quoted in *The Politics of Wilderness Preservation,* by Craig W. Allin (Westport, Conn.: Greenwood Press, 1982), 68.

42. Paul S. Sutter, "'A Blank Spot on the Map': Aldo Leopold, Wilderness, and U.S. Forest Service Recreational Policy, 1909–1924," *Western Historical Quarterly* 29 (summer 1998): 213.

43. Ibid., 189.

Chapter 6 / Employment and Natural Resources

1. BMK, scrapbook, "The MacKaye Reunion. 1916," MKCP; Grover, *Annals of an Era,* 42, 327–337.

2. JHSMK, newspaper clipping book, 1916.

3. LP to Carl Vrooman, 2 June 1916.

4. BMK, "Some Social Aspects of Forest Management," *Journal of Forestry* 16 (Feb. 1918): 212.

5. BMK, "The Soldier, the Worker, and the Land's Resources," *Monthly Labor Review* 6 (Jan. 1918): 51.

6. BMK to MMK, 1, 3 Nov. 1916, SHS.

7. For the history of Everett, Washington, and the events there in 1916, see Norman Clark, *Mill Town* (Seattle: University of Washington Press, 1970); Melvin Dubofsky, *We Shall Be All: A History of the Industrial Workers of the World* (1969; paperback reprint, with a new preface by the author, New York: Quadrangle / New York Times Books, 1973), 337–342; David C. Botting Jr., "Bloody Sunday," *Pacific Northwest Quarterly* 49 (Oct. 1958): 162–172.

8. Clark, *Mill Town,* 90, 143.

9. BMK, looseleaf notes, "IWW Everett Seattle et al. Nov. 1916," 6 Nov. 1916.

10. Ibid.

11. BMK to HMK, 8 Nov. 1916.

12. Botting, "Bloody Sunday," 170–172.

13. BMK, "IWW" notes, 18, 20, 26 Nov., 1916; *Bellingham Herald,* 14 Nov. 1916; "Plan to Stablize Logging Industry," *Seattle Daily Times,* 17 Nov. 1916; "Forest Refuge Is Plan," *Portland Morning Oregonian,* 22 Nov. 1916.

14. BMK to "Friends," 6 Mar. 1954.

15. JHSMK to HMK, 23 Nov. 1916, SHS.

16. BMK to MMK, 18 Feb. 1917, SHS.

17. Emergency Peace Federation circular letter, 22 Feb. 1917, HTC.

18. BMK to PMK, 15 Mar. 1917, 27 Mar. 1917, HTC.

19. PMK to BMK, 31 Mar. 1917.

20. Harold Ickes to Emergency Peace Federation, 2 Apr. 1917, quoted in T. H. Watkins, *Righteous Pilgrim: The Life and Times of Harold L. Ickes, 1874–1952* (New York: Henry Holt, 1990), 157.

21. JHSMK to Marion M. MacKaye, 2 June 1917, HTC.

22. Kennedy, *Over Here*, 123–124, 252–253.

23. BMK, USFS report, "A Plan for Emergency Fuelwood for Washington, D.C.," 1918; BMK, USFS report, "Reconnaissance Plan for Woodlot Development in Locality of Hopkinsburg, Allegan County, Michigan," Jan. 1918.

24. U.S. Forest Service Chief of Investigations [RZ] to "Miss Judson," 31 Dec. 1917.

25. BMK, "Some Social Aspects of Forest Management," 210–211.

26. Ibid., 211–212.

27. Ibid., 212–213.

28. LP to BMK, memorandum, 22 Aug. 1918.

29. BMK, draft U.S. Fuel Administration / Dept. of Labor / USFS report, "Tentative Plan for Operation by U.S. Wood Fuel Corporation," c. 1918; LP to "District Forester," draft letter, 18 Nov. 1918.

30. BMK to MMK, 30 Apr. 1918; BMK to Ethelberta Lill, 23 Apr. 1921.

31. "Common Sense and the I.W.W.," *New Republic*, 27 Apr. 1918, 375–376. MacKaye became a lifelong friend of the author of this unsigned editorial, Robert Bruère. See BMK, "Robert W. Bruère (1876–1964)," *CCB*, Feb. 1965, 2–4.

32. LP, "Living a Long Life," 379.

33. For World War I soldier resettlement programs, see LP, "Living a Long Life," 378–382; Bill G. Reid, "Proposals for Soldier Resettlement during World War I," *Mid-America* 46 (July 1964): 172–186; Bill G. Reid, "Franklin K. Lane's Idea for Veterans' Colonization, 1918–1921," *Pacific Historical Review* 33 (1964): 447–461.

34. Reid, "Franklin K. Lane's Idea," 452–453.

35. Franklin K. Lane to Woodrow Wilson, 31 May 1918, *The Letters of Franklin K. Lane*, ed. Anne Wintermute Lane and Louise Herrick Wall (Boston: Houghton Mifflin, 1922), 285–289.

36. William Kent to Woodrow Wilson, 27 May 1918, HSP.

37. William Kent to Woodrow Wilson, 3 June 1918, HSP.

38. Ibid.

39. Reid, "Franklin K. Lane's Idea," 457.

40. LP, "Living a Long Life," 378–379.

41. H.R. 13415, 65th Cong., 3d sess.

42. BMK, "Making New Opportunities for Employment," *Monthly Labor Review* 8 (Apr. 1919): 131.

43. House Committee on Labor, *Employment for Discharged Soldiers and Sailors: Hearings on H.R. 13415*, 65th Cong., 3d sess., 17 Jan. 1919, 9–39.

44. BMK, looseleaf notes on Canada, Dec. 1918.

45. Ibid., 9, 13 Dec. 1918.

46. Thomas Adams, *Rural Planning and Development* (Ottawa: Commission of Conservation, 1917).

47. BMK to PB, 2 Feb. 1965.

48. BMK, "The First Soldier Colony—Kapuskasing, Canada," *The Public*, 15 Nov. 1919, 1066–68.

49. Ibid.

50. BMK, "Making New Opportunities," 127.

51. H.R. 15672, 65th Cong., 3d sess.

52. H.R. 6426, 66th Cong., 1st sess.

53. BMK, "Making New Opportunities," 139.

54. BMK, "Powell as Unsung Lawgiver," 4.

55. Roy Lubove, *Community Planning in the 1920's: The Contribution of the Regional Planning Association of America* (Pittsburgh: University of Pittsburgh Press, 1963), 93.

56. *ENR*, 11–13.

57. Ibid., 65–77.

58. Ibid., 66–67.

59. Ibid., 70–72.

60. Ibid., 74–75.

61. *GG* (IV), 15; *ENR*, 77.

62. Ibid., 77–78.

63. Ibid., 80, 23.

64. Ibid., 144.

65. PMK to BMK, 15 Sept. 1919.

66. JMK to BMK, 19 Sept. 1919.

67. Filibert Roth to BMK, 29 Sept. 1919.

Chapter 7 / Turning Point

1. SC to Jonathan Daniels, 13 Jan. 1964, SCP.

2. Prince Hopkins to BMK, 19 Nov. 1920; BMK to MMK, 15 May 1920.

3. SC, "My Friend Benton," *LW* 39 (Jan.–Mar. 1976): 17.

4. BMK, "Foreword," 8 Oct. 1922.

5. JHSMK to HMK and MMK, 4 Oct. [1916?].

6. BMK, "A Plan for Cooperation Between Farmer and Consumer," *Monthly Labor Review* 11 (Aug. 1920): 213–233.

7. JHSMK to MMK, 1 Sept. [1919?], MKCP; BMK, "The Lesson of Alaska," *The Public*, 30 Aug. 1919, 930–932; BMK, "The Struggle Between the Nation and Private Interests for Alaska's Riches," *Reconstruction*, Jan. 1920, 27–30.

8. Fred Kerby to BMK, 18, 25 Aug. 1919.

9. *Congressional Record*, 20 Oct. 1919, 7595–7598; SC to Huston Thompson, 15 Dec. 1920, SCP.

10. SC to J. Daniels, 13 Jan. 1964.

11. BMK diary, Feb.–May, 1920.

12. Ibid., 7, 10, 11, 13, 16 Jan. 1920; Citizens' Amnesty Committee letterhead, c. 1920.

13. BMK diary, 27 Jan. 1920 and other Jan.–Mar. entries.

14. BMK diary, 19, 20 Mar. 1920.

15. On Martens and Nuorteva, see: Robert Murray, *Red Scare: A Study in National Hysteria, 1919–1920* (Westport, Conn.: Greenwood Press, 1980), 47, 98–100; Kennedy, *Over Here,* 289; James Weinstein, *The Decline of Socialism in America: 1912–1925* (New Brunswick: Rutgers University Press, 1984), 195; LP, *The Deportations Delirium of Nineteen-Twenty* (Chicago: Charles H. Kerr, 1923), 283–292.

16. BMK diary, 24 Mar. 1920; BMK, Herbert Brougham, and CHW draft letter to Ludwig C. A. K. Martens, 23 Mar. 1920.

17. BMK, JHSMK biographies with BMK et al. to Martens, 23 Mar. 1920.

18. L. Martens to BMK, 2 Apr. 1920.

19. BMK diary, May–June 1920, misc. entries; National Volunteer Committee on Industrial Administration, "The Northwest Program: Preliminary Outline," broadside, 1920; BMK to Herbert Brougham, 14 June 1920.

20. MacKaye Family Association Constitution, 1920.

21. BMK diary, 16, 19, 21 July 1920.

22. BMK, "A Suggested Clearing House for Radicals," c. 1919–1920, 2.

23. Murray, *Red Scare,* 226–229.

24. BMK, editorial, "Ballots vs. Bullets," *ML,* 2 Nov. 1920. Clippings of *ML* editorial columns, marked by MacKaye to indicate his own contributions, are in MKFP, box 182, folders 29 and 30.

25. BMK diary, 14 Aug. 1920.

26. "A Japanese Offer," *ML,* 23 Nov. 1920; "The Latest Moves," *ML,* 26 Nov. 1920.

27. *ML* 1920 editorials: "Airports," 18 Nov.; "Volcano Power," 30 Nov.; "Wave Power," 13 Dec.; "The Census," 4 Oct..

28. "Scenery in Danger," *ML,* 28 Dec. 1920.

29. "Woman's Party Organizer Joins Socialist Ranks," *ML,* 10 Sept. 1920.

30. BMK diary, 30 Sept. 1920.

31. Herbert Brougham obituary, *New York Times,* 19 Apr. 1946; Christopher Lasch, *American Liberals and the Russian Revolution* (New York: Columbia University Press, 1962), 78.

32. HMK to BMK, 6 Oct. 1920; BMK diary, 10 Oct. 1920.

33. "Women and Disarmament," *ML,* 12 Nov. 1920.

34. "Women Must Refuse to Bear Babies for War," *ML,* 10 Nov. 1920.

35. Mabel Irwin, "Birth Control" issue, *Arbitrator* 1 (Aug. 1918).

36. England, "Woman's Is the Responsibility."

37. BMK to MMK, 8 July 1921, SHS; BMK, "Foreword," 8 Oct. 1922.

38. JHSMK to HMK, 3 Nov. 1916, SHS.

39. BMK diary, 27, 28 Sept. 1920.

40. Ibid., 7 Dec. 1920.

41. Ibid., 10 Dec. 1920; "Urges Women To Force Peace," *Chicago Herald-Examiner,* 11 Dec. 1920.

42. BMK diary, 12 Dec. 1920.

43. BMK to Elizabeth Thomas, 12 Dec. 1920.

44. HMK to PMK, 16 Dec. 1920.

45. BMK diary, 25 Dec. 1920.

46. HMK to PMK, 29 Dec. 1920.

47. BMK diary, 1 Jan. 1921; BMK to PMK, 26 Mar. 1921.

48. Thorstein Veblen, *The Engineers and the Price System* (1921; reprint, New York: Viking Press, 1947), 54, 134.

49. SC, "My Friend Benton," 18.

50. BMK diary, 3 Jan. 1921; *The Technical Alliance* (New York: 1921), pamphlet, ATCA; Joseph Dorfman, *Thorstein Veblen and His America* (New York: Viking, 1934), 459–460.

51. *Technical Alliance.*

52. BMK, unpublished reports, "Preliminary Report on Social Wastes in the Lumber Industry," "A Plan for Developing a Program for Industry in the United States," "The Need for Social Engineering," all 1921.

53. BMK diary, 12, 15, 22 Mar., 4, 7 Apr., 1921.

54. *World Disarmament,* Hearings before the House Committee on Military Affairs, 66th Cong., 3d sess., 11 Jan. 1921, 3–7, 25–29, 42–43; *Disarmament,* Hearings before the House Committee on Foreign Affairs, 66th Cong., 3d sess., 14, 15 Jan. 1921, 24–28.

55. JHSMK notebooks, 1921.

56. BMK to E. Lill, 23 Apr. 1921; BMK diary, 2 Apr. 1921.

57. JHSMK to BMK, 5 Apr. 1921.

58. JHSMK to BMK, 6 Apr. 1921.

59. Handbill for Philadelphia peace meeting, 6 Apr. 1921, JHSMK scrapbook.

60. BMK diary, 15, 16 Apr. 1921; BMK to E. Lill, 23 Apr. 1921.

61. "Jessie Hardy MacKaye, Suffragist, Ends Life," *Washington Times,* 19 Apr. 1921; "Noted Suffragist Vanishes in Quest of Death," *New York Evening Journal,* 19 Apr. 1921.

62. "Mrs. MacKaye Gone; Threatened Suicide," *New York Times,* 19 Apr. 1921.

63. *Washington Times,* 19 Apr. 1921.

64. HMK, comments at JHSMK memorial service, 20 Apr. 1921; BMK, "Frog Opera," *LW* 22 (spring 1957): 21.

65. Mary L. Clayton to BMK, 23 Apr. 1921.

66. Helena Hill Weed, letter to editor, *Norwalk (Conn.) Hour,* 22 Apr. 1921.

67. Theresa Russell to BMK, 29 Apr. 1921; Herbert Brougham to BMK, 1 May 1921, 7 May 1921.

68. BMK diary, 11 June 1921.

69. BMK to E. Lill, 23 Apr. 1921.

70. JMK to BMK, 20 Apr. 1921.

71. "Happy New Year," *ML,* 31 Dec. 1920.

72. Martin, "Interview with Benton MacKaye," 12.

Chapter 8 / First Steps along the Appalachian Trail

1. On Charles Harris Whitaker, see LM, ed., *Roots of Contemporary American Architecture* (1952; reprint, New York: Dover Publications, 1972), 434–435; CHW, "The Rejected Stone," unpublished autobiographical ms., Avery Library, Columbia University.

2. CHW to BMK, 19 May 1921.

3. BMK diary, 15, 16, 20, 28, 29 June 1921.

4. BMK, "Memorandum on Regional Planning," 1921, 4–5, 19–25.

5. Ibid., 33, 38, 39–47.

6. Ibid., 48, 49–50.

7. Ibid., 50, 51–52.

8. Ibid., 52–53.

9. Ibid., 55–56, 57.

10. Ibid., 59.

11. BMK diary, 10 July 1921.

12. The late Kermit C. Parsons of Cornell University wrote and edited numerous works about Clarence Stein and the circle of architects, writers, and reformers with whom he worked and collaborated, including: *The Writings of Clarence S. Stein: Architect of the Planned Community,* ed. Kermit Carlyle Parsons (Baltimore: Johns Hopkins University Press, 1998); Parsons, "Collaborative Genius: The Regional Planning Association of America," *Journal of the American Planning Association* 60 (autumn 1994): 462–482; Parsons, "Clarence Stein and The Greenbelt Towns: Settling for Less," *Journal of the American Planning Association* 56 (spring 1990): 161–183.

13. BMK, "Regional Planning and Social Readjustment," 1921, 12–14, 18.

14. MacKaye's article (abbreviated *ATrail* here) was immediately republished in a pamphlet titled *A Project for An Appalachian Trail* (Committee on Community Planning of the American Institute of Architects, [1921]), with an introduction by Clarence S. Stein. (The article is also reprinted in Emblidge, *Appalachian Trail Reader,* 46–57. Stein's introduction is reprinted in *Writings of Clarence Stein,* 112–113.)

15. *ATrail,* 329.

16. Ibid., 326.

17. Ibid., 325.

18. Ibid., 326, 327.

19. Ibid., 327, 328, 330.

20. GP to BMK, 22 Dec. 1921.

21. BMK to MMK, 8 July 1921, SHS.

22. BMK to PMK, 26 Sept. 1921; BMK diary, 27 Aug., 21, 28 Sept., 5 Oct. 1921.

23. CS, "Introduction," in BMK, *A Project for An Appalachian Trail.*

24. BMK to CS, 19 Nov. 1921.

25. BMK diary, Nov.–Dec. 1921, misc. entries.

26. Waterman and Waterman, *Forest and Crag,* 475–488; see also by Laura Waterman and Guy Waterman, "Early Founders of the Appalachian Trail," *ATN* 46 (Sept.–Oct. 1985): 7–11.

27. Hall quoted in Waterman and Waterman, "Early Founders," 10; Waterman and Waterman, *Forest and Crag,* 393.

28. AC to BMK, 20 Oct. 1921, ATCA.

29. Waterman and Waterman, "Early Founders," 10–11.

30. BMK to CS, 11 Dec. 1921.

31. John Nolen to BMK, 12 Dec. 1921, ATCA.

32. Philip W. Ayres to BMK, 1 Feb. 1922, ATCA.

33. *A Trail*, 330.

34. BMK diary, Dec. 1921–Jan. 1922, misc. entries; CS to BMK, 27 Dec. 1921.

35. BMK to MMK, 29 Jan. 1922.

36. BMK to PMK, 22 Jan. 1922.

37. BMK, "The Trail Out: An Outdoor Survey of Our Industrial Wilderness," Mar. 1922.

38. AL, "The Wilderness and Its Place in Forest Recreational Policy," *Journal of Forestry* 19 (Nov. 1921): 718–721; reprinted in AL, *The River of the Mother of God and Other Essays,* ed. Susan L. Flader and J. Baird Callicott (Madison: University of Wisconsin Press, 1991), 78–79.

39. Sutter, "Blank Spot on the Map," 207; AL, "Wilderness and Its Place," 80.

40. Allin, *Politics of Wilderness Preservation,* 69–71; Baldwin, *Quiet Revolution,* 56, 71–77, 108–114, 128–131; David Backes, "Wilderness Visions: Arthur Carhart's 1922 Proposal for the Quetico-Superior Wilderness," *Forest and Conservation History* 35 (July 1991): 128–137.

41. *A Trail*, 326.

42. Ibid.

43. Stephen Fox, *John Muir and His Legacy: The American Conservation Movement* (Little, Brown: Boston, 1981), 159–163.

44. Allin, *Politics of Wilderness Preservation,* 68–69; Nash, *Wilderness and the American Mind,* 198; Robert P. McIntosh, *The Background of Ecology: Concept and Theory* (Cambridge: Cambridge University Press, 1985), 35, 66; Norman T. Newton, *Design on the Land: The Development of Landscape Architecture* (Cambridge: Harvard University Press, 1971), 562–575.

45. BMK to CS, 10 Feb. 1922.

46. BMK, "The Trail Out."

47. Ibid.

48. BMK diary, 15 Mar. 1922.

49. CS to BMK, 2 Mar. 1922.

50. BMK, untitled notes for talk to Boston Society of Landscape Architects, c. Mar. 1922.

51. BMK diary, Mar.–Apr. 1922, misc. entries; BMK, "Some Early A.T. History," 92–93.

52. Waterman and Waterman, *Forest and Crag,* 424.

53. BMK to Frank Place, 6 Dec. 1940.

54. RT, "A Great Trail from Maine to Georgia," *New York Evening Post,* 7 Apr. 1922.

55. BMK diary, Apr. 1922, misc. entries; BMK, "Some Early A.T. History," 92–93.

56. Franklin W. Reed to BMK, 1 Apr. 1922, 23 June 1922, ATCA.

57. Mary Hosmer Lupton, "Halstead Shipman Hedges," *ATN* 41 (May–June 1980): 5–7; Harlan P. Kelsey to BMK, 2 Feb. 1922, ATCA; Paul M. Fink to H. P. Kelsey, 30 Jan. 1922.

58. BMK diary, 19, 21 Apr. 1922.

59. BMK diary, Apr.–June 1922, misc. entries.

60. BMK, "Suggestions for Scouting the Appalachian Trail," 1922.

61. BMK diary, 8–10 June 1922; BMK, "Appalachian Trail, Inc.," 1922.

62. BMK to PMK, 21 June 1922; PMK to BMK, 12 Apr. 1922, ATCA.

63. BMK diary, 15–25 July 1922; BMK to PMK, 26 July 1922.

64. BMK, "Making Geography: A Conservation Survey on the Appalachian Trail," 1922, 2.

65. BMK to PMK, 8 Nov. 1922; BMK diary, 13 Nov. 1922.

66. BMK to CS, 15 Nov. 1922.

67. Ibid.

68. BMK diary, 24–26 Nov. 1922; BMK, "Some Early A.T. History," 93.

69. CS, "Conference with Benton MacKaye re Appalachian Trail," attached to CS to BMK, 7 Dec. 1922.

70. BMK, "Progress Toward the Appalachian Trail," *Appalachia* 15 (Dec. 1922): 244–246.

71. BMK, "Some Early A.T. History," 93–94; "The Appalachian Trail Idea Grows," *New York Evening Post*, 26 Jan. 1923; BMK, "The Job Ahead," excerpt from Jan. 1923 New England Trail Conference speech, attached to BMK to CS, 19 Jan. 1923.

72. BMK, "The Job Ahead."

Chapter 9 / The Regional Planning Association of America and the Appalachian Trail Conference

1. Parsons, "Collaborative Genius," 465; *Writings of Clarence Stein*, 104–105.

2. BMK diary, 29 Jan. 1923; BMK to PMK, 8 Mar. 1923.

3. BMK diary, 17 Feb. 1923. MacKaye noted in his diary the City Club lunch "with group of architects et al," but he did not record the names of those present. In a 19 Feb. 1923 letter to Mumford, apparently the first in their five decades of correspondence, MacKaye described plans for another meeting, to include journalist Lincoln Steffens and economist Thorstein Veblen. "I enjoyed mightily our talk the other day," he concluded. LMP.

4. LM, introduction to *NE*, xiv.

5. BMK to LM, 24 Mar. 1938, LMP. For an extensive personal and intellectual portrait of the friendship between MacKaye and Mumford, see John L. Thomas, "Lewis Mumford, Benton MacKaye, and the Regional Vision," in *Lewis Mumford: Public Intellectual,* ed. Thomas P. Hughes and Agatha C. Hughes (New York: Oxford University Press, 1990), 66–99. Another perspective appears in Donald L. Miller, *Lewis Mumford: A Life* (New York: Weidenfeld & Nicolson, 1989), 207–211.

6. BMK, "Great Appalachian Trail from New Hampshire to the Carolinas," *New York Times*, 2 Feb. 1923.

7. BMK diary, 22 Mar., 20 Apr., 16 May 1923; BMK, fragment of untitled book ms., "Chapter I: The Modern Wilderness—Industry," 20 Apr. 1923, 6.

8. Carl Sussman, ed., *Planning the Fourth Migration: The Neglected Vision of the Regional Planning Association of America* (Cambridge: MIT Press, 1976), 1. Sussman's book is an essential resource concerning the RPAA; but many other scholars have also studied the group and its key figures. Some of their works include: Roy Lubove, *Community Planning in the 1920's: The Contribution of the Regional Planning Association of America;* K. C. Parsons, "Collaborative Genius: The Regional Planning Association of America";

406 · Notes to Pages 171–176

Francesco Dal Co, "From Parks to the Region: Progressive Ideology and the Reform of the American City," in *The American City: From the Civil War to the New Deal*, Giorgio Ciucci et al., (Cambridge: MIT Press, 1979), 293–387; Daniel Schaffer, *Garden Cities for America: The Radburn Experience* (Philadelphia: Temple University Press, 1982); Stanley Buder, *Visionaries and Planners: The Garden City Movement and the Modern Community* (New York: Oxford University Press, 1990), 157–180; Mark Luccarelli, *Lewis Mumford and the Ecological Region: The Politics of Planning* (New York: Guilford Press, 1995); and Edward K. Spann, *Designing Modern America: The Regional Planning Association of America and Its Members* (Columbus: Ohio State University Press, 1996).

9. LM, introduction to CS, *Toward New Towns for America*, 3d ed. (Cambridge: MIT Press, 1966), 15.

10. LM, *Sketches from Life: The Autobiography of Lewis Mumford* (Boston: Beacon Press, 1983), 338.

11. "Minutes of the [18 Apr. 1923] Organization Meeting of the Regional Planning Association," 20 Apr. 1923, CSP.

12. Sussman, *Planning the Fourth Migration*, 6–16.

13. Ibid., 18.

14. LM, in CS, *Toward New Towns*, 13.

15. The minutes and records of the RPAA's activities document numerous payments to MacKaye. For instance, the RPAA's minutes of 5 Feb. 1925 reported that, of the $2,505 the group had raised since its 1923 inception, $750 had been paid to MacKaye "for investigation in connection with the Appalachian Trail." CSP.

16. Sussman, *Planning the Fourth Migration*, 16.

17. "The Regional Planning Association of America," organizational statement, 8 June 1923, CSP.

18. "Proposed Garden City and Regional Planning Association," 7 Mar. 1923, CSP.

19. Ebenezer Howard, *Garden Cities of To-morrow*, ed. F. J. Osborn, with an introductory essay by LM (Cambridge: MIT Press, 1965), 50–57, 150.

20. *ENR*, 106–107.

21. BMK diary, 18–22 May 1923.

22. LM, *Sketches from Life*, 342.

23. Ibid.

24. *GG* (I), 439.

25. Patrick Geddes, "Talks from the Outlook Tower," in *Patrick Geddes: Spokesman for Man and the Environment*, ed. Marshall Stalley (New Brunswick: Rutgers University Press, 1972), 300.

26. Ibid., 325.

27. LM, *The Story of Utopias* (1922; New York: Viking Press, 1962), 279–281.

28. Patrick Geddes, *Cities in Evolution: An Introduction to the Town Planning Movement and to the Study of Civics* (1915; New York: Howard Ferting, 1968), 95–96.

29. "From the Program Committee (Mumford, Chase, MacKaye, Stein)" to RPAA Executive Committee, 12 June 1923, CSP. Reprinted in *Writings of Clarence Stein*, 116–120.

30. BMK diary, June–July 1923, misc. entries; BMK, "Work Progresses on Appalachian Trail Building," *New York Evening Post*, 9 Aug. 1923.

31. LM, "New Trails for Old," *The Freeman*, 4 July 1923, 396–397.

32. BMK diary, 25–28 Oct. 1923; RT, "Great Appalachian Trail Marches Toward Its Goal," *New York Evening Post*, 2 Nov. 1923; BMK, "Some Early A.T. History," 94.

33. *Survey*, editorial, 51 (15 Nov. 1923), 193–194; BMK, "Some Early A.T. History," 94

34. For the relationship between "Giant Power" and regionalist thought in the 1920s, see Thomas P. Hughes, *American Genesis: A Century of Invention and Technological Enthusiasm, 1870–1970* (New York: Viking, 1989), 298–309, 352–359.

35. BMK, "Appalachian Power: Servant or Master?" *Survey Graphic* 51 (1 Mar. 1924), 618–619.

36. Ibid.

37. BMK, "A Suggested Policy for Approaching the Problem of Regional Planning in the United States," May 1924; BMK to CS, 1 Apr. 1924; BMK diary, 11–14 May 1924.

38. Nash, *Wilderness and the American Mind*, 190–191; Arthur Ringland, "National Conference on Outdoor Recreation," *Appalachia* 16 (June 1925): 164–166; John Ise, *Our National Park Policy* (New York: Arno Press, 1979), 252–254.

39. Allin, *Politics of Wilderness Preservation*, 71; Meine, *Aldo Leopold*, 243.

40. BMK diary, 23–31 May 1924; BMK, "Remarks," *National Conference on Outdoor Recreation*, 68th Cong., 1st sess., 1924, S. Doc. 151, 124–127; Sutter, "Driven Wild," 58–59.

41. Sussman, *Planning the Fourth Migration*, 143–144; Schaffer, *Garden Cities*, 43.

42. BMK diary, 1 June–1 July 1924; BMK to CS, 25 July 1964.

43. State of New York, *Report of the Commission of Housing and Regional Planning to Governor Alfred E. Smith* (Albany: J. B. Lyon, 1926), 5.

44. Ibid., 25–38, 49–52, 81; Schaffer, *Garden Cities*, 87.

45. LM quoted in Sussman, *Planning the Fourth Migration*, 230. Sussman treats this episode, including Mumford's 1932 *New Republic* articles and Adams's response, on 221–267.

46. M. Christine Boyer, *Dreaming the Rational City: The Myth of American City Planning* (Cambridge: MIT Press, 1986), 185.

47. For Mumford's "neotechnics," see: LM, *Technics and Civilization* (New York: Harcourt, Brace, 1934), 212–267; LM, *The Culture of Cities* (New York: Harcourt, Brace, 1938), 340; Luccarelli, *Lewis Mumford*, 187–196.

48. LM, *Sketches from Life*, 344.

49. Ibid., 344–346; Sussman, *Planning the Fourth Migration*, 49–51; BMK diary, Aug.–Sept. 1924, misc. entries.

50. LM to BMK, 18 Dec. 1924.

51. LM, "Regions—To Live In," reprinted from *Survey Graphic* 54 (1 May 1925), 151–152, in Sussman, *Planning the Fourth Migration*, 92.

52. Ibid., 90, 92.

53. BMK, "The New Exploration: Charting the Industrial Wilderness," reprinted from *Survey Graphic* 54 (1 May 1925), 153–157, 192, 194, in Sussman, *Planning the Fourth Migration*, 94–110.

54. Ibid., 95, 98.

55. Ibid., 101.

56. LM to Patrick Geddes, 4 Dec. 1924, reprinted in LM, *My Works and Days: A Personal Chronicle* (New York: Harcourt Brace Jovanovich, 1979), 106–107; BMK diary, 11–12 Oct. 1924.

57. BMK diary, Oct.–Dec. 1924, misc. entries.

58. Ringland, "National Conference on Outdoor Recreation," 164.

59. BMK diary, Dec. 1924, misc. entries; Harlean James, "The First Appalachian Trail Conference," *ATN* 11 (Jan. 1950): 9–10; ATC, *Member Handbook* (Harpers Ferry, W.Va.: ATC, 1988), 20–23.

60. "Brief of Proceedings of the Appalachian Trail Conference," 2–3 Mar. 1925, 1–2.

61. Ibid., 2–3.

62. Ibid., 3–5.

63. Ibid., 5–8.

64. BMK diary, 3 Mar. 1925.

65. "Brief of Proceedings," 1.

66. Ronald Foresta, "Transformation of the Appalachian Trail," *Geographical Review* 77 (Jan. 1987): 76–85.

67. BMK, "The RPAA Era, A Reminiscence," 30 Oct. 1969.

68. LM to CS, 18 May 1925, CSP.

69. BMK to HMK, 14 May 1925, SHS.

Chapter 10 / *The New Exploration*

1. RZ to BMK, 26 May 1925.

2. LM, introduction to *NE*, vii.

3. BMK, "Industrial Exploration," parts I–III, *The Nation*, I. "Charting the World's Commodity Flow," 20 July 1927, 70–73, II. "Charting the World's Requirements," 27 July 1927, 92–93, III. "Charting the World's Resources," 3 Aug. 1927, 119–121.

4. Ibid., pt. I, 70.

5. Ibid., pt. III, 119.

6. Ibid., pt. II, 92.

7. Ibid., pt. I, 70–72.

8. BMK diary, 15, 22, 29 Oct., 5, 12, 19 Nov. 1925; BMK notes on New York Civic Club lectures, Oct.–Nov. 1925.

9. Norman Thomas to BMK, 21 Mar. 1925; BMK to N. Thomas, 26 Mar. 1925.

10. CS to BMK, 1 Feb. 1926; Harold M. Ware, "Russian Agricultural Progress," letter to editor, *The Nation*, 25 Nov. 1925, 598; Bruce Bliven, "Mr. Ware and the Peasants," *The New Republic*, 22 July 1925, 232–235; "Americans Operate 15,000-Acre Farming Experiment in Russia," *New York Times*, 18 Apr. 1926, sec. 8, 15.

11. BMK, "Regional Planning Report on the Archangelskoe Rayon, North Caucasus, Russia," Apr. 1926, 1–2, 6–18.

12. Ibid., 18 Apr., 21–23 May, 6–23 July 1926.

13. BMK to "Friends," 6 Mar. 1954.

14. BMK to LM, 13 Nov. 1927.

15. Author interview with Joan Pifer Michaels, 8 Mar. 1988.

16. Waterman and Waterman, *Forest and Crag*, 481; John Flanders, "Afoot in Central

New England: The History of the Wapack Trail," *Appalachia* 49 (15 June 1992): 41–51; BMK to LM, 3 Dec. 1926; BMK, "A Boys' Project for a Balanced Civilization," *Christian Science Monitor,* 13 Jan. 1927; BMK, "The Kidder Mountain Trail," 1926, 10.

17. Introduction, by editors Flader and Callicott, to AL, "Wilderness as a Form of Land Use," as reprinted in AL, *River of the Mother of God,* 134. Leopold's article originally appeared in *Journal of Land and Public Utility Economics* 1 (Oct. 1925): 398–404.

18. AL, "Wilderness as a Form of Land Use," 140.

19. AL to BMK, 3 Feb. 1926.

20. CS to BMK, 15 July 1925.

21. LM to BMK, 22 Dec. 1926.

22. BMK to LM, 3 Dec. 1926; Blair, "Pageantry for Women's Rights," 45.

23. Ibid.

24. BMK to LM, 21 May 1927.

25. BMK to LM, 3 Dec. 1926.

26. Ibid.

27. LM to BMK, 22 Dec. 1926.

28. LM, *Sketches from Life,* 342–343; LM, "Random Notes," 5 Apr. 1968, LMP.

29. BMK to LM, 9 Mar. 1927.

30. BMK to LM, 4 Apr. 1927.

31. LM, *The Golden Day* (New York: Boni and Liveright, 1926), 120.

32. Edward T. Hartman to BMK, 22 May, 12 July, 18 July 1927; BMK diary, 4 June, 4–5 Oct. 1927; BMK to William Roger Greeley, 12 May 1956.

33. BMK, speech to Massachusetts Federation of Planning Boards, "A Cross-Section of Massachusetts: Two Possibilities," 4 Oct. 1927; "Real City and Metropolis Are Contrasted by Benton MacKaye at Greenfield Planning Event," *Fitchburg Sentinel,* 5 Oct. 1927.

34. BMK, notes for speech to Cape Cod Chamber of Commerce, "Cape Cod—The Region. Its Needs and Possibilities," 14 Dec. 1927.

35. Harris A. Reynolds, with BMK and William Roger Greeley, "Zone the State Highways: The Lesson of the Mohawk Trail," Bulletin 146 (Boston: Massachusetts Forestry Association, 16 Jan. 1928).

36. The story of the Governor's Committee on the Needs and Uses of Open Spaces is related in McCullough, *Landscape of Community,* 288–291, and Gordon Abbott Jr., *Saving Special Places: A Centennial History of The Trustees of Reservations: Pioneer of the Land Trust Movement* (Ipswich, Mass.: Ipswich Press, 1993), 29–31.

37. BMK diary, 11 Jan. 1928.

38. BMK, memorandum for Governor's Committee on the Needs and Uses of Open Spaces, "Progress Report on a Survey of Massachusetts for the Purpose of Locating Needed Open Spaces," 4 Apr. 1928.

39. BMK diary, 21 Mar., 4 Apr. 1928; Arthur Comey to BMK, 14 June 1928.

40. BMK diary, May–June 1928, misc. entries.

41. According to MacKaye's annotation on his own copy, he was the author of an appendix titled "Classification of Types of Open Space" in the *Report of the Governor's Committee on Needs and Uses of Open Spaces* (Boston: 1929), 11–15.

42. BMK, "Wilderness Ways," *Landscape Architecture* 19 (July 1929): 237, 240, 244.

43. BMK to Ned Richards, 23 May 1928.

44. BMK to PMK, 8 Jan. 1927.

45. BMK to CS, 29 Dec. 1926.

46. BMK diary, 1 Dec. 1927.

47. BMK to HSMK, 8 Mar. 1928.

48. BMK diary, 3 Nov. 1928.

49. BMK diary, Jan.–Oct. 1928, misc. entries; Harriet (Bridgman) Blackburn to author, 6 Jan. 1990; BMK to LM, 14 Apr. 1928; BMK diary, 18–20 Apr. 1928. For Mumford's relationship with Strongin at this time, see Miller, *Lewis Mumford*, 238, 266–267; LM, *Sketches from Life*, 393–394; LM, *My Works and Days*, 170, 320–324.

50. Author interview with Howard Bridgman, 19 Oct. 1989.

51. Henry Wright to Alfred Harcourt, 12 Dec. 1927; Donald Brace to H. Wright, 5 Jan. 1928; LM to BMK, 3 May 1928; BMK diary, Apr.–May 1928, misc. entries.

52. BMK to LM, 4 Apr. 1927; LM, *Sketches from Life*, 342–343.

53. Lubove, *Community Planning*, 83.

54. *NE*, 148.

55. John Nolen, *New Towns for Old* (Boston: Marshall Jones, 1927).

56. *Epoch*, 2:476.

57. *NE*, 5–13.

58. Ibid., 26, 48–50.

59. Ibid., 48–51.

60. Ibid., 52–55.

61. Ibid., 120, 130, 133.

62. Ibid., 134, 138–139, 203, 205.

63. Ibid., 183–185.

64. Ibid., 192–194, 199, 178–179, 226.

65. BMK diary, 12 Oct. 1928; BMK's Harcourt Brace royalty statements, 1928–1933.

66. Thomas H. Reed, review of *NE*, *American Political Science Review* 23 (Nov. 1929): 1050–1051; John Frederick Lewis Jr., review of *NE*, *Philadelphia Record*, 19 Jan. 1929.

67. RZ, review of *NE*, *Journal of Forestry* 26 (Dec. 1928): 1029–1030.

68. BMK to RZ, 9 Dec. 1928; *NE*, 154.

69. LM, *The City in History* (New York: Harcourt Brace, 1961), 598.

70. LM, *Sketches from Life*, 346.

71. *NE*, 73–74.

Chapter 11 / Trailwork and the "Townless Highway"

1. LM, Introduction to *NE*, xiv.

2. *Encyclopedia Britannica*, 14th ed., 1928, s.v. "regional planning."

3. Austin F. Hawes to BMK, 4 Aug. 1928. See MacKaye's unpublished manuscript, "Fire Trails: A Manual of Forest Fire Prevention for Amateur Woodsmen (in Connecticut)," 1929.

4. BMK to LM, 24 Sept. 1928, LMP.

5. BMK, "Some Early A.T. History," 95–96.

6. Ibid., 95.

7. Waterman and Waterman, *Forest and Crag*, 488 n; BMK, "Memorandum for Mr. Roosevelt: (Re Survey of Taconic Section of the Appalachian Trail," 13 Jan. 1926; FDR to BMK, 15 Jan. 1926; RT to FDR, 14 Feb. 1926, ATCA; William A. Welch to BMK, 28 Dec. 1925; BMK diary, 14, 17 Dec. 1925, 27 May 1926.

8. Waterman and Waterman, *Forest and Crag*, 488–489; ATC, *Member Handbook*, 24–29.

9. ATC, *Member Handbook*, 26–27; RT, "The Long Brown Path," *New York Evening Post*, 1 June 1928; *The Appalachian Trail*, Publication No. 5 (Washington, D.C.: ATC, Jan. 1934), 13.

10. BMK, "Outdoor Culture: The Philosophy of Through Trails," Publication No. 16 (Boston: New England Trail Conference, 1927), 1–9. (Also published in *Landscape Architecture* 17 [Apr. 1927]: 163–174.)

11. Ibid., 6–7.

12. BMK, draft biography for Harvard class of 1900, 1930.

13. BMK, speech to Green Mountain Club, "Why the Appalachian Trail?" 12 Jan. 1929.

14. BMK, "A New England Recreation Plan," *Journal of Forestry* 27 (Dec. 1929): 927–930. (Reprinted in *FGG*, 161–168.)

15. BMK diary, 27 Jan., 1–2, 26 Mar., 10–11 May 1929; BMK, notes for speech to ATC, "The Origin and Conception of the A.T.," 10 May 1929, ATCA; BMK to HA, 20 July 1959.

16. BMK to HSMK, 31 Mar. 1928.

17. BMK to LM, 16 Aug. 1930; BMK, "The Family Locomotive," 12 Aug. 1930.

18. U.S. Bureau of the Census, *Historical Statistics of the United States, Colonial Times to 1970*, bicentennial edition, pt. 2 (Washington, D.C., 1975), 716.

19. Boyer, *Dreaming the Rational City*, 179.

20. BMK, "The Appalachian Trail (A Project in Regional Planning)," 1921, 40–41.

21. *NE*, 224–225.

22. Ibid., 182–200.

23. Henry Seidel Canby to LM, 25 Feb. 1929.

24. BMK, "Our Iron Civilization," *Saturday Review of Literature*, 2 Nov. 1929, 342–343.

25. BMK to CS, 10 Oct. 1929.

26. Ibid.

27. Ibid.; BMK diary, 21 Sept., 5 Oct. 1929.

28. BMK to CS, 10 Oct. 1929.

29. LM to BMK, 22 Oct. 1929; BMK diary, 14 Jan. 1930; BMK, "The Townless Highway," *The New Republic*, 12 Mar. 1930, 93–95.

30. BMK, "The Townless Highway," 93–95.

31. Ibid.

32. BMK, "Super By-Pass for Boston," *Boston Globe*, 31 Oct. 1930.

33. BMK, "Roads vs. Shuttles," *American City* 44 (Mar. 1931): 125–126.

34. BMK, "Highway Approaches to Boston: A Wayside Situation and What To Do About It," Bulletin No. 2 (Boston: Trustees of Public Reservations, 1931).

35. BMK, "Cement Railroads," *Survey Graphic* 68 (1 Nov. 1932): 541–542, 570.

36. See BMK, "Report on a Regional Plan for Washington, Connecticut and Its Environs," 10 Feb. 1931; BMK, "Report on a Regional Plan for the Shepaug River Protective Association, Litchfield County, Connecticut," 24 Aug. 1931.

37. BMK to Robert Crosser, 16 Dec. 1930.

38. Fred Kerby to BMK, 12 Jan. 1931, 21 Feb. 1931.

39. BMK and LM, "Townless Highways for the Motorist: A Proposal for the Automobile Age," *Harper's Monthly*, Aug. 1931, 347–356.

40. Ibid., 355.

41. *Federal-Aid Highway Act of 1956. U.S. Statutes at Large* 70 (1956): 374–387.

42. BMK to CS, 19 May 1930; BMK diary, 10 May 1929.

43. Martin, "Interview with Benton MacKaye," 10.

44. BMK, speech to ATC, "Vision and Reality," 20 May 1930.

45. For profiles of Avery, see Robert A. Rubin, "The Short, Brilliant Life of Myron Avery," *ATN*, Special 75th-Anniversary Issue, July 2000, 22–29; Waterman and Waterman, *Forest and Crag*, 489–493.

46. Ibid., 490.

47. Bill Mersch quoted in David Bates, "Profile: Myron Halliburton Avery," *Potomac Appalachian*, Dec. 1982, 8.

48. HB to BMK, 28 Feb. 1931.

49. BMK, message to 5th meeting of ATC, 12 June 1931.

50. HB to BMK, 20 June 1931.

51. BMK to HB, 29 June 1931.

Chapter 12 / "RP = TH + AT + HT," a Formula for the New Deal

1. LM to HB, 24 Aug. 1932.

2. BMK diary, 2 Mar. 1930.

3. See John L. Thomas, "Lewis Mumford: Regionalist Historian," *Reviews in American History* 16 (Mar. 1988): 162–163.

4. LM to Patrick Geddes, 27 June 1931, in LM, *My Works and Days*, 108.

5. Allen Tate quoted in James D. Hart, *Oxford Companion to American Literature*, 4th ed. (New York: Oxford University Press, 1965), 702.

6. CS to Harris A. Reynolds, 11 Mar. 1931.

7. Sussman, *Planning the Fourth Migration*, 197–198; Louis Brownlow to BMK, 21 May 1931; Charles S. Ascher, "Regionalism Charting the Future," *Survey* 66 (15 Aug. 1931): 460–461.

8. CS to BMK, 25 Mar. 1931.

9. Charles S. Ascher to Peter Dunning, 6 Dec. 1974.

10. BMK, speech at Regionalism Round Table, "Cultural Aspects of Regionalism," 19 June 1931.

11. CS to LM, 17 July 1931, CSP.

12. BMK diary, 3 July 1931.

13. A thorough, amply documented account of the origins and politics of Skyline Drive appears in Darwin Lambert, "Administrative History, Shenandoah National Park,

1924–1936" (Luray, Va.: National Park Service / Shenandoah Natural History Association, 1979), 99–121. See also Harvey P. Benson, "The Skyline Drive: A Brief History of a Mountaintop Motorway," *PATC Bulletin* 9 (July 1940): 67–71.

14. Quoted in Benson, "Skyline Drive," 68.

15. Horace M. Albright as told to Robert Cahn, *The Birth of the National Park Service: The Founding Years, 1913–33* (Salt Lake City: Howe Brothers, 1985): 265–268; Lambert, "Administrative History," 99–100.

16. BMK diary, 8, 11–14 July 1931.

17. BMK to Arno Cammerer, 13 June 1931, quoted in Lambert, "Administrative History," 165.

18. Arthur Demaray (for Cammerer) to HA, 17 June 1931, in Lambert, "Administrative History," 165.

19. Lambert, "Administrative History," 162–163.

20. BMK diary, 17 Sept. 1931.

21. BMK diary, 11 Oct. 1931; RT, "The Long Brown Path," *New York Evening Post,* 16 Oct. 1931.

22. BMK diary, 18–31 Oct. 1931.

23. BMK, "Some Early A.T. History," 96.

24. BMK, "The Appalachian Trail: A Guide to the Study of Nature," *Scientific Monthly* 34 (Apr. 1932): 330–342.

25. BMK, "New England in Genesis," *New England,* Jan. 1931, 3–9; BMK, "New England in Harness," *New England,* Feb. 1931, 5–8.

26. BMK, "The Appalachian Trail: A Guide to the Study of Nature," 342, 330.

27. Ibid., 330.

28. Ibid., 339.

29. BMK to PB, 2 Feb. 1965.

30. BMK, "Memorandum re a Nature Guide Service for the Appalachian Trail," 20 Aug. 1932.

31. William H. Carr, "Two Thousand Miles Along the Crest of the Atlantic Highlands," *Natural History* 33 (July–Aug. 1933): 395–408.

32. BMK, "A Cosmic Occasion," 17 Mar. 1932 (outline), 28 Mar. 1932 (draft article); see also BMK, "Moon's Shadow to Move Across the Long Trail," *Christian Science Monitor,* 22 July 1932.

33. BMK diary, 31 Aug. 1932; BMK to HMK, 4 Sept. 1932.

34. BMK to LM, 30 June 1932, LMP.

35. Arvia MacKaye Ege, *Power of the Impossible: The Life Story of Percy and Marion MacKaye* (Falmouth, Me.: Kennebec River Press, 1992), 330–331, 397–399.

36. BMK to CS, 25 Sept. 1932; BMK diary, Apr.–Dec. 1932, misc. entries.

37. HA to BMK, 14 Oct. 1932.

38. Ibid.

39. BMK, map, "A Project in Public Works: Appalachian America: Highway—Towns—Forestation," c. 1932.

40. BMK to CS, 28 Feb. 1932.

41. BMK to CS, 24 Sept. 1932.

42. BMK, "End or Peak of Civilization?" *Survey* 68 (1 Oct. 1932): 441, 444.

43. BMK to CS, 24 Sept. 1932.

44. BMK to CS, 4 Nov. 1932.

45. BMK to CS, 15 Nov. 1932.

46. BMK, "A Suggestion Re National Planning," 19 Nov. 1932, attached to BMK to CS, 19 Nov. 1932.

47. William E. Leuchtenburg, *Franklin D. Roosevelt and the New Deal, 1932–1940* (New York: Harper Colophon, 1963), 18–19; BMK to CS, 1 Apr. 1933.

48. BMK diary, 18–22 Mar. 1933; BMK, "Memorandum to the Secretary of Labor: Re Forest Communities," 22 Mar. 1933.

49. BMK to CS, 1 Apr. 1933.

50. BMK to CS, 7 Apr. 1933.

51. BMK diary, Feb.-Mar. 1933, misc. entries.

52. Judson King wrote an insider's account of the TVA's origins, *The Conservation Fight from Theodore Roosevelt to the Tennessee Valley Authority* (Washington, D.C.: Public Affairs Press, 1959). MacKaye's 1933 articles promoting the development of the Tennessee Valley were: "The Tennessee River Project: First Steps in a National Plan," *New York Times,* 16 Apr. 1933, sec. 8, p. 3; "The Challenge of Muscle Shoals," *The Nation,* 19 Apr. 1933, 445–446; "Tennessee—Seed of a National Plan," *Survey Graphic* 22 (May 1933): 251–254, 293–294.

53. King, *The Conservation Fight,* 267–276; BMK diary, 19 May 1933.

54. John Collier, *From Every Zenith: A Memoir and Some Essays on Life and Thought* (Denver: Sage Books, 1963), 27, 172–173, 269–284; James M. Glover, *A Wilderness Original* (Seattle: The Mountaineers, 1986), 157–158; BMK diary, 5 May 1933.

55. BMK to CS, 5 May 1933.

56. BMK, "Memorandum to the Commissioner of Indian Affairs: (re Forestation Act)," 24 May 1933.

57. BMK to CS, 9 June 1933.

58. BMK to CS, 20 June 1933.

59. BMK to Col. [George P.] Ahern, 21 June 1933.

60. Ibid.

61. Meine, *Aldo Leopold,* 302–306; BMK diary, 23–24 June, 1933.

62. BMK to HMK, 5 July 1933; AL and BMK, "Sandia Cooperative Flood-Control Project," 26 June 1933.

63. BMK to HMK, 5 July 1933.

64. BMK to HB, 5 July 1933.

65. Meine, *Aldo Leopold,* 302–306; AL, "The Conservation Ethic," *Journal of Forestry,* 31 (Oct. 1933): 634–643. Reprinted and quoted here from AL, *River of the Mother of God,* 183.

66. For a discerning discussion of Leopold's ideas during these years, see Sutter, "Driven Wild," 131–140.

67. BMK diary, July–Aug. 1933, misc. entries.

68. BMK diary, 7 July 1933.

69. BMK to FG, 20 Aug. 1933.

70. BMK to PMK, 27 Aug. 1933.

71. BMK, memorandum for Jay B. Nash, "Progress Report, Indian Emergency Conservation Work, Leupp Reservation, Arizona," 6 Aug. 1933; BMK to John Collier, 20 Aug. 1933.

72. BMK diary, 22, 31 Aug. 1933.

73. CS to LM, 5 Sept. 1933, CSP.

74. John Collier to BMK, 31 Aug. 1933.

75. BMK to HB, 6 Sept. 1933.

76. BMK to HB, 9 Sept. 1933.

77. LM to BMK, 14 Sept. 1933.

78. George Barber to BMK, 2 Mar. 1954; HB to CS, 3 Oct. 1933.

79. BMK diary, 28 Sept. 1933; Eben Alexander to Ernest Hendry, 7 Nov. 1933.

80. BMK to CS, 13 Oct. 1933; CS to HB, 9 Oct. 1933.

Chapter 13 / The Tennessee Valley Authority

1. BMK to CS, 1 Apr. ("Sunday") 1934.

2. Among the multitude of TVA scholars, Daniel Schaffer has most closely examined the fate of regional planning from the perspective of MacKaye and his like-minded colleagues within the agency. Schaffer's articles include: "Benton MacKaye: The TVA Years," *Planning Perspectives* 5 (1990): 5–21; "Ideal and Reality in 1930s Regional Planning: The Case of the Tennessee Valley Authority," *Planning Perspectives* 1 (1986): 27–44; "Environment and TVA: Toward a Regional Plan for the Tennessee Valley, 1930s," *Tennessee Historical Quarterly* 4 (1984): 333–354.

3. RM to E. C. M. ("Ned") Richards, 30 Aug. 1933.

4. Richards to BMK, 19 May 1928.

5. RM, *The Social Management of American Forests* (New York: League for Industrial Democracy, 1930), 22.

6. BMK, "Tennessee—Seed of a National Plan," *Survey Graphic* 22 (May 1933): 294.

7. BMK, "Regional Planning and the Tennessee Valley," 5 Feb. 1934, 36.

8. For accounts of Ned Richards's stormy TVA career, see Philip Selznick, *TVA and the Grass Roots: A Study in the Sociology of Formal Organization* (New York: Harper Torchbooks, 1966), 147–149, and E. C. M. Richards, "The Future of TVA Forestry," *Journal of Forestry* 36 (July 1938): 643–652. For Earle Draper's account of his own career and TVA role, see Joint Committee on the Investigation of the Tennessee Valley Authority, *Investigation of the Tennessee Valley Authority,* 75th Cong., 3d sess., pt. 9, 15–19, 21 Nov. 1938, 3701–3728.

9. ED, memorandum to TVA Board of Directors, "Project R.P.1—Tennessee Valley Section of the National Plan," 20 Feb. 1934, RHP.

10. FDR, message to Congress, 10 Apr. 1933, in *Franklin D. Roosevelt and Conservation, 1911–1945,* vol. 1, ed. Edgar B. Nixon (Hyde Park, N.Y.: General Services Administration / FDR Library, 1957), 151–152.

11. ED, memorandum to TVA Board, "Project R.P.1," 20 Feb. 1934.

12. BMK diary, 27, 31 Mar. 1934; F. W. Reeves to BMK, 3 Apr. 1934.

13. CS to C. L. Richey, 7 Oct. 1933; LM to CS, 4 Oct. 1933.

14. Thomas K. McCraw, *Morgan vs. Lilienthal: The Feud within the TVA* (Chicago: Loyola University Press, 1970), 38.

15. BMK, "Regional Planning and the Tennessee Valley," 5 Feb. 1934, 24.

16. *Tennessee Valley Authority Act of 1933, U.S. Statutes at Large* 48 (1933): 69.

17. Ibid.

18. For the origins of Sections 22 and 23, see a series of TVA memoranda: "Correspondence between Howard K. Menhinick and Tracy B. Augur regarding Origins and History of Sections 22 and 23 of the TVA Act," 21 Dec. 1942, 6 Feb. 1943, 1 Mar. 1943, TVACL. See also Paul K. Conkin, "Intellectual and Political Roots," in *TVA: Fifty Years of Grass-Roots Bureaucracy*, ed. Erwin C. Hargove and Paul K. Conkin (Urbana: University of Illinois Press, 1983), 30–31.

19. "Origin of the Regional Planning and Development Concept in TVA Legislation," 6 Feb. 1943, in "Correspondence between Menhinick and Augur," 3.

20. For a profile of TVA directors Arthur E. Morgan, Harcourt A. Morgan, and David Lilienthal, see McCraw, *Morgan vs. Lilienthal*, 7–23.

21. Selznick, *TVA and the Grass Roots*, 91–93; McCraw, *Morgan vs. Lilienthal*, 31–32; Michael J. McDonald and John Muldowny, *TVA and the Dispossessed: The Resettlement of Population in the Norris Dam Area* (Knoxville: University of Tennessee Press, 1982), 20–24.

22. TA, "Organization of Regional Planning Section for Project R.P.1," TVA memorandum to ED, 26 Apr. 1934, TVACL.

23. BMK, "'Opus One,' Progress Report on Regional Plan No. 1 (Project RP-1)," TVA memorandum to ED and TA, 20 Aug. 1934 (revised 20 Dec. 1934), 2.

24. BMK to SC, 20 July 1935.

25. Selznick, *TVA and the Grass Roots*, 152, 190–193; Schaffer, "Ideal and Reality," 29–31.

26. James E. Moorhead to author, 27 May 1988.

27. Lilienthal quoted in Arthur M. Schlesinger Jr., *The Coming of the New Deal* (American Heritage Library edition, Boston: Houghton Mifflin, 1988), 331.

28. BMK, "Regional Planning and the Tennessee Valley," 5 Feb. 1934, 2.

29. ED, "Your Memo of Apr. 25, Ward Shepard Correspondence," TVA memorandum to BMK, 28 Apr. 1934, TVACL.

30. HB diary excerpt, 25 July 1934, HBP.

31. BMK diary, 27 July 1934.

32. BMK, "Opus One," 5.

33. Ibid., 28.

34. BMK diary, 2–3 May 1934.

35. McDonald and Muldowny, *TVA and the Dispossessed*, 4, 6–7.

36. BMK, "Opus One," 36; Schaffer, "Ideal and Reality," 33; BF, "Forestry Application by the Tennessee Valley Authority," *Journal of Forestry* 33 (Oct. 1935): 851–856.

37. BMK diary, 20 Nov. 1934.

38. RH, "The Knoxville Years," *LW* 39 (Jan.–Mar. 1976), 24.

39. Carroll A. Towne to author, 27 Aug. 1988.

40. RH, "Knoxville Years," 24.

41. Ibid.

42. ED, "Oral History Interview," 24 Apr. 1984, TVA Oral History Collection, TVACL.

43. BMK to CS, 28 Sept. 1934.

44. RH, "Knoxville Years," 25; RH to Paul L. Evans, 3 Feb. 1969.

45. BMK to CS, 28 Sept. 1934.

46. RH, "Knoxville Years," 25; RH to BMK, "Spring Equinox" 1954.

47. HB to BMK, 19 June 1966.

48. MAM, interview with author, Knoxville, Tenn., 4–5 Mar. 1987.

49. BMK to Carrie E. Bliss, 9 Dec. 1934.

50. Georgia Appalachian Trail Club, *Friendships of the Trail: A History of the Georgia Appalachian Trail Club, 1930–1980* (Atlanta: Georgia Appalachian Trail Club, 1981), 70.

51. BMK to Gertrude Beard, 25 Aug. 1934.

52. MAM, interview.

53. Glover, *A Wilderness Original,* 176–180, 184, 198–199, 201.

54. MAM, interview.

55. BMK diary, 14–30 Jan., 23 Aug. 1935; PMK to BMK, 22 Jan. 1935, 5 Feb. 1935.

56. BMK, "Re John Reed and James MK," undated note, quoting Mary ("Peg") MacKaye.

57. BMK to JK, 15 Apr. 1956.

58. BMK to Robert O. Conant, 8 Apr. 1935.

59. BMK, "Physical Planning in the T.V.A. Program," TVA office memorandum to ED, 25 Feb. 1935.

60. BMK and Sam F. Brewster, "The Wild Life Program," TVA office memorandum to ED, 25 Sept. 1934.

61. BMK diary, 1, 19 Nov. 1934.

62. BMK to Leonard Doob, 12 Mar. 1935.

63. BMK, "(Confidential) In Re Park Legislation: Separation of Functions," 24 Mar. 1936.

64. BMK, "Habitability: A Study of Norris Sub-Region," TVA office memorandum to Allen A. Twichell, 22 Oct. 1935.

65. BMK, "Primeval Environment as a Natural Resource," TVA office memorandum to A. A. Twichell, 27 June 1935, 1.

66. Ibid., 20–22.

67. BMK diary, 29 May 1935.

68. BMK, "Primeval Environment," 18–19.

69. Ibid., 22.

70. ED, "Your Memo of June 27, 1935, to Mr. Twichell; Primeval Environment as a Natural Resource," TVA memorandum to BMK, 17 Sept. 1935, TVACL.

71. BMK, "New Name for Late 'Resort and Recreation' Committee," TVA memorandum to TA, 30 Oct. 1935.

72. TA, "Your Unnumbered Opus—New Name for Late 'Resort and Recreation Committee,'" TVA memorandum to BMK, 31 Oct. 1935.

73. BMK, "Your Comments on 'Scenic Resource Committee,'" TVA memorandum to TA, 5 Nov. 1935.

74. BMK, "Suggestion for Instituting a New School in Geography," memorandum to G. Donald Hudson, 15 Nov. 1935.

75. BMK, draft "Scenic Resources Board" bill, 31 Mar. 1936; BMK, "In Re Park Legislation."

76. Glover, *A Wilderness Original,* 167–168; RM to BMK, 28 July 1936.

77. BMK, "In Re Park Legislation," 19.

78. BMK, "Scenic Resources Board" bill.

79. Ibid.

80. BMK to RSY, 7 Apr. 1936, RMP; George Marshall, "Benton as Wilderness Philosopher," *LW* 39 (Jan.–Mar. 1976): 12.

81. Tennessee Valley Authority Department of Regional Planning Studies, *The Scenic Resources of the Tennessee Valley: A Descriptive and Pictorial Inventory* (Knoxville: TVA, 1938), 211–217; RH to Paul L. Evans, 3 Feb. 1969.

82. BMK diary, 7, 10, 16 Apr. 1936.

83. Selznick, *TVA and the Grass Roots,* 187; Schaffer, "Ideal and Reality," 38–41.

84. BMK to LM, 23 Apr. 1936.

85. J. E. Moorhead to BMK, 12 Aug. 1936.

86. Schaffer, "Benton MacKaye," 18.

Chapter 14 / The Wilderness Society

1. Tim Palmer, *Endangered Rivers and the Conservation Movement* (Berkeley: University of California Press, 1986), 52.

2. FDR quoted in Fox, *John Muir,* 200.

3. Lambert, "Administrative History," 130–132; Harley E. Jolley, *The Blue Ridge Parkway* (Knoxville: University of Tennessee Press, 1969), 42–44.

4. Lambert, "Administrative History," 121; RT, "The Long Brown Path," *New York Evening Post,* 19 Aug. 1935; RT, "The Long Brown Path," *New York Evening Post,* 23 Jan. 1936.

5. BMK, "Flankline vs. Skyline," *Appalachia* 20 (1934): 104–108.

6. Ibid., 107.

7. HB, letter to editor, "The Wilds, Parkways, and Dollars," *The Nation,* 11 July 1934, 46–47.

8. Horace Albright to BMK, 9 July 1934.

9. BMK, "Expression of Sentiment Re Skyline Drive," statement for 6th Appalachian Trail Conference, Rutland, Vt., June 1934.

10. MA, statement for Southern Appalachian Trail Conference, 22 May 1934; BMK, address to Southern ATC, 26 May 1934; BMK, resolution adopted at Southern ATC, 27 May 1934, ATCA.

11. Hannah Silverstein, "No Parking: Vermont Rejects the Green Mountain Parkway," *Vermont History* 63 (summer 1995): 133–157; Hal Goldman, "James Taylor's Progressive Vision: The Green Mountain Parkway," *Vermont History* 63 (summer 1995): 158–179; Ruth Gillette Hardy to BMK, 16 July 1934;

12. Hardy to BMK, 16 July 1934

13. RT, "The Long Brown Path," *New York Evening Post,* 6 July 1934.

14. MA to Arno B. Cammerer, 7 Aug. 1934.

15. BMK to Cammerer, 31 Aug. 1934.

16. Cammerer to BMK, 14 Sept. 1934.

17. BMK to Cammerer, 21 Sept. 1934.

18. Sutter, "Driven Wild," 311–313.

19. MA to BMK, 10 Sept. 1934.

20. HA to E. G. Frizzell, 9 Aug. 1934.

21. HB, "Origins of the Wilderness Society," *LW* 5 (July 1940): 10–11.

22. Ibid.

23. Ibid.

24. HB to HA, 22 Aug. 1934, WSP/DEN; HA to BMK, 3 Apr. 1930; RM, "The Problem of the Wilderness," *Scientific Monthly* 30 (Feb. 1930): 148.

25. HA to BMK, 12 Sept. 1934, WSP/DEN.

26. BMK, "(Draft Copy) Invitation to Help Organize a Group to Preserve the American Wilderness," 20 Sept. 1934.

27. BMK to HA, 20 Sept. 1934.

28. Stephen Fox, "We Want No Straddlers," *Wilderness* 48 (winter 1984): 5; HB, "Origins of the Wilderness Society," 11.

29. "Invitation to Help Organize a Group to Preserve the American Wilderness," 19 Oct. 1934, WSP/DEN.

30. Glover, *A Wilderness Original,* 176–177; BMK diary, 19 Oct. 1934.

31. BMK to HS, 22 Oct. 1934.

32. RM to BMK, 24 Nov. 1934, WSP/DEN.

33. BMK diary, 20 Oct. 1934.

34. RSY to John C. Merriam, 4 Apr. 1934, WSP/DEN; RSY to BMK, 26 Oct. 1934, WSP/DEN; RM to BMK, 24 Nov. 1934; HB, "Origins of the Wilderness Society," 11.

35. RSY, editorial, "The Tower of Babel, New Style," *LW* 2 (Nov. 1936): 1–2.

36. AL, "The Wilderness and Its Place in Forest Recreational Policy," in AL, *River of the Mother of God,* 79; RM, "The Problem of the Wilderness," 141; RSY to HB, 31 Oct. 1934, WSP/DEN.

37. HB to HA, 1 Oct. 1934, ATCA; RM, "The Forest for Recreation and a Program for Forest Recreation," in U.S. Forest Service, *A National Plan for American Forestry,* 73d Cong., 1st. sess., 1933, S. Doc. 12, 475.

38. BMK to RSY, 9 Nov. 1934, WSP/DEN.

39. "The Wilderness Society," organizational statement, 21 Jan. 1935.

40. Ibid.

41. BMK to HB, 19 Feb. 1934, HBP; "Editorial: The Cult of the Wilderness," *Journal of Forestry* 33 (Dec. 1935): 955–957.

42. "Minutes of the Wilderness Society from its Beginning, Jan. 20, 1935," WSP/DEN; RM to BMK, 9 Apr. 1935, Paul Oehser personal collection; Fox, "We Want No Straddlers," 9.

43. "The Wilderness Society"; RM to HA, 24 Oct. 1934, WSP/DEN.

44. BMK to RM, 12 Dec. 1935, Paul Oehser personal collection; RM to RSY, 26 Oct. 1935, WSP/DEN.

45. MA, "The Skyline Drive and the Appalachian Trail," *PATC Bulletin* 4 (Jan. 1935): 9–11.

46. Ibid.

47. Harold P. Ickes, "Wilderness and Skyline Drives," *LW* 1 (Sept. 1935): 12.

48. BMK, "Why the Appalachian Trail?" *LW* 1 (Sept. 1935): 7–8.

49. Ibid.

50. MA, resolution on "Skyline Drives" for 7th meeting of ATC, June 1935; RT, "The Long Brown Path," *New York Evening Post,* 27 June 1935.

51. "Trail Conference Resume," *PATC Bulletin* 4 (Oct. 1935): 83–84.

52. RT to BMK, 6 July 1935.

53. BMK to SC, 20 July 1935.

54. Ibid.

55. MA to Richard Mansen, 31 Jan. 1936, ATCA.

56. BMK to MA, 20 Nov. 1935.

57. MA to BMK, 19 Dec. 1935, ATCA.

58. BMK to MA, 4 Feb. 1936.

59. BMK to Eugene C. Bingham, 29 Oct. 1934.

60. See George J. Ellis, "The Path Not Taken," (master's thesis, Shippensburg University, 1993 [copy at ATCA]), for a detailed (and pro-Avery) account of the issues and events surrounding the divisions within the Appalachian Trail Conference and the differences between MacKaye and Avery during the mid-1930s.

61. RM and Althea Dobbins, "Largest Roadless Areas in the United States," *LW* 2 (Nov. 1936): 11–13.

62. AL, "Why the Wilderness Society?" *LW* 1 (Sept. 1935): 6.

63. BMK, "Why the Appalachian Trail?" 8.

Chapter 15 / "Watershed Democracy"

1. BMK diary, July–Aug. 1936, misc. entries; BMK to LM, 16 Aug. 1936, LMP. See also LM, *Culture of Cities,* and SC, *Rich Land, Poor Land: A Study of Waste and Natural Resources* (New York: Whittlesey House, 1936).

2. Parsons, "Clarence Stein and the Greenbelt Towns," 161–179.

3. BMK, "Magna Charta: An Interpretation of the T.V.A. Planning Law," Feb. 1937; William E. Leuchtenburg, "Roosevelt, Norris and the 'Seven Little TVAs,'" *Journal of Politics* 14 (1952): 418–441.

4. BMK to RH, 2 May 1937.

5. BMK to Francis Biddle, 1 July 1938; BMK to HS, 12 July 1938; BMK to HS, 14 July 1938.

6. BMK diary, 28 Jan., 22–23 May 1937; BMK's inscriptions regarding the "Wood-ticks" appear in his personal copy of Allen Chamberlain's *The Annals of the Grand Monadnock* (Concord, N.H.: Rumford Press, 1936), ATCA.

7. BMK diary, 11 Nov. 1936, 27 Jan., Feb. 1, 1937; "Two Damndest Years" scrapbook, ATCA; BMK to "Philosopher's Club," 24 Mar. 1937; BMK to CS, 24 Feb. 1937; Anne Pursel to HB, 28 Jan. 1937, HBP.

8. BMK diary, 28 Mar., 21–29 June 1937.

9. BMK diary, 19 July 1937.

10. BMK diary, 29 Mar. 1937.

11. BMK to Laurence Fletcher, 29 Aug. 1937.

12. CS to BMK, 15 Nov. 1937; BMK to HB, 28 May 1937, HBP.

13. Laurence Fletcher to BMK, 9 Oct., 23 Dec. 1937.

14. Trustees of Public Reservations of Massachusetts, *The Bay Circuit: A Practical Plan for the Extension of the Metropolitan Park System and the Development of a State Parkway through a Number of Reservations in the Circuit of Massachusetts Bay* (Boston, 1937), 14.

15. BMK to CS, 29 Aug. 1937.

16. Bay Circuit Alliance, *Protection Plan for The Bay Circuit Trail and Greenway* (Andover, Mass., 1995).

17. BMK to William L. Stoddard, 27 Mar. 1938; BMK to CS, 3 Apr. 1938; BMK diary, Jan.–Apr. 1938, misc. entries.

18. BMK to LM, 14 Dec. 1937, LMP; Michael F. Mayer to "Folks," 16 Jan. 1938; BMK to HMK, 19 Jan. 1938.

19. BMK diary, 4 Feb. 1938.

20. Ibid., 27 Apr. 1938.

21. Ibid., Apr.–July 1938, misc. entries; BMK to Carol Hartshorne, 18 July 1938.

22. BF to BMK, 25 Aug. 1938, 14 Sept. 1938; BMK to Howard W. Odum, 30 Mar. 1940.

23. Palmer, *Endangered Rivers*, 25.

24. FDR quoted in William E. Leuchtenburg, *Flood Control Politics: The Connecticut River Valley Problem, 1927-1950* (Cambridge: Harvard University Press, 1953), 51.

25. Ibid., 96.

26. *Flood Control Act of 1936, U.S. Statutes at Large* 49 (1936): 1570–1571.

27. RZ quoted in BF and Anthony Netboy, *Water, Land, and People* (New York: Alfred A. Knopf, 1950), 212.

28. BMK diary, 21 Oct., 8 Nov. 1939; BMK, "A Primer of Flood Control (As Applied to the Northeast Region)," 10 Nov. 1938.

29. BMK to HB, 8 Jan. 1939, HBP.

30. BMK, "Flood Control Problems in New England," USFS memorandum to E. N. Munns, 11 Mar. 1939; Leuchtenburg, *Flood Control Politics*, 46–99.

31. BMK, "Flood Control Problems."

32. BMK, "States Rights and Conservation Districts," *USFS Service Bulletin*, 18 Sept. 1939, 2–4.

33. BMK diary, 12 Mar. 1940; BMK, "Democracy in Flood Control: The Lesson of Hill, N.H.," *Survey Graphic* 29 (Sept. 1940): 468–470.

34. BMK diary, Nov.–Dec. 1938–Jan. 1939, misc. entries.

35. BF and Netboy, *Water, Land, and People*, 137–139; BMK, "Flood Control Problems."

36. BMK, "'Little Waters' as a War-Time Measure," REA memorandum to Joseph E. O'Brien, 13 Apr. 1942.

37. BMK, "Democracy in Flood Control," 470.

38. BMK, "Watershed Government," *USFS Service Bulletin*, Nov. 1939, 5–7.

39. BMK diary, 9–10 Dec. 1939; BMK, "Inquiry on flood control methods by City Planning Schools of MIT and Columbia," USFS memorandum to E. N. Munns, 24 Jan.

1940, CSP; Walter L. Creese, *The Search for Environment: The Garden City, Before and After* (New Haven: Yale University Press, 1966), 309–310.

40. *Flood Control Act of 1938, U.S. Statutes at Large* 52 (1938): 1216.

41. BMK, "Region Building in River Valleys: Upstream Community vs. Downstream Slum," *Survey Graphic* 29 (Feb. 1940): 107–108.

42. BMK, "Regional Planning and Ecology," *Ecological Monographs* 10 (July 1940): 349–353.

43. BMK, "Survey report (in process) of Merrimack River Watershed, N.H. and Massachusetts," USFS memorandum for E. N. Munns, 15 Feb. 1940, 14.

44. Robert O'Brien to author, 7 Jan. 1988.

45. Jack J. Preiss, *Camp William James* (Norwich, Vt.: Argo Books, 1978); Calvin W. Gower, "'Camp William James': A New Deal Blunder?" *New England Quarterly* 38 (1965): 475–493; SC, "Young Men in Tunbridge," *Survey Graphic* 33 (May 1942): 229–233.

46. Preiss, *Camp William James,* 97–119; BMK diary, 16–23 Dec. 1940; BF to BMK, 13 Jan. 1941; BMK to Stearns Morse, 30 Nov. 1940; Gower, "Camp William James," 487–493.

47. BMK diary, 6 Mar., 5 Sept., 5 Oct. 1939, 6 June 1940; PMK to BMK, 11, 12, 19 May 1940.

48. BMK diary, 3 May 1937; BMK to CS, 13 Aug. 1939; BMK to LM, 8 June 1943, LMP.

49. BMK diary, 3 Nov. 1936, 29 Oct. 1940.

50. BMK diary, 31 May 1940; BMK to LM, 27 July 1940, LMP.

51. BMK, "'Defense Time' Conservation," *Planners' Journal* 6 (July–Sept. 1940): 71–76.

52. LM to BMK, 25 July 1940.

53. BMK, "Re: The Big Sandy Watershed as Illustration of Possible Modifications of Watershed Conservation Policy to Meet Needs of National Defense," USFS memorandum for E. N. Munns, 22 Aug. 1940.

54. BMK, "A Pattern for Vermont River Basin Development," chap. 4 in *Water Resources and Electrical Energy* (Vermont Commission for the Study of Water Resources and Electrical Energy, 1941), 33–37.

55. BMK to E. N. Munns, 16 Mar. 1941.

56. BMK, "'A.R.M.': A memorandum on Conservation as one of the Three Elements of National Defense," USFS report, Apr. 1941, 1.

57. BMK, "Arsenals of Democracy," 20 May 1941.

58. BMK, "Conservation Units in National Defense," supplement to "A.R.M.," 1–8.

59. E. N. Munns to BMK, 10 June 1941; BMK diary, 2 Apr. 1941.

60. BMK, "War and Wilderness," *LW* 6 (July 1941): 7–8.

61. Avery Means to BMK, 25 Oct. 1941.

62. BMK to William L. Stoddard, 3 Aug. 1941; BMK, "Charting Defense Production via System of Industrial Intelligence," memorandum, Aug. 1941; BMK to HS, 13 Oct. 1941.

63. Samuel T. Dana and Sally K. Fairfax, *Forest and Range Policy,* 2d ed. (New York:

McGraw-Hill, 1980), 147; D. Clayton Brown, *Electricity for Rural America: The Fight for the REA* (Westport, Conn.: Greenwood Press, 1980), 34–98.

64. BMK to CS, 8 Feb. 1942.

65. BMK to SC, 10 Oct. 1943.

66. BMK, "Suggestions for Supplementing the REA Program for 'Appraisal of Rural Resources Available for National Defense,'" memorandum for Joseph E. O'Brien, 23 Mar. 1942.

67. BMK, "An Alaska-Siberia 'Burma Road,'" *The New Republic*, 2 Mar. 1942, 292–294.

68. Laurence Todd to BMK, 18 Mar. 1942.

69. BMK diary, 2 Aug. 1942; BMK to CS, 9 Aug. 1942; BMK, "Mobilization of Rural Resources Adjacent to the Alaskan-Canadian Military Highway," memorandum for J. E. O'Brien, Sept. 1942.

70. BMK to HB, 15 July 1942, HBP.

71. GG (II), 498; BMK diary, 13 Oct. 1944.

72. Brown, *Electricity for Rural America*, 76–98; William J. Neal to BMK, 2 July 1943; BMK to SC, 1 Aug. 1943; B. M. Snoody to BMK, 3 Sept. 1943.

73. BMK, "Field of Work Being Covered by Benton MacKaye under Above Title [Industrial Engineer]," REA memorandum, June 1944; BMK to CS, 1 Aug. 1943.

74. BMK, "'Dairyland' as a Geographic Unit: What Are Its Proper Ultimate Limits?" REA memorandum, Nov. 1943, 25.

75. BMK to LM, 2 Apr. 1944, LMP.

76. BMK diary, 9 Apr. 1944; BMK to HB, 9 Apr. 1944, HBP.

77. BMK, "A Legislative Analysis: The Place of REA and G&T in MVA," REA report, Oct. 1944, 8–12.

78. Donald Worster, *Rivers of Empire: Water, Aridity, and the Growth of the American West* (New York: Pantheon, 1985), 268–269; BF and Netboy, *Water, Land, and People*, 257–259.

79. Worster, *Rivers of Empire*, 268–269.

80. BMK, "The Story of Conservation," book outline, 18 Apr. 1944; BMK diary, Apr.–July 1944, misc. entries; CS to BMK, 24 Apr. 1944; CS and M. M. Samuels to BMK, 15 June 1944; LM to BMK, 30 May 1944.

81. HMK certificate of cremation, 14 Aug. 1944; BMK to Aline M. Stein, 30 Aug. 1944, CSP; BMK to LM, 4 June 1944.

82. CS to LM, 20 Dec. 1944, CSP; LM to BMK, 17 Dec. 1944; BMK to Harry Johnson, 12 Dec. 1944; BMK to HB, 14 Jan. 1945.

Chapter 16 / Wilderness in a Changing World

1. GG (VI), 216; *NE*, 138–139.

2. HB, "The Last Decade, 1935–1945," *LW* 10 (Dec. 1945): 13–17.

3. RM to BMK, 24 Feb. 1937.

4. BMK to RM, 2 Mar. 1937.

5. RM, HA, and RSY to BMK, 25 Mar. 1937.

6. "The Wilderness Society, Minutes of the Organizing Meeting," 24 Apr. 1937, RMP.

7. Ibid.

8. ATC, *Member Handbook,* 36–38; Edward B. Ballard, "Behind the Trail's History," letter to editor, *ATN* 54 (July–Aug. 1993): 3–4.

9. RM to BMK, 18 Nov. 1937.

10. BMK to RM, 3 Dec. 1937.

11. Charles H. W. Foster, *The Appalachian National Scenic Trail: A Time to Be Bold* (Harpers Ferry, W.Va.: ATC, 1987), 172–173.

12. ATC, *Member Handbook,* 36–38.

13. HB, "Last Decade," 16–17; RSY to BMK, 20 Nov. 1939.

14. BMK to RSY, 17 Nov. 1939, WSP/DEN.

15. Glover, *A Wilderness Original,* 272; Fox, "We Want No Straddlers," 10, 14.

16. HB to RSY, 14 May 1940.

17. BMK, "The Gregarious and the Solitary," *LW* 4 (Mar. 1939): 7–8.

18. BMK, "The Spirit of the Wilderness," *LW* 5 (July 1940): 17–18.

19. John C. Hendee, George H. Stankey, and Robert C. Lucas, *Wilderness Management,* USFS Miscellaneous Publication No. 1365 (Washington, D.C.: GPO, Oct. 1978), 35–36, 61–64.

20. Watkins, *Righteous Pilgrim,* 583–584.

21. Ibid., 562–569, 584–591.

22. BMK to RM, 10 May 1935, RMP.

23. Watkins, *Righteous Pilgrim,* 554–555.

24. RM to Harold Ickes, 21 May 1937; RM to RSY, 4 Oct. 1937: both in WSP/DEN.

25. RM to RSY, 12 Dec. 1938, WSP/DEN.

26. HB, "Last Decade," 16–17.

27. RSY to BMK, 6 Feb. 1943; BMK to RSY, 1 Feb. 1943: both in WSP/DEN.

28. BMK diary, 13 July, 8, 14 Aug. 1944, 11, 27 Mar. 1945.

29. HB, "Last Decade," 16.

30. BMK to Wilderness Society Councilors, "Draft Letter #1," 21 May 1945.

31. "Draft Letter #2," 27 May 1945.

32. "Draft Letter #3," 31 May 1945.

33. "Draft Letter #4," 31 May 1945.

34. "Draft Letter #5," c. 1 June 1945.

35. "Draft Letter #6," c. 1 June 1945.

36. BMK to Charles G. Woodbury, 19 June 1949.

37. Ernest S. Griffith to George Marshall, "Events of 1945," 15 Oct. 1945.

38. Fox, "We Want No Straddlers," 10–12; Margaret E. Murie, interview with author, Moose, Wyo., 14 Aug. 1989.

39. E. S. Griffith to G. Marshall, 15 Oct. 1945; HB to BMK, 19 July 1945; BMK diary, 15 Aug. 1945; BMK to HB, 27 Nov. 1945, HBP. For profiles of Murie and Zahniser, see also Fox, *John Muir,* 266–272, and Mark Harvey, "Howard Zahniser: A Legacy of Wilderness," *Wild Earth* 8 (summer 1998): 62–66.

40. BMK diary, 6 Aug. 1945.

41. Paul Boyer, *By the Bomb's Early Light* (New York: Pantheon, 1985), 31.

42. BMK, "Draft Resolution on Atomic Energy," June 1946.

43. HA to BMK, 2 June 1946; HB to BMK, 9 June 1946; BMK diary, 22 June 1946; "News Items of Special Interest," *LW* 11 (Sept. 1946): 28.

44. AL to BMK, 1 May 1946.

45. BF to BMK, 17 Mar. 1946.

46. For a discussion of postwar Forest Service policies, see Paul S. Hirt, *A Conspiracy of Optimism: Management of the National Forests since World War Two* (Lincoln: University of Nebraska Press, 1994), 44–81.

47. BMK diary, 30 Nov. 1945; BMK, "A Wilderness Philosophy," *LW* 11 (Mar. 1946): 4.

48. HB to OM, 1 Aug. 1940; HB journal, 22, 25 June 1945: both in WSP/DEN.

49. H.R. 2142, 79th Cong., 1st sess.

50. House Committee on Roads, *National System of Foot Trails: Hearings on H.R. 2142,* 79th Cong., 1st sess., 24 Oct. 1945, 1–20.

51. BMK, "A Bill to Establish a National System of Wilderness Belts," 31 Jan. 1946.

52. BMK to Daniel K. Hoch, 31 Jan. 1946.

53. BMK, "National System of Wilderness Belts."

54. LM to BMK, 28 Feb. 1946.

55. OM to BMK, 28 Apr. 1946.

56. "News Items," *LW* 11 (Sept. 1946): 23.

57. MA to HZ, 21 Nov. 1946.

58. BMK to HZ, 4 Dec. 1946.

59. HZ to BMK, 14 Jan. 1947.

60. HZ to BMK, 14 Feb. 1947.

61. BMK to WS Councilors, 2 June 1947; Meine, *Aldo Leopold,* 497; "News Items," *LW* 12 (autumn 1947): 29.

62. BMK to Joan S. Wright, 15 Jan. 1968.

63. See, for example, "Nature Sanctuaries in the United States and Canada," *LW* 15 (winter 1950–51), an entire issue devoted to an inventory compiled by committees of the Ecolological Society of America and the Ecologists Union; and "Report of the [Society of American Foresters] Committee on Natural Areas," *Journal of Forestry* 47 (Feb. 1949): 137–146.

64. "News Items," *LW* 12 (autumn 1947): 29.

65. BMK to OM, 7 Jan. 1947.

66. BMK, memorandum on "wildland patches" to WS members, 9 Nov. 1947; "On Preserving Our 'Wildland Patches,'" *LW* 13 (autumn 1948): 9–12, (winter 1948–49): 11–14.

67. Meine, *Aldo Leopold,* 477–478; Hirt, *Conspiracy of Optimism,* 62; BMK to William Vogt, 31 July 1948.

68. BMK, editorial, *LW* 13 (autumn 1948), inside front cover.

69. BMK, "Primeval Security: The Import of the Pan-American Wildlife Treaty," *LW* 13 (spring 1948): 1–4.

70. CS to LM, 25 July 1947, CSP.

71. BMK diary, Mar.–Apr. 1947, misc. entries.

72. OM to BMK, 27 Aug. 1948; Margaret E. Murie interview, 14 Aug. 1989.

73. BMK to WS Councilors, draft letter, c. June 1948.

74. BMK to HZ, 29 Jan. 1950, WSP/DEN.

75. House Committee on Merchant Marine and Fisheries, Subcommittee on Fisheries and Wildlife Conservation, *The Preservation of Wilderness Areas: An Analysis of Opinion on the Problem,* report prepared by C. Frank Keyser, 81st Cong., 1st sess., 1949, Committee Print 19; BMK to Ernest Griffith, 6 Nov. 1949; BMK to HZ, 6 Nov. 1949, WSP/DEN.

76. Morris L. Cooke to BMK, 24 Jan. 1950.

77. BMK to HZ, 4 Feb. 1950.

78. BMK to M. Cooke, 29 Jan. 1950; M. Cooke to BMK, 26 Apr. 1950.

79. BMK to M. Cooke, 28 June 1950.

80. M. Cooke to BMK, 18 July 1950. For the full three-volume report of the President's Water Resources Policy Commission, see *A Water Policy for the American People* (Washington, D.C.: GPO, 1950).

81. BMK, "Dam Site vs. Norm Site," *Scientific Monthly* 71 (Oct. 1950): 241.

82. Ibid., 243.

83. Ibid., 244.

84. Ibid., 245–247.

85. Samuel P. Hays, *Beauty, Health, and Permanence: Environmental Politics in the United States, 1955–1985* (Cambridge: Cambridge University Press, 1987), 2–3.

86. James Glover and Regina Glover, "The Natural Magic of Olaus Murie," *Sierra* 72 (Sept.–Oct. 1987): 73.

87. OM to BMK, 29 Apr. 1950.

88. "The Second Wilderness Conference," *LW* 16 (spring 1951): 26–33.

89. BMK to WS Councilors, 2 June 1947.

Chapter 17 / "Geotechnics of North America"

1. BMK, "Life History of Benton MacKaye for 50th Anniversary Report, Harvard 1900," 1950.

2. In 1947, for example, MacKaye collaborated with Bernard Frank and a PATC member, Ivan Tarnowsky, to promote what became a popular educational nature trail, linked to the Appalachian Trail, in Shenandoah National Park's White Oak Canyon. See Darwin Lambert, *The Undying Past of Shenandoah National Park* (Boulder, Colo.: Robert Rhinehart, 1989), 266; BF to BMK, 2 May 1947; BMK to Edward D. Freeland, Superintendent, Shenandoah National Park, 13 May 1947.

Two years later he worked with Laurence Fletcher, William Wharton, William Roger Greeley, and Connecticut ecologist Frank Egler on a less successful effort to develop a network of "ecologic areas" in southern New England. See BMK, memorandum to Frank Egler, et al., "Questions (and suggested answers) re Establishment and Uses of Ecologic Areas ('specimens of environment')," 5 July 1949; BMK to L. B. Fletcher, 29 June, 11 July 1949.

3. Boyer, *Bomb's Early Light,* 33–45.

4. BMK to LM, 4 Mar. 1946, LMP.

5. BMK, notes for talk on world government referendum, c. Oct. 1946; BMK diary, 21 Oct. 1946.

6. James T. Patterson, *Grand Expectations: The United States, 1945–1974* (New York: Oxford University Press, 1996), 165–205.

7. Boyer, *Bomb's Early Light,* 34.

8. BMK (signed "Federalist"), *The Montachusett Federalist* (Boston: United World Federalists, 1949), comprising a series of five 1949 letters in *Fitchburg Sentinel;* BMK, "Raising the Pyramid," *Fitchburg Sentinel,* 15 July 1949.

9. BMK, "'More Perfect' than What?" *Fitchburg Sentinel,* 24 Jan. 1949.

10. BMK diary, 23 Sept. 1949; Boyer, *Bomb's Early Light,* 349; BMK, "Washington and Lee, Patriots," *Fitchburg Sentinel,* 22 Feb. 1950.

11. Boyer, *Bomb's Early Light,* 327–328; K. C. Parsons, in *Writings of Clarence Stein,* 503–504.

12. CS to LM, 16 Jan. 1948.

13. CS to BMK, 25 July 1947, CSP; LM to BMK, 27 July 1947.

14. Catherine Bauer Wurster to CS, 13 Apr. 1948, CSP.

15. "Minutes of the Regional Planning Association of America," 22 Apr. 1948, CSP; CS, "Memorandum on Purpose and Future Organization of R.D.C.A.," 11 Feb. 1949, CSP.

16. RDCA, minutes, 20 Oct. 1950; RDCA, executive meeting minutes, 12 Dec. 1950: both in CSP; BMK, "Fields of Administration of the United States Government," RDCA report, 15 May 1951.

17. "Federal Administration: By Subject and by Region," RDCA minutes, 18 May 1951, 2–4.

18. Ibid., 6.

19. Frederic N. Cleaveland, memorandum to John A. Parker, 2 Nov. 1951, CSP.

20. HZ to CS, 23 July 1951, CSP.

21. BMK to CS, 9 July 1951.

22. BMK, "The Kitimat Museum," memorandum to CS, 5 May 1952.

23. BMK, "Non-Gregarious Recreation for Kitimat, B.C.," memorandum to CS, 6 Nov. 1951.

24. Kenneth T. Jackson, *Crabgrass Frontier: The Suburbanization of the United States* (New York: Oxford University Press, 1987), 203–218, 226–228, 293; BMK diary, 23 May 1954.

25. BMK, winter solstice postcard to friends, 1954; BMK to AB, 27 May 1954, HBP.

26. BMK to HB, 14 Jan. 1945.

27. SC, "Confidential Report" to Guggenheim Foundation, 22 Nov. 1944, attached to CS to BMK, 20 Dec. 1944.

28. LM to BMK, 27 July 1947.

29. BMK to HB and AB, 6 Feb. 1946, HBP.

30. BMK to Paul Kellogg, 28 Sept. 1948.

31. GG (V), 172–173.

32. GG (III), 558–559.

33. *GG* (IV), 15.

34. *GG* (VI), 216–217.

35. *GG* (VII), 285.

36. *GG* (VI), 217.

37. BMK to Lambert Davis, 4 Feb. 1950.

38. Kenneth Ross to BMK, 3 Mar. 1969.

39. BMK to LM, 20 Feb. 1954, CSP.

40. BMK to LM, 21 Nov. 1963, LMP.

41. AB to BMK, 30 May 1975.

42. C. J. S. Durham, "A View from the Earldom," *LW* 39 (Jan.–Mar. 1976): 27.

43. HZ to CS, 22 May, 22 Aug. 1950, CSP; Paul Oehser to BMK, 5 Dec. 1950.

44. BMK to LM, 8 June 1964, LMP; Paul H. Oehser, interview with author, Washington, D.C., 11 June 1986.

45. BMK to Robert Bruère, 23 Oct. 1961.

46. BMK, "Word Estimate, Geotechnics of North America," 4 July 1960.

47. BMK to LM, 12 May 1963, LMP.

48. BMK to CS, 29 June 1964, CSP.

49. BMK to CS, 5 Aug. 1965, CSP.

50. BMK to LM, 12 Nov. 1960.

51. LM to BMK, 15 Nov. 1960.

52. Paul T. Bryant, *The Quality of the Day: The Achievement of Benton MacKaye* (Ann Arbor: University Microfilms, 1981); BMK to PB, 6 Feb. 1962; BMK to PB, 26 July 1964; PB, comments at symposium, "Benton MacKaye and the Appalachian Trail," University at Albany, State University of New York, 22 Nov. 1996.

53. Donald Jackson to BMK, 17 Jan. 1966.

54. LM to D. Jackson, 10 Aug. 1966.

55. BMK to CS, 3 Aug. 1968.

56. LM to BMK, "Memorandum on the Geotechnics of North America," 2 Nov. 1970.

57. Ibid.

58. Roy P. Basler to BMK, 13 July 1972; BMK to R. Basler, 22 July 1972; BMK, *A Two-Year Course in Geotechnics* (Shirley Center, Mass.: n.p., 1972).

59. LM to HB, 24 Aug. 1932.

60. MacKaye completed several typed drafts of "Geotechnics of North America." The draft on deposit at the Manuscripts Division of the Library Congress is identified as "Wilderness Society Edition, Copy No. 2"; it was prepared during 1964–1965. The draft cited below, which includes MacKaye's revisions through 1968, is on deposit among the MacKaye Family Papers at the Dartmouth College Library. It is identified and catalogued as "Wilderness Society Edition, Copy #3," box 196, folders 24–39.

61. For a discussion of the open space movement of the 1950s and 1960s and its "key role in the evolution of the environmental movement," see Adam W. Rome, "William Whyte, Open Space, and Environmental Activism," *Geographical Review* 88 (April 1998): 259–274.

62. *GNA,* ch. 1, 10–21.

63. Ibid., ch. 5, 3.

64. Ibid., 66, 76.

65. Ibid., ch. 8, 1, 7, 108.

66. Ibid., ch. 7, 40–57, 60, 66.

67. Ibid., ch. 8, 102.

68. Ibid., ch. 9, 18.

69. Ibid., 8, 26B-27, 47, 52.

70. Ibid., 90, 92.

71. Ibid., 95.

Chapter 18 / Linking Action with Prophecy

1. *NE*, 133.

2. LM, typescript, "Benton MacKaye," 24 Mar. 1969, attached to LM to BMK, 6 Mar. 1974.

3. HPH, "The Shirley Influence," 22.

4. BMK, "A Pageant of Her Deeds," 11 Oct. 1953, scene 5, 1.

5. BMK to Helen Northup, 18 Oct. 1953.

6. LM, citation, Trustees of Public Reservations Conservation Award, Jan. 1954.

7. BMK, series of maps and charts, "Cape Cod to Berkshire," 17 Jan. 1954; BMK diary, 12, 20 Jan. 1954.

8. BMK to "Friends," 6 Mar. 1954.

9. LM, *Sketches from Life,* 341.

10. BMK to Mary ("Peg") MacKaye, 29 Apr. 1950.

11. BMK to "Friends," 6 Mar. 1954.

12. Glover, *A Wilderness Original,* 244–246.

13. House Special Committee on Un-American Activities (HUAC) to Attorney General Francis Biddle, 17 Oct. 1941, BMK Federal Bureau of Investigations file (Freedom of Information Act request no. 292865), in author's possession.

14. BMK FBI file.

15. BMK diary, misc. entries, 1950–1964.

16. BMK diary, 16–19 (including "Douglas Expedition" memo) Mar. 1954.

17. BMK to HZ, 17 Feb. 1954, WSP/DEN; BMK and Kenneth W. Ross to HZ, 27 Feb. 1954, HBP.

18. Jack Durham, "The C & O Canal Hike," *LW* 48 (spring 1954): 1–26; Fox, *John Muir,* 239–242.

19. Nash, *Wilderness and the American Mind,* 219. For a comprehensive study of the Echo Park struggle, see Mark W. T. Harvey, *A Symbol of Wilderness: Echo Park and the American Conservation Movement* (Albuquerque: University of New Mexico Press, 1994).

20. BMK diary, 5 Nov. 1952.

21. Allin, *Politics of Wilderness Preservation,* 105–108; T. H. Watkins, "Untrammeled by Man: The Making of the Wilderness Act of 1964," *Audubon* 91 (Nov. 1989): 86–88.

22. BMK, letter to the editor, "Saving the Wild," *Washington Post,* 26 May 1960.

23. BMK, "If This Be Snobbery," *LW* no. 77 (summer–fall 1961): 3–4.

24. BMK diary, 4–5 May 1964; "1906 . . . Howard Clinton Zahniser . . . 1964," *LW* no. 85 (winter–spring 1964): 3.

25. BMK to SB, 11 Sept. 1974.

26. *Wilderness Act, U.S. Statutes at Large* 78 (1964): 890–896.

27. George Marshall to BMK, 26 Feb. 1959.

28. BMK to Joan Safford Wright, 15 Jan. 1968.

29. Martin, "Interview with Benton MacKaye," 35.

30. BMK, "Wilderness as a Folk School," *LW* 17 (winter 1952–1953): 6.

31. Frederick Gutheim, interview with author, Washington, D.C., 10 June 1986.

32. BMK to Ruth Blackburn, 8 Dec. 1967.

33. AB to Walter Wright, 12 Jan. 1979, HBP.

34. BMK to "Trusting Twelve," 6 Mar. 1959; BMK to Aline M. Stein, 19 Dec. 1962; CS to BMK, 21 May 1959; CS to BMK, 25 Feb. 1959: all in CSP.

35. Regarding Stein's episodic mental difficulties, see BMK to CS, 24 Feb. 1937; CS to BMK, 7 Mar. 1937; CS to BMK, 18 May 1937; CS to BMK, 6 Dec. 1942; BMK diary, Jan.–Feb. 1953. See also K. C. Parsons, in *Writings of Clarence Stein*, xxviii, 359–361.

36. LM to BMK, 21 June 1966.

37. Ege, *Power of the Impossible*, 672, 677; BMK diary, 30–31 Aug. 1956; BMK to LM, 31 Aug. 1956, LMP; BMK to Ernest Peabody, 4 Sept. 1956.

38. BMK to LM, 6 Sept. 1956, LMP.

39. BMK to CS, 11 Apr. 1963, CSP.

40. BMK to Cora H. Thomas, 12 June 1956.

41. BMK diary, 1 Aug. 1962.

42. BMK to LM, 5 May 1958, LMP; BMK to HB, 11 Nov. 1963; BMK diary, 17 Jan. 1965.

43. BMK to Arthur Drinkwater, 4 July 1965.

44. BMK to Aline Stein, 1 May 1966.

45. Lucy Johnson to HB and AB, undated (c. autumn 1967), HBP.

46. BMK to Michael Nadel, 8 June 1965, WSP/DEN.

47. Citation for Department of the Interior Conservation Service award, attached to Stewart L. Udall to BMK, 16 Feb. 1966.

48. BMK to HB, 11 June 1966.

49. BMK, "Of Wilderness Trails and Areas: Steps to Preserve the Original America," 30 May 1966.

50. S. Udall to BMK, 20 July 1966; Department of the Interior, Bureau of Outdoor Recreation, *Trails for America* (Washington, D.C., 1966), 67–73.

51. BMK to PB, 8 June 1966.

52. BMK, "Benton MacKaye's Message," *LW* 30 (autumn 1966): 9; Michael Frome, *Strangers in High Places: The Story of the Great Smoky Mountains*, expanded ed. (Knoxville: University of Tennessee Press, 1994), xxvii–xxxiii.

53. ATC, *Member Handbook*, 48–49; Foster, *Appalachian National Scenic Trail*, 14–16.

54. Foster, *Appalachian National Scenic Trail*, 15–16.

55. Cliff Gaucher to BMK, 29 Mar. 1966; BMK to M. Nadel, 8 Apr. 1966.

56. BMK to Ruth Blackburn, 22 Mar. 1967.

57. BMK to C. Gaucher, 16 May 1968.

58. BMK to Stanley Murray, 20 Oct. 1968.

59. BMK to S. Murray, 20 Oct. 1970.

60. James R. Jones to BMK, 1 Oct. 1968; *National Trails System Act, U.S. Statutes at Large* 82 (1968): 919–926; *Wild and Scenic Rivers Act, U.S. Statutes at Large* 82 (1968): 906–918; BMK, "Recreational Possibilities of Public Forests," 29.

61. George Marshall to BMK, 13 May 1968; BMK, letter to editor, "Adopted Son Dies," *Fitchburg Sentinel*, 18 Mar. 1968.

62. HB to BMK, 9 Sept. 1957.

63. BMK to Freeman Tilden, 28 Nov. 1966.

64. BMK to S. Udall, 16 June 1968; BMK to Sen. Edward M. Kennedy, 16 Jan. 1969.

65. Stewart Brandborg to BMK, 23 Dec. 1969.

66. Hays, *Beauty, Health, and Permanence,* 52–57, 62.

67. BMK, "Memorandum on Environment," 4 May 1970; *NE,* 144.

68. BMK to CS, 20 June 1966.

69. BMK to Marion Stoddart, 4 Dec. 1966; BMK diary, 15 Dec. 1966; John Berger, *Restoring the Earth* (New York: Alfred A. Knopf, 1986), 9–25.

70. BMK to Paul Oehser, Oct. 29, Nov. 29 1966.

71. BMK, "The Appalachian Trail: A Guide to the Study of Nature," 342.

72. *Ex9,* 49.

73. Philip Chase to BMK, 25 Mar. 1968; Lucy Johnson to AB, 8 May, 29 Oct. 1968, HBP.

74. BMK diary, July–Aug. 1969, misc. entries; Christy Barnes to Lucy Johnson, 17 Aug. 1969, HBP.

75. BMK to LM, 11 Sept. 1969, LMP.

76. HPH, "The Shirley Influence," 23.

77. Ibid.

78. BMK to Michael Nadel, 29 Oct. 1971.

79. Waterman and Waterman, *Forest and Crag,* 559.

80. Lester Holmes to BMK, 6 Jan., 13 Nov. 1973; ATC, *Member Handbook,* 54.

81. BMK diary, 6–7 Aug. 1953; Shaffer, *Walking with Spring,* 110.

82. Waterman and Waterman, *Forest and Crag,* 641.

83. Ibid., 563.

84. BMK to Robert Wirth, 7 June 1966.

85. ATC, *Member Handbook,* 54.

86. Jane Jacobs, *The Death and Life of Great American Cities* (1961; New York: Vintage Books, 1992), 20–21.

87. Ibid., 374.

88. LM to BMK, 4 Mar. 1968.

89. BMK diary, 7 Feb., 20 Feb.–8 Mar. 1975.

90. Ibid., 27 Apr. 1975.

91. Ibid., 1–2 June, 2 July 1975; Sally K. Fairfax interview, 30 May 1985, quoted in Foster, *Appalachian National Scenic Trail,* 204 n. 494.

92. BMK diary, 17 May 1975; Constance L. Stallings, "Elder of the Tribe: (1879–1975), The Last Interview with Benton MacKaye," *Backpacker* 14 (Apr. 1976): 54–57, 81–85.

93. BMK diary, 10, 26 Sept. 1975.

94. Lucy Johnson to SC, 25 Jan. 1976, WSP/DC.

95. LM et al., "Benton MacKaye: A Tribute," *LW* 39 (Jan.–Mar. 1976): 3, 6–34.
96. L. Johnson to SC, 25 Jan. 1976.

Epilogue: A "Planetary Feeling"

1. BMK to Nancy Durham, 13 June 1964.
2. *NE*, 202.

NOTE ON SOURCES

A delightful article by Crista Rahmann Renza, "The MacKaye Family Papers," in the *Dartmouth College Library Bulletin* (22 [November 1981]: 15–24), first led me to the Dartmouth College Library, in Hanover, New Hampshire. The well-organized MacKaye Family Papers in the Rauner Special Collections Library at Dartmouth were a fundamental source for this biography.

Other manuscripts, correspondence, and materials that relate principally to the MacKayes are in the Marion and Percy MacKaye Collection, Harvard Theatre Collection, Houghton Library, Cambridge, and the Shirley Historical Society, Shirley, Massachusetts. The Manuscripts Division of the Library of Congress holds a copy of Benton MacKaye's manuscript "Geotechnics of North America" and an autobiographical manuscript and some correspondence of Col. James Morrison McKaye.

Some Benton MacKaye correspondence, manuscripts, maps, and photographs are at the Appalachian Trail Conference (ATC), headquartered in Harpers Ferry, West Virginia. The ATC is also the repository of MacKaye's personal library and some of the contents of his "Sky Parlor."

Other manuscript collections and archives I consulted were: Lewis Mumford Papers, Rare Book and Manuscript Library, University of Pennsylvania, Philadelphia; Lewis Mumford Collection, Guggenheim Library, Monmouth University, West Long Branch, N.J.; Clarence S. Stein Papers, Rare Book and Manuscript Collections, Carl A. Kroch Library, Cornell University, Ithaca, N.Y.; Harvey and Anne Broome Papers, McClung Historical Collection, East Tennessee Historical Center, Knoxville; Robert Marshall Papers, Bancroft Library, University of California, Berkeley; Harry A. Slattery Papers, Rare Book, Manuscript, and Special Collections Library, Duke University, Durham, N.C.; Charles Harris Whitaker autobiographical manuscript "The Rejected Stone," Avery Library, Columbia University, New York; Harvard University Archives, Cambridge; Tennessee Valley Authority Corporate Library, Knoxville; Wilderness Society Papers, Western History and Genealogy Department, Denver Public Library; Wilderness Society Papers, Wilderness Society, Washington, D.C.; Potomac Appalachian Trail Club, Vienna, Virginia; and the Stuart Chase Papers, Louis L. Post Papers, Gifford Pinchot Papers, and National Woman's Party Papers, all at the Library of Congress, Manuscripts Division.

♣

Many publications used in preparation of this book are cited in the endnotes; a few works proved especially useful as starting points for further research. Percy MacKaye's sprawling two-volume biography of his father, *Epoch: The Life of Steele MacKaye, Genius of the Theatre, in Relation to His Times and Contemporaries* (New York: Boni & Liveright, 1927), in fact chronicles several generations of MacKayes. Another rich if disorderly compendium of family history and documents is *Annals of an Era: Percy MacKaye and the MacKaye Family, 1826-1932,* edited and with an introduction by Edwin O. Grover (Washington, D.C.: Dartmouth College/The Pioneer Press, 1932). Percy's daughter, Arvia MacKaye Ege, continues the family's tale from her parents' perspective in *The Power of the Impossible: The Life Story of Percy and Marion MacKaye* (Falmouth, Me.: Kennebec River Press, 1992).

Paul T. Bryant prepared "A Benton MacKaye Bibliography" (*The Living Wilderness* 39 [January–March 1976]: 33–34), which includes most of MacKaye's published works. Bryant's excellent 1965 University of Illinois dissertation, *The Quality of the Day: The Achievement of Benton MacKaye* (Ann Arbor: University Microfilms, 1981), was based in part on the author's personal conversations and correspondence with MacKaye. Bryant was also editor of MacKaye's collection of articles and essays, *From Geography to Geotechnics* (Urbana: University of Illinois Press, 1968).

Personal remembrances of MacKaye by Lewis Mumford, Stuart Chase, George Marshall, Paul H. Oehser, Frederick Gutheim, Harley P. Holden, Paul T. Bryant, Robert M. Howes, and C. J. S. Durham appear in *The Living Wilderness* 39 ([January–March 1976]: 3–33). Mumford's introduction to the 1962 and 1991 paperback reprint editions of *The New Exploration: A Philosophy of Regional Planning* (Urbana: University of Illinois Press, 1962; Urbana-Champaign: University of Illinois Press and Appalachian Trail Conference, 1991) also provides an engaging personal and biographical portrait of MacKaye.